世界龙胆荟萃

GENTIANS OF THE WORLD

袁永明　著
YUAN Yong-Ming

山东科学技术出版社
·济南·

图书在版编目（CIP）数据

世界龙胆荟萃：汉英对照 / 袁永明著. -- 济南：山东科学技术出版社，2024. 11. -- ISBN 978-7-5723-1984-6

Ⅰ. ①世… Ⅱ. ①袁… Ⅲ. ①龙胆科－世界－图集 Ⅳ. ① Q949.776.4-64

中国国家版本馆 CIP 数据核字 (2024) 第 050736 号

世界龙胆荟萃

SHIJIE LONGDAN HUICUI

责任编辑：陈 昕 庞 婕 徐丽叶

主管单位：山东出版传媒股份有限公司
出 版 者：山东科学技术出版社
地址：济南市市中区舜耕路 517 号
邮编：250003 电话：（0531）82098088
网址：www.lkj.com.cn
电子邮件：sdkj@sdcbcm.com
发 行 者：山东科学技术出版社
地址：济南市市中区舜耕路 517 号
邮编：250003 电话：（0531）82098067
印 刷 者：山东临沂新华印刷物流集团有限责任公司
地址：山东临沂高新技术产业开发区龙湖路
邮编：276017 电话：（0539）2925659

规格：16 开（210 mm × 278 mm）
印张：52.25 **字数：**600 千
版次：2024 年 11 月第 1 版 **印次：**2024 年 11 月第 1 次印刷
定价：880.00 元
审图号：GS 鲁（2024）0202 号

内容简介

龙胆科植物花色绚丽多彩，是地球上最为美丽的野花。它们的生境特殊，生长在地球的最高处，主要分布于世界各地的高山地区，备受人们喜爱。

从攻读博士学位开始，本书作者 30 多年来长期追寻拍摄世界各地的龙胆科植物及其生长环境。现从积累的 3 万多幅照片中，精选出 1600 多幅，经过严谨、认真的科学鉴定和对相关物种的形态分类、生态地理等特征的综合分析，重点介绍 300 余种（含种下单元）龙胆科龙胆族相关植物的学名、形态特征、地理分布和生长环境等属性。每个物种都配有从多角度和在不同生长状态下拍摄的多幅彩色照片，以期全面描述它们的形态和美学特征。本书包括的种类覆盖了全球龙胆科龙胆族的全部 22 个属中的 21 个属及超过 30% 的种类，尤其龙胆属的种类覆盖接近 57%。

本书是系统描述龙胆科植物多样性的科学图册，全面而通俗地记述了它们的形色之美、生长环境之美。本书力求科学严谨与通俗易读相结合，让人赏心悦目。

本书可作为植物学工作者、园艺工作者、自然爱好者和摄影爱好者的参考书，有助于他们对龙胆科植物全面、准确地识别和深入理解，并真正了解龙胆花的多样美丽。本书在手，美丽的龙胆花就不再遥不可及，相信它能唤起人们对龙胆、对高山大川、对大自然生物多样性和地球这颗蓝色星球更加炙热的爱。

About the book

Plants of Gentianaceae have bright and colorful flowers, being the most beautiful flowers on the Earth. They have also unique habitats, occupying the highest habitats on the earth, mostly growing at high elevation plateaus and mountains, making them the most attractive wild flowers on this blue planet Earth.

Started with doctoral dissertation researches, the author has been pursuing photographing of plants of Gentianaceae family and their environments around the world for over 30 years. This book now selected over 1600 photos out of a collection of over 30000 photos accumulated. After rigorous scientific identification and comprehensive analysis of the morphological, ecological and geographical features of the relevant species, the book focuses on over 300 species (including intraspecific taxa) of tribe *Gentianeae*, introducing their scientific names, morphological characteristics, geographical distribution, and habitats. Each species is illustrated with multiple color photos taken from different angles and at different growth states, in order to comprehensively describe their morphological characteristics and their beauty. This book covered 21 out of the 22 genera and over 30% species of the tribe worldwide, in particular, the coverage of species of the genus *Gentiana* had reached nearly 57%.

As a scientific atlas that describes the diversity of Gentianaceae plants and their environments, this book provides a comprehensive and yet popular demonstration of the diversity and its beauty. Seeking careful balancing among readability, pleasures, completeness in coverage, and scientific accuracy, this book is an easy yet a serious reading on the biodiversity of Gentianaceae plants, based on a scientific and systematic summary of decades of field observations of wild Gentianaceae plants around the world done by the author.

This book can serve as a good reference for botanists, horticulturists, nature photographers, and nature enthusiasts, helping them to have a comprehensive and in-depth understanding of Gentianaceae plants, accurate identification of species, and meanwhile to truly experience the beauty of Gentianaceae biodiversity. This book brings the beautiful gentians from remote and high mountains hard to access to reader's finger tips. The author believes it can evoke people's stronger love for gentians, for mountains and rivers, for the nature and our blue planet Earth.

作者简介：

袁永明，理学博士。1962 年出生于中国甘肃南部一个生长着陕南龙胆花的山村，在龙胆花的陪伴下长大，在兰州大学生物学系植物学专业完成本科和硕士阶段学习，1996 年在瑞士纳沙泰尔大学植物研究所取得博士学位。曾先后在中国科学院、兰州大学、瑞士纳沙泰尔大学和瑞士企业等不同单位从事与植物学专业有关的科研、教学和产品研发工作。现退休赋闲。他酷爱大自然，喜欢山野旅行和摄影，半生与花花草草打交道，这既是他的专业，也是他的爱好。

About the author

Yong-Ming Yuan, Doctor in Science, was born in 1962 in a mountain village in southern Gansu Province, China. He grew up with the companion of the flowers of *Gentiana piasezkii* which is wildly growing in the mountains. He completed his undergraduate and graduate studies in botany at the Department of Biology of Lanzhou University. Later, in 1996, he obtained a doctoral degree from the Institute of Botany at the University of Neuchatel in Switzerland. He has been engaged in scientific researches, teaching, and the development of natural products related to botany, in different institutions in Chinese Academies of Sciences, Lanzhou University, University of Neuchatel and Swiss enterprises for many years, and just retired recently. He has a great passion for nature, enjoys traveling in the mountains and taking photos, and spends half of his life dealing with flowers and plants, which is not only his profession, but also his passion.

递我一支龙胆花，有如给我一把火炬，
让我用这蓝色的、光焰四射的花朵之炬来引领自己……
——D. H. 劳伦斯

Reach me a gentian, give me a torch,
Let me guide myself with the blue, forked torch of this flower ...
- D. H. Lawrence

Foreword 1

L'ouvrage que Yuan Yong-Ming offre aux lecteurs témoigne du profond attachement de l'auteur à ses racines, à sa terre natale, à ses parents, Madame CAI Long-E, Monsieur YUAN Xue-Yi, à l'ensemble de sa famille. Issu d'une famille d'agriculteurs modestes, vivant sur des terres difficiles du comté de Tanchang, Yuan a sans doute participé aux travaux des champs. A ce titre, il s'est imprégné de la nature, a compris la fragilité des écosystèmes pastoraux et forestiers. L'essentiel des ressources énergétiques reposant sur l'utilisation du bois, Yuan a vu la limite de la forêt s'éloigner du village. Il a vécu les efforts accrus de sa famille pour assurer leurs ressources. Il a pris conscience de l'affaiblissement de la biodiversité. Parmi les plantes familières, *Gentiana piazeskii*, une étoile bleue, à gorge finement ponctuée de noir allait devenir sa motivation constante dans la poursuite de ses recherches. L'ethnobotaniste Pierre Lieutaghi, rapporté par Alain Hervé, se plaignait : *La mention de la beauté a disparu des revues scientifiques contemporaines*. Elle est omniprésente dans le travail de Yuan.

Après quelques travaux de cytologie, Yuan Yong-Ming a compris que le nombre chromosomique n'avait pas de réelle valeur à l'échelle du genre, sinon au niveau de certaines sections. Des erreurs d'interprétation du critère cytologique avaient déjà conduit à la description de genres tels *Favargera* et *Kuepferella* dont la validité, jamais confirmée, a été encore infirmée par les travaux de Yuan. Dans le deuxième moitié du 20e siècle, les botanistes européens pensaient disposer des éléments leur permettant d'esquisser des phylogénies. Ils imaginaient que le genre s'était diversifié à partir de l'ancêtre de *Gentiana lutea*, une espèce à longévité exceptionnelle parmi les espèces herbacées. Par évolution divergente, elle aurait produit des espèces à cycle plus court, annuel ou bisannuel. Les arguments étaient très faibles. La question restait ouverte. Yuan a compris que la question devait être reprise sur la base d'études moléculaires. Il est rapidement devenu un pionnier dans l'application des ITS (Internal transcribed spacer) à l'étude de la systématique et de la biogéographie des *Gentianaceae*.

Dans la deuxième moitié du XXe siècle, le genre *Gentiana* était compris dans un sens très large. Pour les botanistes européens, il était superflus de multiplier les genres pour quelques espèces facilement identifiables à des Gentianes *sensu lato*. C'était oublié que l'essentiel de la diversité des gentianes se trouvaient dans les montagnes d'Asie centrale. Les travaux de Yuan apportent d'une manière incontestable des lettres de noblesse aux genres *Comastoma*, *Gentianella* et *Gentianopsis* que les flores européennes dédaignent encore aujourd'hui. Nul doute que la synthèse de Yuan trouvera un écho dans les flores à venir.

Le présent ouvrage ne résulte pas seulement de patients travaux de laboratoires et de réflexions scientifiques. Aventurier souvent solitaire dans ses excursions les plus périlleuses, Yuan a prospecté intensément les montagnes chinoises et complété ses recherches dans les Alpes, les Balkans, le Caucase. Il y a constamment recherché le vrai, le perfectionnement. Avec son extraordinaire expérience du terrain, Yuan a ressenti la responsabilité de raconter, de partager ses émerveillements, mais aussi de dénoncer la fragilité des écosystèmes alpins exposés aux aménagements touristiques, puis plus récemment menacés par les perturbations climatiques. Partout dans le monde, la montagne est martyrisée par l'extension du réseau routier, l'urbanisation, l'exploitation des ressources hydrauliques et minières, le développement du tourisme et du sport. Les montagnes asiatiques échapperont-elles à cette péjoration ? Jusqu'à aujourd'hui, le maintien de la qualité des pelouses alpines, dans le Qinghai en particulier, résulte sans doute du pâturage extensif par les yacks, conduit avec intelligence par les populations indigènes sur des territoires immenses. Les images des périphéries de ville telles Zeku (fig. 2-13, p.053) ou Aba (*Gentiana straminea* figure on p.306) sont instructives ; les contrastes sont saisissants. Deux sociétés s'y affrontent: la citadine, qui construit, aménage, innove, s'étend ; la pastorale, patiente qui ménage les sols, exploite sans compromettre, utilise sans détruire. Les pelouses encore préservées résisteront-elles à l'urbanisation croissante? Nous l'espérons avec Yuan Yong-Ming. Malheureusement, le réchauffement climatique entraînant la fonte du permafrost risque de perturber ce fragile équilibre et d'affecter la survie des milieux riches en Gentianacées.

Soucieux de mettre à disposition des amateurs et des professionnels des documents correctement identifiés, Yuan a réuni en 30 années d'expériences 30000 photos. Elles donnent une image inégalée de la diversité des *Gentianeae*, tout en illustrant l'évolution de la photographie. Toutes les images prises au-delà des années 2000 sont d'une précision remarquable. Précédemment, le rendu des bleus avec les émulsions argentiques était souvent décevant. Il est parfait avec l'avènement des procédés numériques.

Dans les chapitres descriptifs, la définition des genres est fondée sur la synthèse de nombreux travaux personnels et une connaissance exhaustive de la bibliographie. Yuan distingue bien les caractères liés à la fonction qui ont des origines multiples et peuvent induire en erreur la classification. Par exemple, les gentianes de petites taille doivent trouver un moyen de disperser les graines. Après l'anthèse, les unes allongent la tige, d'autres le carpophore. Chacun de ces caractères apparaît indépendamment plusieurs fois, dans des sections non apparentées. Au contraire, des traits comme la structure du tégument séminal ou la forme de la corolle donnent de précieuses informations sur les liens de parenté.

Yuan Yong-Ming retient aussi les noms vernaculaires chinois. Dans quelques cas, il propose de nouveaux noms pour toujours associer science et vulgarisation et rapprocher la science du public. En revanche, il ne figure pas les noms d'auteurs des taxons cités. Sa nomenclature rigoureuse, fondée sur les recherches les plus actuelles, permet d'éviter toute confusion. Chaque lecteur trouvera dans International Plant Names Index (IPNI) les informations complémentaires, éludées pour ne pas alourdir inutilement le texte.

De nouvelles perspectives de recherches sont aussi esquissées sur les interactions étroites entre les gentianes et leurs pollinisateurs. La diversité des coloris ornant la gorge des corolles, les contrastes de couleur, pointillés, lignes foncées ou claires, doivent jouer un rôle dans l'attraction des insectes. Ils déterminent la précision de la pollinisation, garante de la stabilité de chaque espèce. Observateur attentif, Yuan mentionne la variabilité intraspécifique notant des variants dans la coloration, comme chez *G. choanantha* et *G. leucomelaena*. Ebauche d'une spéciation en cours ?

L'intérêt du présent ouvrage dépasse évidemment le cadre des montagnes centrasiatiques, chinoises en particulier. Il souligne les principaux centres de spéciation et met en évidence les grands courants migratoires qui ont permis l'installation des *Gentianeae* sur tous les continents. Bien entendu, il intéressera l'ensemble des botanistes mondiaux, mais son intérêt touchera aussi tous les amateurs de fleurs de montagne. Le soussigné avait été fasciné par l'ouvrage publié en 1955 par Henri Gaussen et Paul Barruel intitulé « Montagnes ». Il offrait des images de *Saussurea*, *Meconopsis*, *Rheum* rapportées d'expéditions dans l'Himalaya. Nul doute que le livre de Yuan connaisse le même succès éditorial et suscite de nombreuses vocations de botanistes. Il encouragera les lecteurs à entreprendre des voyages dans les principaux centres de diversité des *Gentianeae*, en particulier dans les montagnes centrasiatiques. Il rend aussi hommage à sa famille, ses parents en particulier, dont la fierté serait légitime s'ils avaient pu en connaître l'achèvement.

Philippe Kuepfer

[Prof. Dr. Philippe Kuepfer is professor emeritus from the University of Neuchatel. He was the doctoral supervisor of the author of this book. For many years, Prof. Kuepfer has been teaching botany and studying systematics, cytology, and evolution of the plant families including Gentianaceae, Brassicaceae, and Ranunculaceae with important publications and wide international impact. There are quite a few plant Latin names given after his name.]

序一

本书作者对祖国、家乡、父母（母亲蔡龙娥女士和父亲袁学义先生）及整个家庭有着深深的依恋。他出生于甘肃省宕昌县一个淳朴的农民家庭，在那贫瘠的土地上，他时常帮助家里干农活。正因如此，他得以沉浸在大自然中，并了解畜牧和森林生态系统的脆弱性。由于大部分能源供应都来自木材，他看到了森林线从村庄向外远去的过程，也目睹了家人们为保护当地资源所做的努力，并深刻意识到生物多样性的削弱和减退问题。在他熟悉的植物中，陕南龙胆（*Gentiana piazeskii*）就像蓝色的星星，有着精致的蓝黑色花冠，这或许成了他持续研究的动力。阿兰·埃尔韦（Alain Hervé）曾报道过民族植物学家皮埃尔·利乌塔吉（Pierre Lieutaghi）的抱怨：当代科学期刊中对美的提及已经消失。但这种美，在这本书中却无处不在。

在细胞学方面做了一些工作后，作者意识到染色体数目在龙胆属的水平上没有真正的价值，只是在一些组的水平上有一定参考意义。利用细胞学标准的解读，使 *Favargera* 和 *Kuepferella* 等属一级的描述出现错误，其有效性从未得到证实。本书作者的工作进一步印证了这一点。在 20 世纪下半叶，欧洲的植物学家们确信他们已经掌握了系统演化的骨架要素。他们认为，龙胆属是从类似于黄龙胆（*Gentiana lutea*）的祖先分化而来；黄龙胆是一种寿命特别长的草本物种。通过分化进化，得以形成较短生命周期（一年或两年）的物种。当然，支持该说法的证据非常薄弱，问题仍然悬而未决。作者认为，这个问题需要依据 DNA 分子研究再次深入探讨，他很快就成为利用 ITS 序列研究龙胆科系统学和生物地理学的先驱。

在 20 世纪下半叶，龙胆属的概念是很广义的。对于欧洲的植物学家们来说，将一个很容易识别的只有少数几个物种的广义属进行拆分是多余的。人们忘记了，龙胆的多样性大部分体现在亚洲的山区。作者的研究毫无疑问地给欧洲植物志中较罕见的喉毛花属 *Comastoma*、假龙胆属 *Gentianella* 和扁蕾属 *Gentianopsis* 等带来了它们的“贵族身份证书”。他的综合结论也将在未来的植物志中得到回应。

作者是一位富有探险精神的野外工作者，经常独自进行冒险旅行，他深入勘察了中国的许多山脉，并完成了阿尔卑斯山、巴尔干半岛和高加索地区的野外考察。他一直追求真实，追求完美。凭借着丰富的野外工作经验，他觉得有责任讲述科考故事，分享他的奇遇，同时也提醒世人，旅游业发展，当然，还有气候变化对高山生态系统带来的新近威胁。在世界各地，由于道路网的扩展、城市化、水和矿产资源的开发及旅游和体育业的发展，山区正在遭受折磨。亚洲的山区能躲过这种恶化吗？目前为止，高山、草原的生态保护主要依靠当地政府和人民的努力，尤其是青海等广泛放牧牦牛的地区。青海省泽库县（第 053 页图 2-13）和四川省阿坝藏族羌族自治州（第 302 页麻花艽图）等城镇近郊的图像很具有启发性：城市居民们致力于建设、发展、创新和扩张；而以土地为生的田园派们则珍惜土地，坚持有节制地开发、不破坏地使用。反差如此强烈。自然生态能否抵抗日益增长的城市化？我们和作者一样，都希望如此。不幸的是，导致永久冻土层融化的全球变暖可能会破坏这一脆弱的平衡，从而影响龙胆科植物的生境。

为了尽量给植物学专业人士和业余爱好者提供经过准确鉴别的素材，作者在过去 30 多年里收集了 3 万多张照片。它们不仅提供了一幅幅展示龙胆科植物无可比拟的生物多样性画卷，同时也展示了摄影技术的演变。2000 年之前，一般用胶片的银乳液来渲染龙胆的蓝调，结果常常令人失望；而 2000 年以后拍摄的所有照片都非常清晰准确。蓝色显然更适合用数字技术表达。

在本书的描述性章节中，属的定义是作者基于大量个人研究结果和大量文献知识的整合。他很好地区分了具有多重起源和可能误导分类的与功能相关的形态特征。例如，小龙胆们需要找到一种有效扩散种子的方法，开花结果后，有些种类的茎会快速增长，而另外一些则是雌蕊柄快速增长。所有这些性状在不相关的组中多次重复起源。相反，像种皮的形状结构或花冠形状这样的特征则提供了有关家族内进化关系的重要信息。

作者还谨慎地选用或新拟了中文植物名称，将科学性与普及性联系起来，使本书更易于公众阅读。另一方面，为了避免混淆，书中所有植物学名中的命名人都被省略了。关于龙胆科命名方面的研究文献有很多，读者都可以从国际植物名称索引（IPNI）这类在线数据库中找到有关信息。

书中对新的研究视角，如龙胆花与其传粉者之间密切的相互作用等，也做了简要概述。花冠喉部的颜色多样性、颜色对比、颜色深浅、纹饰的形状如斑点或线条等，在吸引昆虫方面发挥着重要作用，它们决定了授粉的精准度，从而保证了每个物种的稳定性。作为一个敏锐的观察者，作者提到了有关花色和纹饰丰富的种内变异，比如反折花龙胆（*G. choanantha*）和蓝白龙胆（*G. leucomelaena*）丰富的种内变异现象。这是正在进行中的物种分化的草案？

这本书的内容显然超出了中亚山区的框架，尤其是超出了中国山区。它强调了主要的物种分化中心，并强调了龙胆族植物在各大洲定居的主要传播散布路线和趋势。当然，这将引起全世界植物学家的兴趣，也将引起所有山花爱好者的兴趣。

本人曾经痴迷于亨利·高森（Henri Gaussen）和保罗·巴鲁尔（Paul Barruel）于 1955 年出版的名为《大山》（*Montagnes*）的书，书中呈现了喜马拉雅山探险队拍摄的风毛菊属

（*Saussurea*）、绿绒蒿属（*Meconopsis*）和大黄属（*Rheum*）的一些图像。毫无疑问，本书将同样成功，它将吸引许多植物学家，将鼓励读者前往龙胆族重要的多样性中心，特别是中亚山区。书中作者还向他的家人，特别是他的父母致敬，如果他们在天有灵，一定会为这本书的完成感到无比骄傲。

Philippe Kuepfer

Philippe Kuepfer 博士是瑞士纳沙泰尔大学退休教授，是本书作者的博士生导师。Kuepfer 教授长期从事植物学教学及对龙胆科、十字花科、毛茛科等植物类群的系统演化、细胞学和分类学研究，有广泛的国际影响，许多植物类群都是以他的名字命名的。

序 二

龙胆是龙胆科植物的简称。在高山地区分布的龙胆主要是龙胆族物种。龙胆与报春、杜鹃被誉为高山三大名花。这些高山龙胆生长在崇山峻岭之间，常常隐藏于阴影之中，仿佛是一种不可捉摸的幻影。然而，当它开放时，却绽放出一种夺目的蓝紫色光彩，宛如天空中的闪电在大地上投下的影子。自古以来，龙胆就被视为一种与天地灵气相通的神奇之物，被用于祭祀、医疗乃至仪式之中。它那独特的颜色和形态，更让诗人们无数次被其所迷恋，将之描绘得淋漓尽致。

袁永明博士撰写的《世界龙胆荟萃》集科普性、可读性与科学性于一体。作者通过 30 多年的野外科学考察，从收集的 3 万多张精美照片中精选了 1600 张左右的代表图片，力求全面展现龙胆各物种的姿态和形态特征，并对每个照片都进行了准确鉴定。该书图片优美、文字简洁，具有极强的可读性和观赏性；而建立在准确鉴定基础上的分类和分布描述，又展现了其极高的科学性！图片和文字一起，为龙胆爱好者和研究者等人员快速鉴定龙胆种类提供了参考。

该书展现的龙胆物种或矗立在峭壁之巅，或隐藏于溪流之畔，或被云雾环绕，或沐浴在阳光之下……通过这些图片，读者能够一窥龙胆的独特魅力，感受大自然的神奇与美丽。

让我们一起，踏上探寻龙胆的神奇之旅吧！

刘建全

刘建全，博士，兰州大学生态学院院长、教授、博士生导师，早年师从并共事于刘尚武教授和何廷农教授。他长期从事包括龙胆花在内的许多青藏高原特有动植物的起源、演化和生态学研究，成就斐然，著述颇多，影响深广。

Foreword 2

Gentian is the common name for plants in the Gentianaceae family. The gentians found in high-altitude areas are primarily from the tribe *Gentianeae*. Gentian, primrose, and rhododendron are known as the "three famous alpine flowers". These gentians growing on the mountains often hide in the shadows between the towering peaks, almost like an elusive phantom. Yet when they bloom, they display a dazzling blue-purple luster, like the shadow of lightning cast upon the earth from the sky. Since ancient times, gentians have been seen as a mystical plant connected to the spiritual essence of heaven and earth, used in sacrifices, medicine, and rituals. Its unique colors and forms have captivated poets, who have described them vividly in their works.

The book "Gentians of the World" written by Dr. Yong-Ming Yuan combines scientific rigor and readability. The book not only has beautiful, moving images, but also concise and fluid writing, showcasing the author's in-depth research on gentian plants. Through over 30 years of field research and a collection of 30 000 exquisite photos, the author has carefully selected representative images to comprehensively present the various forms of each species, while also providing accurate identification and detailed descriptions. The book is not only aesthetically pleasing with its images and writing, but also highly scientific in its taxonomic identification and descriptions, providing a quick reference guide for researchers and users to identify gentian species in the future.

The gentian species featured in the book stand atop precipices, hide by streams, are shrouded in mist, or bask in sunlight...Through these images, readers can glimpse the unique charm of gentians and experience the wonders and beauty of nature.

Let us embark on a journey to explore the marvels of the gentian together!

Jianquan Liu

[Dr Jianquan Liu, professor at the Colleage of Ecology, Lanzhou University, was formerly postgraduate student and collaborator of Prof. Shang-Wu Liu and Prof. Ting-Nong Ho. Dr. Liu have been studying the origin, evolution and ecology of some animals and plants endemic to the Qinghai-Tibet Plateau, including the gentians, and have achieved great success with lots of publications which had wide and significant impacts.]

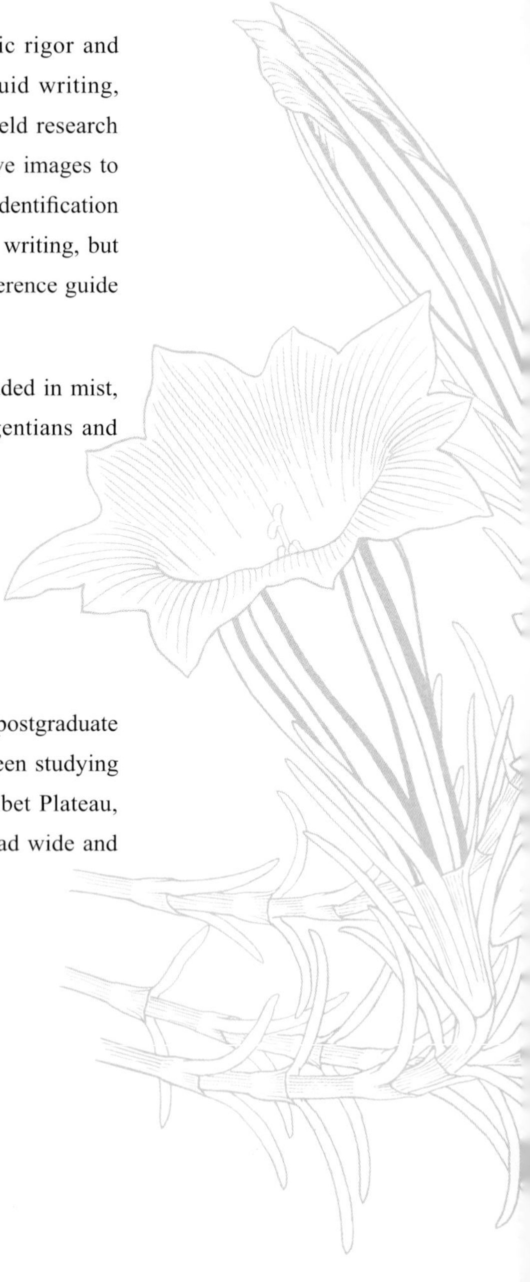

Gentiana lawrencei var. *farreri*; 中国青海省河南蒙古族自治县
Henan, Qinghai, China. 11.09.2023. 摄影 /Photo：马惠

龙胆之路　高阻且长
行则将至　无限风光

Roads to gentians, long and winding,
whenever you go, endless sceneries...

前 言

高山是一种情怀，而于我，高山更是一种印记。我的家乡在中国的陇南山区，这里是秦岭和青藏高原的结合部，这里的山兼具了秦岭的俊秀挺拔和青藏高原的恢弘广阔。生在山里，长在山里，再加上祖辈山民的言传身教，山在我的骨头和血液中留下了各种各样的印记，有山野的巍峨沉默，有山路的艰难曲折，当然还有山花的烂漫美丽和鲜艳夺目。其中，最为绚烂的山花莫过于山顶上的各色龙胆花了，有金黄的华龙胆，像号角向蓝天召唤；有蓝中带粉的秦艽花，成簇成串，招蜂引蝶；有湛蓝湛蓝、透着织锦般光泽的扁蕾花，总爱在风中摇曳，翩翩起舞；更有成片成丛的陕南龙胆花，在山坡上绽放，在草丛中绽放。每朵蓝紫色的花都张着五只角，鲜艳亮丽、轻盈活泼，就像一群璀璨的星星，在开心地嬉戏。龙胆花留在了记忆里，也融在了血液中，龙胆花的蓝成了我记忆深处最浓厚的色彩。

命运眷顾，32 年前我得以跨过高山越过平原去瑞士求学，来到欧洲，密切接触同样秀美雄伟的阿尔卑斯山。视野得以拓展，见识得以扩展，见到更多山外的群山，也见识到更多秀美靓丽的山花，尤其是同样艳光四射的蓝色龙胆。也许是命中注定，更为有幸的是，我的博士生导师 Philippe Kuepfer 教授与我心有灵犀，一拍即合，共同选择全球龙胆属植物作为我的博士论文研究对象，使我从此与龙胆结下了毕生的不解之缘。博士论文的研究工作是在全球范围内，在形态学、染色体和 DNA 分子等不同水平上，比较世界各地的龙胆属植物，以期探明众多龙胆属植物物种的生物多样性形成机制，了解不同地区的物种间的进化关系，为科学分类建立依据，并进一步推断世界不同地区的龙胆类群的起源、散播和分化历史。

龙胆花备受世界各地人们的喜爱。龙胆花色彩绚烂夺目，是地球上最为美丽的野花。它们的生境特殊，主要生长在世界各地的高山地区。龙胆属植物种类繁多，广泛分布在北半球几乎所有的高大山区，尤其是喜马拉雅山和中国的横断山、青藏高原以及欧洲的阿尔卑斯山地区。这项工作对于我来说，充满了挑战，更充满了诱惑。其中，最大的挑战、最大的诱惑，就是去世界各地的高山地区进行野外考察，见识更多更美的山川，也见识更多更美的龙胆。

得益于对山的天然亲近喜好，也得益于对喜马拉雅山和青藏高原的天然联系和对阿尔卑斯山的便捷后天联系，在接下来的数年间，我得以一次次地深入龙胆花在世界各地的故乡，多次进入喜马拉雅山腹地，从希夏邦马的冰湖到珠峰脚下的农庄，从帕里古镇的荒原到错那的深谷山坡，从多雄拉山口的冰川槽谷里攀爬到墨脱湿热河谷的徒步，更有了高黎贡山的露宿，天山的明月和阿尔泰的篝火，也有了甘南草原的繁花似锦，有了贡嘎山的壮丽巍峨、青海湖的草原广阔，也有了祁连山的油菜花海、长白山的天池和连天瀑布；当然还有法国中央高原上美丽的火山地貌上连片的黄龙胆和德国巴伐利亚深处的神秘城堡，也有了瑞士阿尔卑斯山最大冰川的壮丽美景，有了福尔卡山口和圣哥达山口的野餐，也有了北美落基山山间小道上的漫步。野外工作充满了艰辛，更充满了快乐和幸福。最为重要的是，这些年的野外工作让我得以见到更多更美的龙胆花，有黄的、白的和红的，但最令人迷醉的仍然是那些龙胆的蓝色。与龙胆花一次次邂逅，被龙胆花一次次折服，我至今都清晰地记得几乎每一次与以前没见过的新龙胆花的相遇，和相遇时自己从内心到四肢、从灵魂到肉体的悸动和颤烁。

真的很幸运，也真的很幸福。我们的地球是浩瀚宇宙里的一颗孤独星球，一颗唯一的蓝色星球。为什么它是蓝色的？因为那里有蓝色的天空，有蓝色的海洋，也许还因为那里有蓝色的龙胆花。龙胆花花色缤纷，红、黄、蓝、白皆有，却以蓝色最为突出。为什么龙胆花是蓝色的？因为它们生长的地方山高水高，离天最近，蓝是天空的颜色，也是大海的颜色。龙胆花生长的地方，不但山高地阔，还远离尘世，没有拥挤和喧嚣；龙胆花不但生长得又高又远，还不与百花争春，只爱在秋天里绽放，在寒风中起舞，超凡脱俗。龙胆花坦然地姹紫嫣红，却努力地与蓝天大海同色，只因为蓝是爱的颜色。龙胆花就是地球高原上的蓝色精灵，是高原的爱之所托，自足自安，自信自在，自然自乐。

光阴如梭，从我 1990 年开始拍摄龙胆花，几十年一晃而过，那些藏在我幻灯片夹和硬盘中的龙胆花一直在我的心底发酵。当时的胶片开始泛黄，而胶片上的龙胆花有些还在，有些已经变了模样，有些已经不知所踪，尽管她们都曾经存在过，美丽过，短短几十年，却已然是沧海桑田，天翻地覆，龙胆花尤其如此。前些日子有暇去了云南，重返香格里拉和丽江市区周边的山坡草地，那里曾是我初识许多龙胆花的地方，当年曾有很多龙胆花唾手可得，可现在却已经踪影全无。原来甘海子的松树林和草地上满目皆是龙胆花，现在已经变成了缆车站和停车场。随着交通的便利，人类活动的加剧给龙胆花造成了无形的胁迫，气温逐年升高和降雨减少更是造成了更加全局性的环境变化。那些龙胆花在那儿，它们的美丽都曾经被我目睹过，我便有责任把她们的美丽记叙下来，呈现出来，好让她们的美好存在供后来的人们知晓和赞叹，从而可以与大家一起共享那些我曾经见证过的美丽。是时候把它们整理出来了！

除了我的主观愿望，一本比较系统完整又科学可靠的龙胆花图册也是许多人的客观需求，因为被龙胆花所吸引的人远远不止我一个，实际上许许多多热爱自然的人，很多都是“龙胆控”。他们也热爱龙胆，被龙胆花所困，欲罢不能。然而，龙胆花不止难以企及，更难以准确辨识和鉴定，因为龙胆花的种类太多了，变化太大了。同时，由于旅行的便捷、拍照的便捷和互联网的便捷，网络上的植物图片铺天盖地，其中不乏形形色色的龙胆花。然而，首先，网络上的龙胆花图片资源十分庞杂混乱，大部分图片由拍摄者直接上传分享，缺乏严谨的对物种种类的科学鉴定，张冠李戴或指鹿为马的“作品”司空见惯；其次，由于网络高效便捷，较小的局部鉴定错误得以迅速传播和放大，以讹传讹。因此，一本专业的、系统的鉴定和描述龙胆花种类的图册就愈发显得必要。

基于以上认识，我从自己30多年来拍摄的3万多幅龙胆花及其生境的照片中精选整理出1600多幅图片，仅以亚欧龙胆科龙胆族植物为主结集。在此基础上，为了更完美地弥补一些重要类群和地区的缺口，也特别征集了近年来十分活跃的Adrien Favre博士、付鹏程博士等龙胆科专家的一些照片。此外，还征集了一些世界其他地区的专家和龙胆爱好者拍摄的龙胆科植物照片，提高了本书在种类和地理区域两方面的覆盖度。这样做的目的是让大家更全面地欣赏龙胆科植物生物多样性的繁复多样和美丽。所有这些照片均已征得拍摄者的许可，并在照片图注中注明了拍摄作者，其余所有没有标注拍摄者的照片均为作者本人拍摄。本书最后共选定了300余种代表种（含种下单元），涵盖了龙胆科龙胆族的全部22个属中的21个属及超过30%的种类，其中龙胆属共选取了200种，约占属内总种数的57%。此外，笔者刻意选择较多的照片，尤其是比较独特和漂亮的种类，从不同角度和不同季节或不同生长状态“描述”一些特定的对象，希望能从物种的多样性和这些多样性的美，这两个不同角度给大家呈现龙胆花，相信能够在给读者增知的同时，也能使大家赏心和悦目。

本书中族、属的概念，属下单元（组）、种类的划分和定义及描述主要参照《中国植物志》《欧洲植物志》及近几年发表的研究论文和专著，种类的排列则以种的学名字母为序。另外，需要说明的是，中文名称的取舍是本着继承传统、清晰简约、避免重复的原则进行的，除了沿用《中国植物志》的中名外，新拟了个别中文名称，比如以“耳褶属”来命名新近建立的*Kuepferia*属，以“狭蕊属”来命名*Metagentiana*属，以“匙萼龙胆”来对应*Gentiana spathulisepala*（而非原来的*Gentiana stragulata*），以“维纳斯龙胆”来替代原“喜马拉雅龙胆”作为*Gentiana venusta*的中名（以区别*Gentiana himalayaensis*），等等。妥当与否，请大家注意并指正。

本书覆盖的地域广阔，资料的积累过程艰辛漫长。受植物学野外考察的各种条件，比如交通、天气和季节等方面的限制，在地域覆盖、种类的代表性和图片艺术性等诸多方面尚有不少缺憾。在物种鉴定方面，虽然再三斟酌，力求准确，但基于图片和活植物的鉴别鉴定与基于蜡叶标本的传统分类学命名与鉴定之间还是有一定距离的，不当之处，欢迎广大读者批评斧正。

袁永明
2023年10月于上海

家乡的蓝色龙胆花就是我的幸运之星

Blue flowers of gentians from home village are my lucky stars.

陕南龙胆 *Gentiana piasezkii*

中国甘肃省宕昌县 Tanchang, Gansu, China. 08.08.2012

Preface

High mountains are a kind of sentiment, and for me, high mountains are a kind of imprint. My hometown is located in the mountainous areas in southern Gansu, China. This is the junction of the Qinling Mountains and the Qinghai-Tibet Plateau, where the mountains combine the beauty and straightness of the Qinling Mountains with the grandeur and vastness of the Qinghai-Tibet Plateau. Born and raised up in the mountains, combined with the words and deeds of my parents and grandparents, the mountains have left various marks on my bones and in my memory, including the towering silence of the mountains, the curving trials on the steep slopes, and of course, the beautiful and vibrant flowers of the mountains. The most splendid mountain flowers among them are none other than the various colored gentian flowers on the mountaintop, the golden flowers of *Sinogentiana striata* summoning to the blue sky like horns, the blue pink flowers of *Gentiana macrophylla*, clustered in strings and clusters, attracting bees and butterflies, and the deep blue flowers of *Gentiana piasezkii*, shining as silk and dancing gracefully in the wind, while blooming on the slopes and in the grass. Each blue purple flower of *Gentiana piasezkii* has five angles, bright and lively. When many of its flowers start to bloom, they just like a group of brilliant stars blinking in the sky. The gentian flowers remained in my memory and also blended into my blood. The blue color of the gentian flowers has become the strongest color in my memory.

Fate favors me. Thirty-two years ago, I was able to across the mountains and plateaus and went to Switzerland to study, coming to Europe to have close contact with the equally beautiful and majestic Alps. My vision was expanded, my knowledge was expanded, and I could see more mountains outside the mountains, as well as more and beautiful mountain flowers, especially the equally brilliant blue gentians. Perhaps it was fate, but fortunately, my doctoral supervisor Professor Philippe Kuepfer and I had a highly consensus: to study the global gentians as the object for my doctoral dissertation, which led me to form a lifelong happy bond with gentians. The research work of the doctoral dissertation was to compare gentian plants from all over the world at different levels including morphology, chromosomes, and DNA molecular markers, in order to understand the mechanisms of diversification of numerous gentian species, to understand the evolutionary relationships between species in different regions, to establish a basis for scientific classification, and further more to infer the origin, dispersal, and differentiation history of gentian taxa in different regions of the world.

Gentians are beloved by people around the world. Gentian flowers have dazzling colors and are the most beautiful wild flowers on the Earth. Their habitat is unique, mainly growing in high mountain areas around the world. The genus *Gentiana* has numerous species, widely distributed in almost all high mountainous areas of the Northern Hemisphere, especially in the Himalayas, the Hengduan Mountains and Qinghai-Tibet Plateau in China, and the Alps in Europe. This job was full of challenges and temptations for me. The biggest challenge and also temptation was to travel to high mountain areas around the world, to see more and more beautiful mountains and rivers, and also to see and collect more and more beautiful gentians.

Thanks to my natural affinity for mountains, the natural connection between me and the Himalayas and the Qinghai-Tibet Plateau, as well as the convenient access to the Alps from Switzerland, I was able to travel deeper into the hometowns of gentian flowers around the world in the following years, and have visited the hinterland of the Himalayas multiple times, from the icy lakes of Xixiabangma to the farms at the foot of the Everest, from the wild fields nearby the Pari ancient town to the deep valley slopes of Cuona, Climbing from the glacier edge at the Duoxiongla Pass to the hot and humid valley of Yaluzangbu river by Motuo, camping in Gaoligong Mountain, bathing in the bright moon in Tianshan Mountain and by the bonfire in Altai, walking through the blooming flowers in the grassland of southern Gansu, watching the magnificent and majestic peaks of the Gongga Mountain, the vast grasslands of Qinghai Lake, the sea of colza flowers in Qilian Mountain, the Tianchi Lake and the sky high waterfall in Changbai Mountain. Of course, there are also the contiguous yellow gentians growing in the field of the beautiful volcanic landforms on the Massif Central in France, and the mysterious castle deep in Bavaria, Germany. There are also magnificent views of the largest glacier in the Swiss Alps, picnics at the Furka Pass and the Saint Gotthard Pass, as well as strolling along the paths between the Rocky Mountains in North America. Field work is full of hardships, but also full of joy and happiness. Most importantly, these years of wilderness work have allowed me to see more and more beautiful gentian flowers, including yellow, white, and red, but the most intoxicating ones are still the blue ones of those gentians. Encountering the gentian flowers time and time again, being captivated by the gentian flowers time and time again, I still vividly remember almost every encounter with a new gentian flower that I had never seen before, and the throbbing and trembling of myself from the heart to the limbs, from the soul to the body when I met.

I am really lucky and also very happy. Our Earth is a lonely planet in the vast universe, the only blue planet. I am wondering why it is blue? Because there is a blue sky, a blue ocean, and perhaps also because there are blue gentian flowers. Gentian flowers are colorful, with red, yellow, blue, and white colors, but blue is the most prominent. Why are gentian flowers blue? Because they grow in high mountains and waters, closest to the sky. Blue is the color of the sky and also the color of the sea. The place where gentian flowers grow is not only wide and mountainous, but also far away from the civilized world, without crowds and noise. Gentian flowers not only grow high and far away, but also do not compete with other flowers for spring. They only love to bloom in autumn and dance in the cold wind, extraordinary and outstanding. Gentian flowers are calm and colorful, but they strive to be the same color as the blue sky and sea, just because blue is the color of love. Gentian flower is the blue creature on the plateau of the Earth, entrusted by the love of the plateau, self-satisfied, self-confident, and naturally happy.

Time flies like a shuttle. I started shooting gentians in 1990, and decades passed by in a flash. The gentian flowers hidden in my slide folder and hard drive have been fermenting and maturing in my heart. Some films started to turn yellow. Some of the subjects on my films were still there, some had changed their appearance, and some had gone. Although they had all existed before and were beautiful, in just a few decades, they were already a world of dramatic change, especially for the gentian flowers. A few months ago, I had time to go to Yunnan and returned to the hillside grasslands around Xianggelila and Lijiang, where used to be the place when I first met many gentian flowers. At that time, there were many gentian flowers that were easy to get with a random walk, but now they have disappeared completely. The pine forests and grasslands at Ganhaizi used to be covered with gentian flowers, but now they have become a cable station and the parking lots for tourists. With the convenience of transportation and the intensification of human activities, gentian flowers have been invisibly threatened, and the increasing temperature and decreasing rainfall have caused more global environmental changes. Those gentian flowers were once there, and their beauty had been witnessed by me before. Once I met them, I feel I had a responsibility to narrate and present their beauty, so that their beautiful existence can be known and admired by other people, so that we can share the beauty I had witnessed with everyone. It is the time to get them organized and presented!

In addition to my subjective desires, there is also an objective need for a relatively systematic, complete, and scientifically reliable picture book of gentian flowers, because I am not the only one attracted by the gentian flowers. In fact, many people who love nature became "gentian controlled", also love gentians and are trapped by gentian flowers, and unable to escape. However, gentian flowers are not only difficult to reach, but also difficult to be accurately identified with their species, because simply there are too many species and the differences between them are subtle. Meanwhile, due to the convenience of travel, photography, and access to the internet, there are countless plant images on the internet, including various types of gentian flowers. However, the image resources of the gentian flowers on the internet are very diverse and chaotic, with most of the images uploaded and shared directly by photographers, lacking rigorous scientific identification of species. Cases assigning oranges as apples or referring to deer as horses are also common. Secondly, due to the high efficiency and convenience of the network, smaller local identification errors can quickly spread and amplify, leading to the spread of errors. Therefore, a comprehensive summary of gentians with professional and systematic identification and description of the species becomes increasingly necessary.

Based on the above understanding, the author of this book has selected and organized more than 1600 pictures from over 30 000 photos of gentian flowers and their habitats taken over the past 30 years, mainly focusing on the plants of tribe *Gentianeae* of the Gentianaceae family in Asia and Europe. On this basis, in order to better fill the gaps in some important taxa and regions, we have also specially collected some photos from Dr. Adrien Favre and Dr. Peng-Cheng Fu, who have been very active experts in the Gentianaceae family in recent years. In addition, we also collected photos of Gentianaceae plants taken by experts and enthusiasts from other parts of the world, enriching the coverage of this book in terms of species and geographical regions. The purpose of doing so is to enable everyone to fully appreciate the complexity, diversity, and beauty of biodiversity of the Gentianaceae family. All these photos have been taken with the permission of the photographer, and the author of the photo has been indicated in the caption. All other photos without indication of the author have been taken by the author himself. As a result, over 300 representative species (including intraspecific taxa) were selected, covering 21 out of the 22 genera of tribe *Gentianeae* of the Gentianaceae family and nearly 30% of the species. Among them, 200 species belonging to the genus *Gentiana* were selected, accounting for approximately 57% of the total species within the genus. In addition, the author deliberately chooses a large number of photos, especially unique and beautiful species, to "describe" specific objects from different perspectives, seasons, or growth states, hoping to present gentian flowers from two different perspectives: species diversity and the beauty of these diversity. I believe doing so can not only increase readers' knowledge, but also entertaining their hearts and eyes.

The concepts of tribe and genus, as well as the division, definition, and description of infrageneric taxa (sections) and species in this book were mainly based on the *Flora of China*, *Flora of Europe*, and research papers and monographs published recently. The species were arranged in alphabetical order based on their scientific names. Furthermore, it should be noted that the use and selection of Chinese names was based on the principles of following tradition, being clear and concise, and avoiding repetition. In addition to using the Chinese names adopted in the *Flora of China*, some new Chinese names have been proposed, such as naming the newly established *Kuepferia* genus after "Ear Zhe", *Metagentiana* genus after "Xia Rui", and *Gentiana spathulisepala* (instead of the original *Gentiana stragulata*) with "Chi E Longdan", replacing the original "Ximalaya Longdan" with "Weinasi Longdan" as the Chinese name for *Gentiana venusta* (to distinguish *Gentiana himalayaensis*), and so on. Whether it is appropriate or not, please pay attention and make your comments and suggestions.

This book covers a vast area and the process of accumulating information is arduous and lengthy. Due to various limitations in field investigations such as transportation, weather, and seasons, there might be still many shortcomings in terms of regional coverage, representativeness of species, and artistic feature of some photos. In terms of species identification, although careful consideration has been made to strive for accuracy, mistakes may still occur due to the fact that identifications based on pictures and living plants may not perfectly match the identifications based on herbarium materials in classic taxonomy. Any critical comment and correction are very much appreciated.

YUAN Yong-Ming, Shanghai, October 2023

蓝色的爱 Love is blue.
蓝玉簪龙胆 *Gentiana veitchiorum*
中国青海省河南蒙古族自治县 Henan, Qinghai, China. 09.09.2023

欧洲中部到南部特产的坷路龙胆通常生长在陡峭的高山钙质岩为基质的石质山坡草地上，时常被称作号角龙胆，它们的花朵确实就像大地的号角，在称颂和呼唤蓝天。

Gentiana clusii, endemic to central to southern Europe, is commonly known as Trumpet Gentian. They grow in calcium soil on alpine rocky slopes and meadows. Their flowers do look like trumpets of the Earth praising and calling the blue sky.

瑞士伯尔尼州莎斯哈尔山 Chasseral, Bern, Switzerland. 06.06.2004

目 录

中国四川省康定市 Kangding, Sichuan, China. 05.07.2021

1

认识龙胆

To Understand Gentians

龙胆科是被子植物中一个比较进化的类群，隶属于唇形类龙胆目，包含了 100 属、1700 多个种，可进一步划分为 7 个族，广泛分布在世界各地。其中，龙胆族有 22 属，约 980 种，构成了龙胆科的主体，是龙胆科内最常见、最美丽和最醒目的类群。大部分龙胆族植物花色绚丽多彩，是地球上最美丽的野花。它们的生境特殊，主要生长在世界各地的高山地区，备受人们青睐。龙胆花与报春花、杜鹃花一起，并称为“三大高原名花”。需要说明的是，本书只专注于讨论龙胆族植物，不涉及科内其他族群，在提到龙胆科、龙胆族或龙胆花时，若非特别说明，均特指龙胆族植物，敬请读者注意。

虽然龙胆科在被子植物中只是个中等大小的科，但绝对称得上是我们这个蓝色星球上最独特、最有范儿、最有风格的一个类群。

首先，龙胆花的范儿体现在它们的高，就是它们生长地的海拔高。龙胆花大多数种类集中在地表的高海拔地区，是离天空最近的植物类群，个别种类，像乌奴龙胆，分布海拔超过了 6000 米，达到了植物生长的上限，是地球上分布海拔较高的为数不多的几个种类之一。

龙胆花的第二个很有范儿的特征体现在它们的艳丽。龙胆花的色泽鲜艳夺目，五彩缤纷，尤其是龙胆花独特的蓝色，在“三大高原名花”中最为独特，令人迷醉；龙胆花开之时，灿然如织锦，靓丽如云霞，总能让人怦然心动。

龙胆花还有一个很有范儿的气质，在于它们的“出类拔萃”和与众不同。龙胆科植物的根内富含独特的环烯醚萜苷类（iridoid glycosides）苦味物质，使得许多龙胆种类都是苦口的良药；龙胆科植物的体内构造也比较独特，除了正常的维管束构造，还常常有内生韧皮部，比许多植物多生了一个韧皮部；龙胆科植物虽属于被子植物的双子叶植物类，却兼有某些典型的单子叶植物特征，比如平行的叶脉，有些种类容易与相差甚远的植物混淆，比如欧洲常见的黄龙胆（*Gentiana lutea*），在苗期仅有叶子的它常常让人与一起生长的单子叶植物藜芦混淆；另外，不像大部分植物在春夏开花，“三大高原名花”中，唯独龙胆的花多开放在深秋，秋风萧瑟之时，高处不胜寒，高原上百草枯黄，唯有龙胆花在秋风中盛开；龙胆花开在高山顶，颜色绚丽、霜重色愈浓，这是不是表明它们具有某种与众不同的英雄气质？

龙胆花鲜艳夺目，深受大众喜爱，也备受植物学家的重视。许多中外植物学家为龙胆科的研究做出了杰出贡献，发表了大量研究论文，出版了不同地区的植物志及个别属的专著。中国的龙胆种类是世界上最为丰富的，已故中国植物分类学家何廷农先生和刘尚武先生等人为龙胆科的研究做出了重要贡献，他们主编的《中国植物志》（龙胆科卷）中文版和英文版、英文版的《世界龙胆属专著》和《獐牙菜属和近缘属的世界性分类修订》是迄今为止最全面的龙胆科植物分类学文献，为我们奠定了认识龙胆科，尤其是龙胆族植物的基础。

此外，美国学者 Lena Struwe 和 Victor Albert 在 2002 年曾主编《龙胆科植物系统学与自然历史》一书，在科的水平上对全球龙胆科植物的系统分类和进化历史进行了总结和讨论。波兰学者 Jan Rybczynski 等人也曾于 2014 年和 2015 年主编过《龙胆科植物　第 1 卷：特征与生态学》和《龙胆科植物 第 2 卷：生物技术与应用》等著作，也在科的水平上讨论和总结了有关专题。另外，捷克学者 Josef Halda 还曾在 1997 年出版过龙胆属线描图专著，也同时讨论了龙胆属的属下分类。

这些研究论文和著作在不同层面对于深入理解龙胆科植物的系统分类和进化关系十分重要，对于龙胆科植物物种的鉴别与鉴定也十分重要。但是，对于物种特征的描述却非常抽象和“学术”，非植物分类学专业的人不但理解受限，更难体会到龙胆花的生动鲜活和美学属性。因此，本书力图用精美的照片来诠释和描述龙胆花。为了更好地理解专业的植物志书，使用更规范和统一的术语，在照片与文字文献间建立起正确的对应和关联，让我们先从形态特征和术语开始来认识龙胆科植物。

The Gentianaceae are a relatively advanced family of angiosperms belonging to the order Gentianales of Lamiids. The family includes about 100 genera and over 1700 species. It can be further divided into 7 tribes and is widely distributed around the world. Among them, 22 genera and about 980 species belong to tribe *Gentianeae*, which constitutes the main part of the Gentianaceae family. The plants of the tribe are the most common, beautiful, and prominent members within the family. Most species of tribe *Gentianeae* have colorful and showy flowers, making them the most beautiful wild flowers on the Earth. Their habitats are unique, mainly growing in high mountain areas around the world, and thus they are highly favored by people around the world. Gentians, along with primrose and rhododendrons, are well known as the three famous alpine flowers from the high plateaus. It should be noted that this book only focuses on discussing tribe *Gentianeae*, and does not deal with any other tribes within the family. So readers should be advised that when mentioning the gentian family, gentians, or gentian flowers in the book, unless otherwise specified, it is always referring to the plants of the tribe *Gentianeae*.

Although Gentianaceae is only a medium-sized family among angiosperms, it can definitely be called the most unique, special, and stylish group on our blue planet. The first feature of gentian flowers can be described as "high", that refers to the altitude at which they grow. Most of the gentian species are concentrated in high-altitude areas on the surface of the Earth, making them the closest plants to the sky. Some species, such as *Gentiana urnula*, grows at altitude over 6000 meters, reaching the upper limit of plant growth and is one of the few species with the highest altitude distribution on the Earth. The second distinctive feature of gentian flowers is their splendor. The color of gentian is dazzling and colorful, especially its unique blue color, which is the most unique and intoxicating among the three famous alpine flowers. When gentian flowers bloom, they shine like brocade and are as beautiful as rosy clouds and rainbows, always breathtaking. Gentian flowers also have a very stylish temperament, which lies in their outstanding and unique qualities. The roots of Gentianaceae plants are rich in unique bitter substances iridoid glycosides, making many species bitter thus good medicines. The anatomical structures of Gentianaceae plants are also quite unique. In addition to the normal vascular bundle structure with one phloem as in most plants, there is often an inner phloem in vascular bundles of gentians. Although plants in the Gentianaceae family belong to the dicotyledon group of angiosperms, they also possess certain typical characteristics of monocotyledons, such as the parallel leaf veins. Some gentian species are easily confused with other plants that differ greatly, such as the European common yellow gentian *Gentiana lutea*, which often has only leaves during the seedling stage and is often confused with the monocotyledonous plants *Veratrum* that often grows together. In addition, unlike most plants that bloom in spring and summer, among the three famous alpine flowers, only gentian flowers mostly bloom in late autumn. When the autumn wind is howling, the high altitude region is extremely cold, and the grasses on the plateau wither and turn yellow, whereas, gentian flowers start to bloom in the autumn wind. Gentian flowers bloom on high mountaintops, shining like brocade, and the frost and snows could only make their colors brighter and stronger, and thus more colorful and showy. Doesn't this indicate that they have a unique heroic temperament?

Gentian flowers are bright and eye-catching, loved by the public, and also highly valued by botanists. Many botanists from both China and abroad have made outstanding contributions to the study of Gentianaceae, published numerous research papers, monographs of some genera, and flora for different regions/countries. China is extremely rich in gentian species, the most abundant in the world. The late Chinese plant taxonomists Prof. Ting-Nong Ho and Prof. Shang-Wu Liu have made important contributions to the study of Gentianaceae. They led the compilation of both Chinese and English versions of the *Flora of China* (Gentianaceae). They had also published monographs in English, including "A worldwide monograph of *Gentiana*" and "A worldwide monograph of *Swertia* and its allies", the most comprehensive taxonomic literature on Gentianaceae plants to date. These studies have established the foundation for our understanding of Gentianaceae, especially for the tribe *Gentianeae*.

In addition, American scholars Lena Struwe and Victor Albert edited and published the book "Gentianaceae: Systematics and Natural History" in 2002, which summarized and reviewed the classification and evolutionary history of the Gentianaceae family global wide. The Polish scholar Jan Rybczynski et al. also edited and published works such as "The Gentianaceae. Volume 1: Characterization and ecology" and "The Gentianaceae, Volume 2: Biotechnology and Applications" in 2014 and 2015 respectively, and discussed the relevant topics at the family level. In addition, Czech scholar Josef Halda also published an illustration monograph on the genus *Gentiana* in 1997 and also discussed classification of the genus.

These research papers and works are crucial for a deeper understanding of systematics, classification and evolutionary relationships of Gentianaceae at different levels. They can be helpful for the identification of species of Gentianaceae as well. However, the description of species characteristics in those references is very abstract and "academic". For non-professionals in plant taxonomy, understanding of those references is limited. It is also difficult to experience the vividness and aesthetic properties of gentians. Therefore, this book strives to use exquisite photos to interpret and describe gentian flowers. In order to better understand professional botanical records, use more standardized and unified terminology, and establish correct correspondences and associations between photos and written literature, let's start by understanding the morphological characteristics and terminology of plants in the Gentianaceae family first.

龙胆族植物均为草本，一年生、两年生或多年生，个别种类为多年生一次性开花，即生长多年后一次性开花结果，然后个体即告死亡。

1.1 龙胆风格：形态特征与术语

龙胆科龙胆族植物均为草本，一年生、两年生或多年生，个别种类为多年生一次性开花，即生长多年后一次性开花结果，然后个体即告死亡，如美丽轮叶菜。

龙胆科植物（尤其是龙胆属植物）的幼苗其实挺有意思，但尚未得到足够的重视和研究。龙胆种子一般在成熟的当年萌发，萌发是子叶出土的方式，下胚轴伸长，子叶叶状；形成的幼苗初期会生成一个莲座叶丛，叶丛的大小和形态因种类而异；幼苗以这种形态越冬，来年春夏会继续生长，抽薹开花，来自幼苗的莲座叶丛得以保留或枯萎，消退或宿存。龙胆科的许多一年生植物也是以这种方式萌发生长的，所以严格地说是越年生植物或二年生植物。图 1-1 就是通常一年生的马耳山龙胆的幼苗与成年植株。

龙胆科植物的根细弱或粗壮，肉质或须状，通常有不定根，有时也伴生菌根（图 1-2）。

茎大多直立、斜升、平卧或匍匐，一些属包括蔓龙胆、双蝴蝶和翼萼蔓为缠绕藤本；多数多年生种类具地下根状茎或匍匐茎，有单轴和合轴两种分枝方式，有些种类具有明显的根头，也称为茎基，即茎的基部部分，局部膨大，节间密集，宿存有残存的叶柄（图 1-3）。

叶对生或轮生，个别种类互生，具明显的平行叶脉或弧形叶脉，无托叶；幼苗、不育的营养枝或有时主枝顶端的叶常密集地生长在一起，节间很短，形成的构造被称为莲座叶丛（图 1-4）。

花单生、簇生或排列成各式聚伞花序，通常为圆锥状聚伞花序，也被称为聚伞圆锥花序。花两性，极个别单性（如黄秦艽属），4 基数或 5 基数，偶尔 6~8 基数。花萼下部或基部联合成萼筒，上部分离为花萼裂片，也被称为萼齿；萼筒多为倒锥形，也有管状或钟状，花萼筒有时

图 1-1　马耳山龙胆的幼苗与成年植株

Fig. 1-1　Seedling and adult plants of *Gentiana maeulchanensis*

A. 幼苗下胚轴和莲座叶丛远轴面（底面）观
B. 幼苗生长状态与莲座叶丛近轴面（顶面）观
C. 成年花果期的植株
A. Hypocotyl and the rosette leaves abaxial view
B. Normal growing status and adaxial view of rosette leaves of seedling
C. Adult plant at flowering/fruiting stage

图 1-2　龙胆属植物的形态特征范例
Fig. 1-2　Morphological examples of gentians

沿一侧开裂；花萼裂片，即萼齿的数目和形态多变，是重要的识别和分类特征，有丝状、钻状、线形、匙形、三角形、卵形、心形等不同形状，直立、平展、反曲或卷曲等不同姿态（图 1-5）。花冠基部或下部也联合成管状、倒锥状、高脚杯状、漏斗状、钟状或辐状，偶尔近全裂（如黄龙胆）或有距（如花锚属）（图 1-6）；花冠上部分离成花冠裂片，在花蕾期花冠裂片通常按逆时针方向相互重叠旋转排列（从顶部向下俯视方向）；很多种类，尤其是龙胆属的花冠裂片之间有自花冠筒延伸而来、形态与花冠裂片不同的裂片状构造，被称为“褶”，褶在调节花冠的开合中起重要作用；花冠裂片和褶的形态、大小同样是非常重要的识别特征（图 1-7）。在花冠筒、花冠裂片基部或子房基部通常生有不同形状的蜜腺，有些种类无蜜腺（图 1-8）。雄蕊着生于花冠筒上与花冠裂片互生或偶尔着生在花冠裂片间的缺刻处；花药基部或背部着生，2 室。子房上位，通常 1 室或至少上部 1 室，有时胎座隔膜向内突入成 2 室；花柱通常较子房短，个别种类花柱细长；柱头顶生，通常 2 裂，有时盘状，在肋柱花属则沿两侧缝线下延。

果实通常为 2 瓣开裂的蒴果，偶尔呈浆果状；果实通常呈柱状、卵状、倒卵状或椭球状。果实成熟时，子房柄通常会明显伸长，有助于种子的有效散播（图 1-9）。

种子通常多数，较小，多为椭球形，偶尔四面体形，有些种类的种子周边具翅，有助于种子传播；翅的形态大小多变，可作为识别特征；种皮表面有复杂多变的附属结构，有海绵状、蜂窝状、手指状、片状等，是重要的识别特征，可借助显微镜观察。

A. 十字秦艽，单轴分枝、莲座叶丛
B. 大花龙胆，单轴分枝、莲座叶丛
C. 黄龙胆，肉质根及弧形叶脉
D. 小齿龙胆，合轴分枝，莲座叶丛
E. 草甸龙胆，圆锥状肉质根
F 川东龙胆，肉质须根

A. *Gentiana cruciata*, monopodial branching, rosette leaves
B. *Gentiana szechenyii*, monopodial branching, rosette leaves
C. *Gentiana lutea*, fleshy & stout roots, parallel veins
D. *Gentiana microdontha,* sympodial branching, rosette leaves
E. *Gentiana praticola*, cylindric fleshy roots
F. *Gentiana arethusae*, fleshy fibrous roots

图 1-3　长萼龙胆之根、根状茎及分枝构造细节
（中国青海省久治县，2023 年 9 月 11 日）

Fig. 1-3　Details of roots, rhizomes and stem branches of *Gentiana dolichocalyx*
(Jiuzhi, Qinghai, China. 11.09.2023)

图 1-4　龙胆科植物的形态特征范例

Fig. 1-4　Morphological examples of gentians

A. 矮獐芽菜，垫状植物
B. 东俄洛龙胆，平卧茎单轴分枝
C. 大钟花，直立茎及平行脉
D. 中国龙胆，匍匐茎合轴分枝
E. 翼萼蔓，缠绕茎
F. 马利筋龙胆，斜升茎弧形叶脉

A. *Swertia handeliana*, cushion-like plant
B. *Gentiana togolensis,* prostrate stems monopodial branching
C. *Megacodon stylophorus*, erect stem, parallel veins
D. *Gentiana chinensis*, stolon & sympodial branching
E. *Pterygocalyx volubilis*, twisting stems
F. *Gentiana asclepiadea*, ascending stems & parallel veins

图 1-5　龙胆科植物花萼形态特征范例

Fig. 1-5　Examples of calyx forms in gentians

A. 长梗秦艽，佛焰苞状萼筒
B. 黄山龙胆，萼裂片三角形
C. 福丽奇龙胆，卷曲的萼裂片
D. 弯叶龙胆，卵形萼裂片
E. 星萼龙胆，针状开展的萼裂片
F. 陕南龙胆，具翅的花萼筒
A. *Gentiana waltonii,* spathe-like calyx
B. *Gentiana delicata*, triangle calyx lobes
C. *Gentiana froelichii*, curly calyx lobes
D. *Gentiana curviphylla*, ovate calyx lobes
E. *Gentiana asterocalyx,* needle-like calyx lobes
F. *Gentiana piasezkii*, winged calyx tube

图 1-6　龙胆科植物花冠形态特征范例

Fig. 1-6　Examples of corolla forms in gentians

A. 大花肋柱花，辐状花冠，背着花药
B. 黄龙胆，全裂花冠，基着花药
C. 七叶龙胆，钟状花冠筒
D. 六叶龙胆，管状花冠筒
E. 蓝玉簪龙胆，漏斗状花冠筒
F. 高杯喉毛花，高脚杯状花冠筒

A. *Lomatogonium macranthum*, rotate corolla & dorsifixed anthers
B. *Gentiana lutea*, corolla divided to base & basifixed anthers
C. *Gentiana heptaphylla*, campanulate corolla tube
D. *Gentiana hexaphylla*, tubular corolla tube
E. *Gentiana veitchiorum*, funnelform corolla tube
F. *Comastoma traillianum*, salverform corolla tube

图 1-7 龙胆科植物花冠裂片和褶的形态特征范例

Fig. 1-7 Examples of corolla lobes and plicae in gentians

A. 耳褶，椭圆形裂片和耳状褶
B. 大花蔓龙胆，三角形裂片和齿状褶
C. 达乌里秦艽，倒卵形裂片和卵形褶
D. 福建蔓龙胆，卵形裂片和啮齿状褶
E. 华龙胆，卵形裂片和偏斜圆形褶
F. 开张龙胆，椭圆形裂片和 2 裂褶
G. 大花扁蕾，椭圆形裂片和流苏褶
H. 春龙胆，椭圆形裂片和舌状褶
I. 刺芒龙胆，三角形裂片和平截锯齿状褶
J. 无茎龙胆，三角形裂片和圆形褶
K. 丝瓣龙胆，平截裂片和丝状流苏褶
L. 高雅狭蕊，卵形裂片和啮齿状褶

A. *Kuepferia otophora*, elliptic lobe & auricular plicae
B. *Crawfurdia angustata*, triangular lobe & denticulate plicae
C. *Gentiana dahurica*, obovate lobes & ovate plicae
D. *Crawfurdia pricei*, ovate lobe & erose plicae
E. *Sinogentiana striata*, ovate lobe & oblique round plicae
F. *Gentiana aperta*, elliptic lobe & 2-clifted plicae
G. *Gentianopsis grandis*, elliptic lobe & fimbriate plicae
H. *Gentiana verna*, elliptic lobe & ligulate plicae
I. *Gentiana aristata*, triangular lobe & truncate denticulate plicae
J. *Gentiana acaulis*, triangular lobe & round plicae
K. *Gentiana exquisita*, truncate lobe & fimbriate plicae
L. *Metagentiana gentilis*, ovate lobe & erose plicae

A. 辐花，片状附属物
B. 德国假龙胆，流苏状毛
C. 高杯喉毛花，流苏状鳞片
D. 椭圆叶花锚，花冠裂片距
E. 观赏獐牙菜，线形蜜腺
F. 毛萼獐牙菜，鳞片状蜜腺
G. 獐牙菜，腺斑

A. *Lomatogoniopsis alpina*, membrane
B. *Gentianella germanica*, fimbriate hairs
C. *Comastoma trillianum*, fringed scales
D. *Halenia elliptica*, corolla spur
E. *Swertia decora*, linear nectaries
F. *Swertia hispidicalyx,* scale-like nectaries
G. *Swertia bimaculata*, nectary patches

图 1-8　龙胆科植物不同花冠附属物范例

Fig. 1-8　Examples of corolla appendant in gentians

图1-9　龙胆科植物果实形态特征范例

Fig. 1-9　Examples of fruit forms in gentians

A. 阿尔泰龙胆，卵状革质果皮
B. 膜果龙胆，柱状膜质果皮
C. 叶萼龙胆，卵状膜质果皮
D. 五岭龙胆，椭圆状革质果皮
E. 匙萼龙胆，长柱状膜质果皮
F. 昆明龙胆，具长果柄的卵状膜质果皮

A. *Gentiana grandiflora*, ovoid leathery carpel
B. *Gentiana hyalina*, cylindric membrane carpel
C. *Gentiana phyllocalyx*, ovoid membrane carpel
D. *Gentiana davidii*, ellipsoid leathery carpel
E. *Gentiana spathulisepala*, long cylindric membrane carpel
F. *Gentiana duclouxii,* ovate membrane carpel with a long stipe

花果形态特征在多数植物中是可靠的分类学特征，能充分体现不同植物类群间的系统演化关系。研究表明，有些龙胆科植物的花果特征是可靠的分类学特征，但有些种类的花果形态特征，比如獐牙菜属花的蜜腺位置及形态、龙胆属的成熟蒴果的形状等性状都与它们的功能高度相关，呈现出明显的多次起源和平行进化现象。因此，如果仅依靠这类个别特征来定义类群的话，有可能误导分类学处理。

接下来，让我们再看看龙胆科整体的分类概况。

The plants of tribe *Gentianeae* are herbs, annual, biennial or perennial. Some species are monocarpic, i.e. perennial plants die off after flowering and fruiting.

1.1 The Styles of Gentians: Morphology & Terminology

The plants of tribe *Gentianeae* (Gentianaceae) are herbs, annual, biennial or perennial. Some species such as *Frasera speciosa* are monocarpic, i.e. perennial plants die off after flowering and fruiting.

The seedlings of plants in the Gentianaceae family, especially those in the genus *Gentiana*, are quite interesting, but have not yet received sufficient attention and research. Seeds of gentians generally germinate in the same year of maturity. Germination style is usually epigeal, with hypocotyl elongation and cotyledons leafy. In the early stage of seedling formation, a leaf rosette is formed. The size and shape of the rosette leaves vary depending on species. The seedlings overwinter in this form and will continue to grow, sprout and bloom, in the following spring and summer. The rosette leaves from the seedlings can be withered or preserved, disappear or persistent. Some of the so-called annual plants of the Gentianaceae family also germinate and grow in this way, so strictly speaking, they are either perennial or biennial plants. Figure 1-1 shows the seedlings and adult plants of *Gentiana maeulchanensis*, commonly known as annual.

The roots are thin or robust, fleshy or fibrous, often adventitious and sometimes mycorrhizal (Fig. 1-2).

The stems mostly erect, ascending, creeping, prostrate, or twisting as vines in some genera including *Crawfurdia*, *Tripterospermum*, and *Pterygocalyx*. Most perennial species have rhizomes or stolon, with two branching modes: monopodial and sympodial. Some species with distinct caudex, also known as stem base, which are the enlarged, extremely shortened base of the stem with dense internodes and covered with remnant of persistent petioles (Fig. 1-3).

The leaves opposite or whorled, occasionally alternative, with parallel or curved veins, without stipules. Seedlings, sterile vegetative branches, or sometimes the leaves at the apex of the main branch often grow densely, forming a structure named rosette (Fig. 1-4).

The flowers are solitary, clustered, or arranged into various cymose inflorescences, usually a thyrse. Flowers are bisexual, rarely unisexual (such as *Veratrilla*), with 4- or 5-merous and occasionally 6-8-merous. The lower part or basal part of the calyx connate to form a calyx tube, while the upper part free as calyx lobes, also known as calyx teeth. The calyx tube is mostly obconic, tubular or campanulate, sometimes split along one side. The number and morphology of calyx lobes are variable and important diagnostic features, with different shapes such as filamentous, subulate, linear, spathulate, triangular, ovate, and cordate, as well as different postures such as upright, flat, reflexed, or curled (Fig. 1-5). The base or lower part of the corolla is also combined into a tubular, obconic, salverform, funnelform, campanulate tube, or radiate, or occasionally fully divided (e.g. *Gentiana lutea*), or with a spur (e.g. *Halenia*) (Fig. 1-6). The upper part of the corolla divided into corolla lobes, which usually overlap and rotate with each other counterclockwise in the bud stage (when seen from the top). Nearly all species in the genus *Gentiana* have corolla structures named plicae, which are extension segments of corolla tube alternative to the corolla lobes. Plicae may play important roles in regulating the opening and closing of the corolla. The shape and size of corolla lobes and plicae are also very important diagnostic features (Fig. 1-7). Nectary often present on corolla tube, at the base of corolla lobes, or at the base of ovary, usually with different shapes, and some species do not have nectary (Fig. 1-8). Stamens are inserted on the corolla tube and alternate with the corolla lobes, or occasionally at sinus between the corolla lobes. Anthers are basifixed or dorsifixed, 2-locular. The ovary is superior, usually 1-loculed or at least 1-loculed in the upper part, and sometimes the placental septum protrudes into 2-loculed. The style is usually shorter than the ovary, and sometimes styles are slender. The stigmata are terminal, usually 2-lobed, sometimes discoid, or sometimes styles absent and stigmata lobes decurrent along carpel sutures as in *Lomatogonium*.

The fruit is usually a 2-valved capsule, occasionally berrylike, usually cylindrical, ovoid, obovoid or ellipsoidal. When the fruit mature, the ovary stalk usually elongates significantly, which would help with dispersal of seeds (Fig. 1-9).

The seeds are usually numerous and small, often ellipsoidal, occasionally tetrahedral. Some seeds with wings of variable shapes and sizes around them, which helps with seed dispersal and are good diagnostic characters. The surface structure of the seedcoat is complex and variable, being sponge like, honeycomb like, finger like, flake like, etc., which are important diagnostic features, and can be observed by using microscopes only.

Morphological characters of flowers and fruits are usually reliable features for classification and can reflect truly evolutionary history and phylogenetic relationships of plants in many plant taxa. However, researches have revealed that in Gentianaceae while some characters of flowers and fruits are reliable some others such as nectary types, positions and their structures in the genus *Swertia*, the shapes of mature capsules in the genus *Gentiana*, are highly function-related features that had multiple origins and significant parallel evolution, and thus can mislead classification if it is based on single character of this type.

Next, let's take a look at the overall classification overview of the family Gentianaceae.

1.2 龙胆家族：龙胆科分类概要

自从 Gilg 于 1895 年定义了龙胆科的范畴和建立科下分类后，100 多年间几乎没有人在科的层面研究和修订龙胆科的分类。直到 2002 年，才有美国学者 Lena Struwe 和 Victor Albert 组织一些专家编写了《龙胆科植物系统学与自然历史》一书，在科的水平上对全球龙胆科植物的系统分类和进化历史进行了总结和讨论，对科下分类系统进行了修订。结合近代的研究结果，有些属，如睡菜属（*Menyanthes*）、荇菜属（*Nymphoides*）等被移出龙胆科，原来放在马钱科等其他科的个别类群则被归入龙胆科，如灰莉属（*Fagraea*）、星花莉属（*Anthocleista*）等。这样定义的龙胆科和科下分类系统基本体现了这个科的系统演化历史，得到了 DNA 分子系统演化研究结果的支持，因而也基本上得到了众多植物学家的认可。

按照这个比较受认可的分类系统，龙胆科可以划分为 7 个族，约 100 个属，共包括约 1700 个种。其中，龙胆族有 22 个属，包括约 980 种。关于龙胆科科下分类及其他龙胆族以外的族的分类超出了本书的范畴，就不做展开讨论。这里仅就这个分类系统的基本情况，进行简要总结和介绍（表 1-1）。

从表 1-1 中可以看出，从种类数的多寡来说，以温带和高山种类为主的龙胆族占了该科一半以上，但从属的数目来看，热带的属数目比较多，且以包含种类数比较少的小属为主，这个现象很可能是不同学者在定义属时，采用了不同的分类学标准所造成的。

龙胆族（*Gentianeae*）的范畴长期以来保持相对稳定，仅对特产于加那利群岛的绮皇花属（*Ixanthus*）一个单种属的归属有不同见解。早期的植物学家，如 Ho & Liu（1990）遵从更早的

Gilg（1895）的传统，把绮皇花属包括在龙胆族内，因为它的植物体宏观形态特征与龙胆族有些类群（如獐牙菜属）比较相似。但现代的系统发育分析，尤其是基于 DNA 分子序列的系统发育分析表明它与地中海地区的特有属绮莲花属（*Blackstonia*）的关系最为密切，应该归属于绮龙花族（*Chironieae*），而非龙胆族。本书在龙胆族外特意收录了绮皇花，以供读者参考比较，权当彩蛋。此外，龙胆族（*Gentianeae*）植物的外部形态和不同的基因序列分析均表明它包含了两个支系，即龙胆亚族（*Gentianinae*）和獐牙菜亚族（*Swertiinae*）。

龙胆族内有几个大属，如龙胆属（*Gentiana*）、假龙胆属（*Gentianella*）、獐牙菜属（*Swertia*）等，有关它们的属下分类长期以来充满争议。细分派把它们又分为若干个小属，归并派则把不同的小属合并成较庞杂的大属，当然还有各种折中派走不同的中间路线。这个分与合的争议仍在继续，本书不做讨论。随着研究的深入，尤其是随着大量基因组信息的引入，在不久的将来可能趋向明朗和统一。

关于龙胆属的分类学研究最为活跃，属下分类系统也争议极大。最狭义的龙胆属定义，尤其是以染色体数目为依据的分类系统，只涵盖了欧洲本土所产的与黄龙胆近缘的 5~7 个种，都是高大的多年生草本，而将其余的 350 多个一年生和多年生种类分别划分为 10 多个大小不等的属；而另一个极端——最广义的龙胆属定义，则把龙胆族的许多属于獐牙菜亚族的属，如假龙胆属（*Gentianella*）、扁蕾属（*Gentianopsis*）、喉毛花属（*Comastoma*）等都统统归入龙胆属。更多的学者采取不同的折中路线，将獐牙菜亚族的属都排除在外，将龙胆属的大部分种类都仍放在龙胆属内。同时，在龙胆属内采用属下等级“组”，来区别不同的属下小类群。Favre 等（2020）最新的基于多位点的 DNA 分子系统学研究基本支持这种折中分类系统，将龙胆属的约 354 个种划分为 13 个组，本书采用这个系统。关于这些组的划分、每个组所包括的种类数目、各组的主要分布区及本书涵盖的种类数等信息总结在表 1-2 中。

龙胆科的模式属是龙胆属，龙胆属的模式种是黄龙胆（*Gentiana lutea*）。但有意思的是，黄龙胆其实是最不像龙胆的，尤其是花的形态（图 1-10），而蓝玉簪龙胆（*Gentiana veitchiorum*）虽不是模式种，却也许更像典型意义上的龙胆（图 1-11）。

表 1-1　龙胆科植物分类概要

Table 1-1　Outlines of Classification on Gentianaceae Family

族 Tribe	亚族 Subtribe	属 Genus	中文属名 Chinese Name	大约种数 / 个 species number	主要分布区域 Major Distribution Region
Chironieae 绮龙花族	Chironiinae 绮龙花亚族	*Bisgoeppertia*		2	西印度群岛 West Indies
		Blackstonia	绮莲花属	4	地中海周边 Mediterranean
		Centaurium	百金花属	20	地中海周边 Mediterranean
		Chironia	绮龙花属	15	非洲 Africa
		Cicendia	黄绮草属	2	地中海周边 Mediterranean
		Eustoma	洋桔梗属	3	北美洲 North America
		Exaculum	小红管属	1	地中海周边 Mediterranean
		Geniostemon		5	墨西哥 Mexico
		Gyrandra		5	中美洲 Central America
		Ixanthus	绮皇花属	1	加那利群岛 Canary Islands
		Orphium	海玫木属	2	南非 South Africa
		Sabatia	绮玫花属	20	北美洲 North America
		Schenkia		5	地中海周边 Mediterranean
		Zeltnera	美金花属	25	美国 USA
		Zygostigma		1	南美洲 South America
	Canscorinae 穿心草亚族	*Duplipetala*		2	东南亚 Southeast Asia
		Canscora	穿心草属	9	热带非洲、亚洲和澳大利亚 Tropical Africa, Asia and Australia
		Cracosna		3	东南亚 Southeast Asia
		Hoppea		2	印度 India
		Microrphium		2	东南亚 Southeast Asia
		Phyllocyclus		5	东南亚 Southeast Asia
		Schinziella		1	热带非洲 Tropical Africa
	Coutoubeinae 克托白亚族	*Coutoubea*		5	南美洲 South America
		Deianira		5	热带美洲 Tropical America
		Schultesia		20	热带美洲 Tropical America
		Symphyllophyton		2	巴西 Brazil

（续表 continued）

族 Tribe	亚族 Subtribe	属 Genus	中文属名 Chinese Name	大约种数 / 个 species number	主要分布区域 Major Distribution Region
Chironieae 绮龙花族	Coutoubeinae 克托白亚族	*Xestaea*	微盏草属	1	尼加拉瓜 Nicaragua
Exaceae 藻百年族		*Cotylanthera*	杯药草属	4	印度洋盆地周边 Indian Ocean Basin
		Exacum	藻百年属	65	印度洋盆地周边 Indian Ocean Basin
		Gentianothamnus	龙胆木属	1	马达加斯加 Madagascar
		Klackenbergia		2	马达加斯加 Madagascar
		Ornichia		3	马达加斯加 Madagascar
		Sebaea	小黄管属	100	非洲 Africa
		Tachiadenus	啤酒树属	11	马达加斯加 Madagascar
Gentianeae 龙胆族	Gentianinae 龙胆亚族	*Crawfurdia*	蔓龙胆属	16	中国 - 喜马拉雅 Sino-Himalayan
		Gentiana	龙胆属	354	亚欧大陆及北美洲 Eurasia, North America
		Kuepferia	耳褶属	12	中国 - 喜马拉雅 Sino-Himalayan
		Metagentiana	狭蕊属	12	中国 - 东南亚 China, Southeast Asia
		Sinogentiana	华龙胆属	2	中国 China
		Tripterospermum	双蝴蝶属	25	东亚、东南亚 East Asia, Southeast Asia
	Swertiinae 獐牙菜亚族	*Bartonia*	巴脱泥属	3	北美洲 North America
		Comastoma	喉毛花属	15	亚欧大陆及北美洲 Eurasia, North America
		Frasera	轮叶菜属	15	北美洲 North America
		Gentianella	假龙胆属	256	亚欧大陆、北美洲、南美洲、澳大利亚、新西兰 Eurasia, North America, South America, Australia, New Zealand
		Gentianopsis	扁蕾属	24	亚欧大陆及北美洲 Eurasia, North America
		Halenia	花锚属	80	亚欧大陆、北美洲、南美洲 Eurasia, North & South Americas
		Jaeschkea	口药花属	3	喜马拉雅西部 West Himalayas
		Latouchea	匙叶草属	1	中国 China
		Lomatogoniopsis	辐花属	3	中国 China
		Lomatogonium	肋柱花属	20	亚欧大陆及北美洲 Eurasia, North America
		Megacodon	大钟花属	2	中国 - 喜马拉雅 Sino-Himalayan

（续表 continued）

族 Tribe	亚族 Subtribe	属 Genus	中文属名 Chinese Name	大约种数 / 个 species number	主要分布区域 Major Distribution Region
Gentianeae 龙胆族	Swertiinae 獐牙菜亚族	*Obolaria*	铜币草属	1	北美洲 North America
		Pterygocalyx	翼萼蔓属	1	东亚 East Asia
		Sinoswertia	异型花属	1	中国 China
		Swertia	獐牙菜属	135	亚欧大陆、北美洲及非洲 Eurasia, North America, Africa
		Veratrilla	黄秦艽属	2	中国 - 喜马拉雅 Sino-Himalayan
Helieae 荷莉族		*Adenolisianthus*		1	巴西 Brazil
		Aripuana		1	巴西 Brazil
		Calolisianthus		6	热带美洲 Tropical America
		Celiantha		3	热带美洲 Tropical America
		Chelonanthus		7	热带美洲 Tropical America
		Chorisepalum		5	委内瑞拉、圭亚那、苏里南 Venezuela, Guyana & Suriname
		Helia		2	巴西 Brazil
		Irlbachia		9	热带南美洲 Tropical South America
		Lagenanthus		1	哥伦比亚、委内瑞拉 Columbia & Venezuela
		Lehmanniella		2	哥伦比亚、秘鲁 Colombia & Peru
		Macrocarpaea		90	热带美洲 Tropical America
		Neblinantha		2	巴西、委内瑞拉 Brazil & Venezuela
		Prepusa		5	巴西 Brazil
		Purdieanthus		1	哥伦比亚、委内瑞拉 Columbia & Venezuela
		Roraimaea		2	巴西、委内瑞拉 Brazil & Venezuela
		Rogersonanthus		3	圭亚那、委内瑞拉 Guyana & Venezuela
		Senaea		1	巴西 Brazil
		Sipapoantha		1	巴西、委内瑞拉 Brazil & Venezuela
		Symbolanthus		30	热带美洲 Tropical America
		Tachia		10	热带南美洲 Tropical South America
		Tetrapollinia		1	热带南美洲 Tropical South America
		Wurdackanthus		2	圭亚那、委内瑞拉 Guyana & Venezuela
		Yanomamua	琴瑶草属	1	巴西 Brazil

（续表 continued）

族 Tribe	亚族 Subtribe	属 Genus	中文属名 Chinese Name	大约种数 / 个 species number	主要分布区域 Major Distribution Region
Helieae 荷莉族		*Zonanthus*		1	古巴 Cuba
Potalieae 坡塔莉族	Potaliinae 坡塔莉亚族	*Anthocleista*	星花莉属	14	热带非洲 Tropical Africa
		Fagraea	灰莉属	70	热带亚洲 Tropical Asia
		Potalia	风轮莉属	9	热带南美洲 Tropical South America
	Faroinae 法若尼亚族	*Congolanthus*		1	热带非洲 Tropical Africa
		Djaloniella		1	热带非洲 Tropical Africa
		Enicostema	玉塔草属	3	泛热带间断 pantropical disjunctive
		Faroa		19	热带非洲 Tropical Africa
		Karina		1	扎伊尔 Zaire
		Neurotheca		3	南美洲、热带非洲 South America, Tropical Africa
		Oreonesion		1	热带非洲 Tropical Africa
		Pycnosphaera		5	热带非洲 Tropical Africa
		Urogentias		1	坦桑尼亚 Tanzania
	Lisianthiinae 丽色花亚族	*Lisianthius*	丽色花属	30	热带美洲 Tropical America
Saccifolieae 勺叶木族		*Curtia*		11	圭亚那、乌拉圭 Guayana & Uruguay
		Hockinia		1	巴西 Brazil
		Saccifolium	勺叶木属	1	圭亚那 Guayana
		Tapeinostemon		7	热带南美洲 Tropical South America
		Voyriella		1	热带南美洲 Tropical South America
Voyrieae 微丽兰族		*Voyria*	微丽兰属	19	热带美洲、热带非洲 Tropical America, Tropical Africa

表 1-2　龙胆属属下分类系统、所含种数（含种下类群）及其分布区

Table 1-2　Infrageneric classification, species numbers (including intraspecific taxa) and distribution of the genus *Gentiana*

组 Section	总种数 / 个 Total Species numbers	分布 Distribution	本书包括种数 Species included in this book
盘柱组 *Calathianae*	15	欧洲、中亚、北美洲 Europe, Central Asia, North America	9
小龙胆组（广义） *Chondrophyllae s.l.*	182	亚欧大陆、北美洲、南美洲 Eurasia, North America, South America	73
大花组 *Ciminalis*	7	欧洲 Europe	6
秦艽组 *Cruciata*	22	亚欧大陆 Eurasia	19
高山龙胆组 *Frigida*	18	亚欧大陆、北美洲 Eurasia, North America	15
龙胆组 *Gentiana*	7	欧洲 Europe	6
匍茎组 *Isomeria*	10	亚欧大陆、北美洲 Eurasia, North America	9
多枝组 *Kudoa*	25	中国 - 喜马拉雅地区 Sino-Himalayas	17
微籽组 *Microsperma*	8	中国 - 喜马拉雅地区 Sino-Himalayas	7
单轴组 *Monopodiae*	15	中国 - 东南亚地区 Sino-Southeast Asia	14
叶萼组 *Phyllocalyx*	1	中国 - 喜马拉雅地区 Sino-Himalayas	1
龙胆草组 *Pneumonathe*	42	亚欧大陆、北美洲 Eurasia, North America	22
四数组 *Tetramerae*	2	中国西南部 Southwest China	2
总计 Total	354		200

欧洲常见的由林奈命名的黄龙胆是龙胆属的命名模式，但也是龙胆属内长得最不像龙胆的物种。

Gentiana lutea commonly seen in Europe and named by Linnaeus is the type of the genus *Gentiana*, while it is also the only species in the genus that the least resembles the rest species in the genus.

图 1–10　龙胆属的命名模式种：黄龙胆
（瑞士纳沙泰尔拉土恩）

Fig. 1-10　*Gentiana lutea*, the type species of the genus *Gentiana*
(La Tourne, Neuchatel, Switzerland. 15.06.2018)

图 1-11　蓝玉簪龙胆，虽然不是模式种，但其特征几乎可以代表龙胆花的全部特征（中国甘肃省玛曲县）

Fig. 1-11　*Gentiana veitchiorum*, although not a type, can present nearly all characters of a gentian (Maqu, Gansu, China. 10.09.2023)

中国青藏高原比较常见的蓝玉簪龙胆，长得就像比较典型的龙胆。

Gentiana veitchiorum, commonly seen in Qinghai-Tibet Plateau of China, is somewhat a more typical gentian.

Gentianaceae family can be divided into 7 tribes, with approximately 100 genera and approximately 1700 species.

1.2 The Gentianaceae Family: Outline of Classification

Since Gilg defined the scope of Gentianaceae family and the intrafamilial classification in 1895, almost no one has studied and revised the classification at the family level for over 100 years. It was not until 2002 that Lena Struwe and Victor Albert organized some experts and wrote the book *Gentianaceae: Systematics and Natural History*, which summarized and discussed the systematics and evolutionary history of Gentianaceae worldwide at the family level, and revised the classification of the family by combining multiple recent research results, some genera such as *Menyanthes* and *Nymphoides* were removed from Gentianaceae. Some other groups that were originally placed in Loganiaceae were moved into Gentianaceae, such as the genera *Fagraea*, *Anthocleista* and so on. This revised classification of Gentianaceae basically reflects the evolutionary history of the family, which was supported by the results of DNA molecular phylogenetic studies, and thus has been widely adopted by many botanists.

According to this well accepted classification, the Gentianaceae family can be divided into 7 tribes, with approximately 100 genera and approximately 1700 species. Among them, the tribe *Gentianeae* has 22 genera, approximately 980 species. The classification of the family and other tribes other than *Gentianeae* is beyond the scope of this book and will not be discussed further. Here is a brief summary and introduction of the basic situation of this classification, as shown in Table 1-1.

It can be seen from the table that in terms of the number of species, the tribe *Gentianeae*, mainly composed of temperate and alpine species, accounts for more than half of the family. However, in terms of the number of genera, there are more tropical genera, mostly small genera including relatively a few species, rather than bigger genera with lots of species. This phenomenon is likely caused by different botanists adopting different taxonomic criteria for classification.

The delimitation of the tribe *Gentianeae* has remained relatively stable for a long time. The only different opinions perhaps are about the attribution of the monotypic genus *Ixanthus*, which is endemic to the Canary Islands. Early botanists (such as Ho & Liu, 1990) followed earlier tradition of Gilg (1895) to include it within tribe *Gentianeae*, as its morphological characteristics were similar to some groups of the tribe (such as the genus *Swertia*). But modern phylogenetic analysis, especially based on DNA molecular sequences, suggested that it was closely related to Blackstonia from the Mediterranean areas, and should be better included within the tribe *Chironieae*. This book included this genus as a bonus for comparison and reference purpose only. In addition, the tribe *Gentianeae* consists of 2 lineages, as subtribe *Gentianinae* and subtribe *Swertiinae*, well supported by both gross morphology and DNA molecular data.

There are several big genera within tribe *Gentianeae*, such as *Gentiana*, *Gentianella*, *Swertia*, etc. Their infrageneric classifications have been controversial for long. The splitters divided them into many smaller genera, while the lumpers merged different small genera into larger heterogeneous genera. And various compromising schoolers take different middle options. The dispute may continue, however, controversial process of separation versus integration may become clearer and reach a more consensus conclusion with the deepening of researches, especially with the introduction of a large amount of genomic information in the coming decades.

The taxonomic study on the genus *Gentiana* was the most active in recent decades, and the infrageneric classification of the genus was also highly controversial. The narrowest definition of the genus *Gentiana*, based on chromosome numbers, included only 5-7 species that closely related to the yellow gentian found in Europe, while the remaining -350 species were divided into over 10 genera of varying sizes. On the opposite side, the most generalized definition of the genus *Gentiana* included many genera belonging to the subtribe *Swertiinae*, such as *Gentianella*, *Gentianopsis*, *Comatoma* etc., within the genus *Gentiana*. Most taxonomists adopted different compromise, excluding the genera of the subtribe *Swertiinae* and still including most related species within the genus *Gentiana*. At the same time, the infrageneric category "section" is used to distinguish different subgroups. The phylogentic studies done by Favre et al. (2020) based on multiple DNA segments basically supports this compromise classification, dividing the approximately 354 species of *Gentiana* into 13 sections. This book here adopts this system. The division of these sections, the number of species included in each section, the main distribution areas of each section, and the number of species covered in this book are summarized in Table 1-2.

The type genus of the Gentianaceae family is the genus *Gentiana*, and the type species of genus *Gentiana* is *Gentiana lutea*. However, it is interesting to note that *Gentiana lutea* is actually the least similar to other species in the genus in many aspects especially floral morphology (Fig. 1-10), while *Gentiana vetchiorum*, although not type, is perhaps more typical of the genus *Gentiana* (Fig. 1-11).

滇西龙胆 *Gentiana georgei*
中国青海省河南蒙古族自治县 Henan, Qinghai, China 09.09.2023

线叶龙胆 *Gentiana lawrencei* var. *farreri*
中国青海省河南蒙古族自治县 Henan, Qinghai, China. 09.09.2023.

2

龙胆地球

Gentians on the Earth

与地球上所有的植物一样，龙胆科植物在地球表面既不是随机分布的，也不是均匀分布的，而是以龙胆科植物特有的格局在水平和垂直两个维度上分布的。造成地表不同地区间龙胆科植物的多寡及种类组成差异的原因，既有不同地区的地质和气候因素，也有地理、土壤和其他动植物组成差异等因素，更有地质和气候演进历史及龙胆科植物本身演化历史方面的因素。我们观察到龙胆花当下的分布格局是以上历史和现存自然因素综合作用的结果。理解龙胆花的分布格局不仅有助于我们快速识别它们的种类、更好地理解不同物种间的进化关系，更有利于我们理解和明了这些生物多样性的形成历史和形成机制，以便我们更好地欣赏和保护它们。

龙胆科植物呈全球性分布，但多数种类集中在温带和亚热带的高山地区，是世界各大山系，包括喜马拉雅山、横断山、阿尔卑斯山、落基山和安第斯山等山脉中最为醒目的高山植物。其中，尤以喜马拉雅山和中国的横断山及青藏高原地区的种类最为丰富，居世界之首。下面，就让我们先来看看龙胆科植物的水平分布式样，即龙胆科植物在地球表面的经度和纬度两个方向上在不同大陆间的分布格局吧。

Like all plants on the Earth, plants of the Gentianaceae family are neither randomly nor uniformly distributed on the surface of the Earth. They are distributed in two dimensions on the Earth, horizontally and vertically, in a pattern specific to the Gentianaceae family. The reasons for the differences in the number and composition of species in the family between different regions on the surface include not only geological and climatic factors in different regions, but also geographical, soil, and other differences in biological compositions. There are also historical factors related to geological and climatic history of the regions, as well as the phylogenetic and evolution history of the Gentianaceae family itself. The current distribution pattern of gentians that we observed today is the result of a combination of historical and existing natural factors. Understanding the distribution pattern of gentians not only can help us quickly identify the species, but also can help us better understand the evolutionary relationships between different species. It is also beneficial for us to understand and comprehend the formation history and mechanisms of these biodiversity, so that we can better appreciate and protect them.

The Gentianaceae family are globally distributed, but most species are concentrated in high mountains in temperate and subtropical areas. They are the most prominent alpine plants in various mountain systems around the world, including the Himalayas, the Hengduan Mountains, the Alps, the Rocky Mountains, and the Andes. Among them, the Himalayas, the Hengduan Mountains and the Qinghai-Tibet Plateau region in China, are particularly rich in species, ranking the number one in the world. Now, let's first take a look at the horizontal distribution pattern of Gentianaceae family, that is, the distribution pattern of Gentianaceae plants between different continents in the longitude and latitude directions on the surface of the Earth.

2.1 多样性的热点地区

笼统地讲，龙胆科，甚至龙胆族都是全球性分布的。从亚洲、欧洲、非洲、南北美洲到大洋洲，各大洲的大陆和岛屿上都有它们的身影。然而，无论是物种类别的组成还是物种数目的多寡，不同类群（族、属或种）在世界各地的分布格局是截然不同的。如前面表 1-1 中所列出的主要分布区所体现的，粗略来说，绮龙花族（*Chironieae*）以环大西洋和地中海为中心分布；藻百年族（*Exaceae*）以环印度洋盆地为中心分布；荷莉族（*Helieae*）、坡塔莉族（*Potalieae*）、勺叶木族（*Saccifolieae*）和微丽兰族（*Voyrieae*）一起以泛热带间断分布或以中南美洲为主的新热带为中心分布；只有龙胆族（*Gentianeae*）是以北半球温带地区为中心，同时向南美洲和大洋洲扩散的方式分布的。

龙胆族（*Gentianeae*）作为一个整体，粗看呈全球分布，世界各大洲都有。从物种多样性的丰富程度来看，可以发现龙胆族（*Gentianeae*）的物种主要集中在世界各地的高山地区，包括亚洲东部的横断山、青藏高原和喜马拉雅山，非洲的乞力马扎罗山和埃塞俄比亚高原，欧洲的阿尔卑斯山和喀尔巴阡山，北美的落基山和阿帕拉契亚山，南美的安第斯山以及澳大利亚的大分水岭等。这些山区构成了龙胆族物种多样性的热点地区。

如果从种以上分类等级来考察这些热点地区的龙胆族特有类群分布情况，它们的情形各不相同。在龙胆族的 22 个属中，以横断山 – 喜马拉雅山为中心的东亚热点地区拥有最多的特有属，有 12 个，包括蔓龙胆属（*Crawfurdia*）、耳褶属（*Kuepferia*）、狭蕊属（*Metagentiana*）、华龙胆属（*Sinogentiana*）、双蝴蝶属（*Tripterospermum*）、口药花属（*Jaeschkea*）、匙叶草属（*Latouchea*）、辐花属（*Lomatogoniopsis*）、大钟花属（*Megacodon*）、翼萼蔓属（*Pterygocalyx*）、异型花属

（*Sinoswertia*）和黄秦艽属（*Veratrilla*）。北美洲东部地区拥有 3 个特有属，分别是巴脱泥属（*Bartonia*）、轮叶菜属（*Frasera*）和铜币草属（*Obolaria*）。世界其他地区包括欧洲均没有特有属。亚欧大陆和北美洲有 3 个共有属，分别是喉毛花属（*Comastoma*）、扁蕾属（*Gentianopsis*）和肋柱花属（*Lomatogonium*）。剩余的 4 个属，即龙胆属（*Gentiana*）、假龙胆属（*Gentianella*）、花锚属（*Halenia*）和獐牙菜属（*Swertia*），都是广布属，它们的分布超出了亚欧大陆和北美洲，进一步扩展到南美洲、澳大利亚或非洲。仔细观察这 4 个广布属的分布格局，尤其是结合各属的系统演化历史，可以发现很有意思的生物地理格局。

龙胆属的约 354 个种，可划分为 13 个组，如前面表 1-2 所示。种的多样性也是以横断山－喜马拉雅山为中心的东亚热点地区最高，有 5 个组为这一地区所特有，包括多枝组、微籽组、单轴组、叶萼组和四数组；欧洲的阿尔卑斯山和喀尔巴阡山热点地区次之，有 2 个特有组，即大花组和龙胆组；另有 1 个组即盘柱组的多样性中心在欧洲，仅个别种扩散到北美洲；世界上其他地区均没有特有组。有 1 个组即秦艽组的分布仅限于亚欧大陆；有 3 个组是亚洲、欧洲和北美洲所共有，即高山龙胆组、匍茎组和龙胆草组，其中前两者的多数种仅限于东亚地区，仅个别种扩散到欧洲和北美洲，但后者龙胆草组只有个别种分布在亚欧大陆，大多数种类则分别限于北美洲西部的落基山或东部的阿帕拉契亚山；龙胆属只有 1 个组即小龙胆组（广义）的个别种类进一步扩散到南美洲、北非和澳大利亚北部局部地区。有关该属物种的多样性分布及可能的起源和散布历史的简化图见图 2-1。Favre 等（2016）较深入地研究了龙胆属的起源和散布历史，根据 DNA 分子序列信息推测龙胆属可能是 3400 万年前左右起源于青藏高原的，随后不同的分支向欧洲、北美洲和其他地区散布。印度板块与亚欧大陆板块的碰撞造成的青藏高原隆升是龙胆属植物分化和散布的主要驱动力，他们据此也提出了该属各组比较详细的散布线路图假说。

假龙胆属（*Gentianella*）是龙胆族中种类数第二多的属，仅次于龙胆属，约有 256 种，包括花冠有流苏（约 27 种）、花冠无流苏（约 218 种）和双蜜腺类（约 11 种）三大类型。按照地区来划分的话，南美洲的安第斯山区种类最为丰富，有近 173 种；澳大利亚和新西兰，共有约 32 种；亚洲约 26 种；欧洲约 15 种（含小种）；北美洲约 12 种；北非西部仅有 1 种。其中，个别种是跨区域分布的。仅从种的数量看，南半球似乎在本属的起源与演化中更为重要，然而，基于多个 DNA 分子序列的系统演化分析却表明，该属中的原始类群仅分布于东亚和部分欧洲地区，至少花冠无流苏的大部分种类是从东亚起源，在第三纪末才传播到北美洲，再进一步传到南美洲、澳大利亚和新西兰。南美洲的大部分种类都是当地二次分化的结果，而澳大利亚和新西兰的种类也只是一次单一远距离传播事件带来的二次分化的结果（图 2-2）（von Hagen & Kadereit, 2001）。

花锚属（*Halenia*）的情况与假龙胆属有些类似，总种数 40~80 种。该属大部分种类分布在南美洲的安第斯山区。具体种数的统计有截然不同的观点。细分派认为有 60 多种；而归并派认为只有约 20 种，中美洲地区约有 15 种，北美洲有 1 种本土特有，东亚地区有 2 种，其中 1 种与

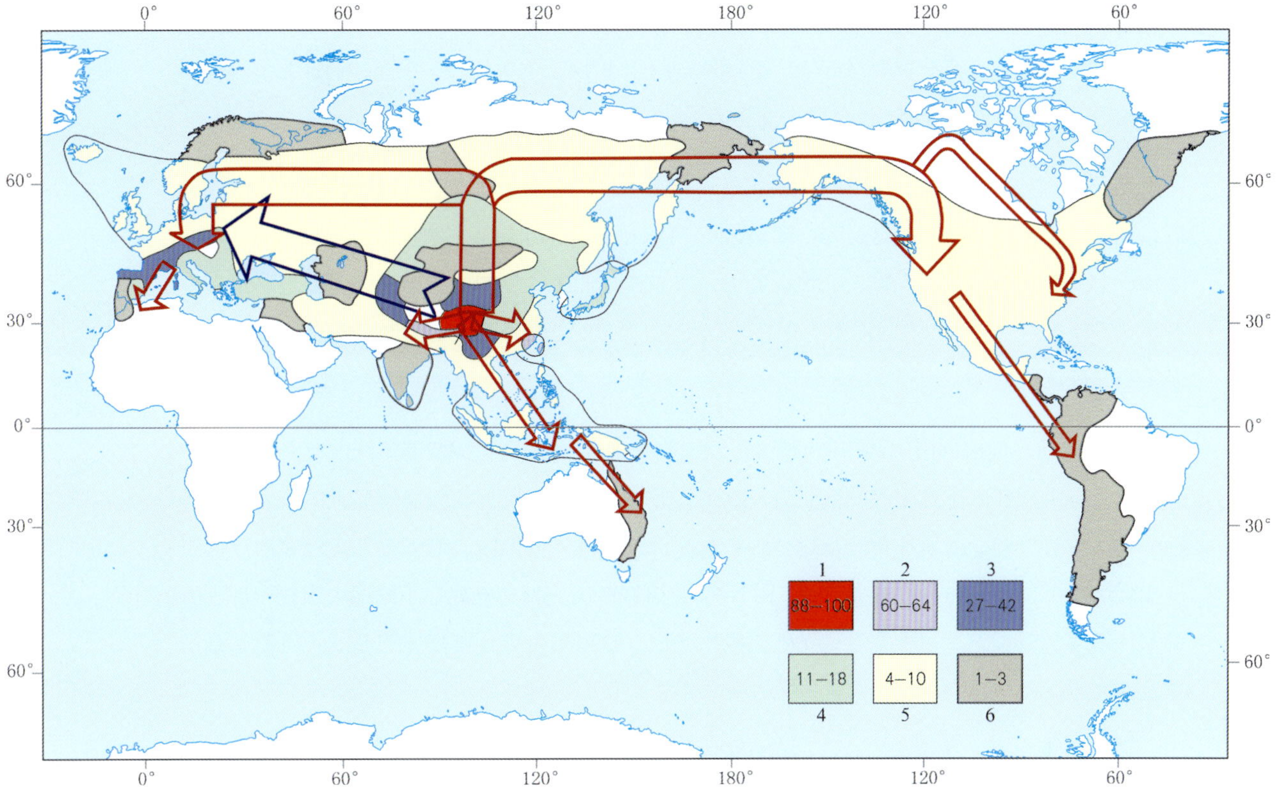

图 2-1　龙胆属植物物种多样性的地理分布及不同分布区间的起源与散布关系示意（根据 Ho & Liu, 1990 改编）

Fig. 2-1　Schematic diagram showing species numbers of different regions and the possible origin place and dispersal routes of those species in the genus *Gentiana* (Modified after Ho & Liu, 1990)

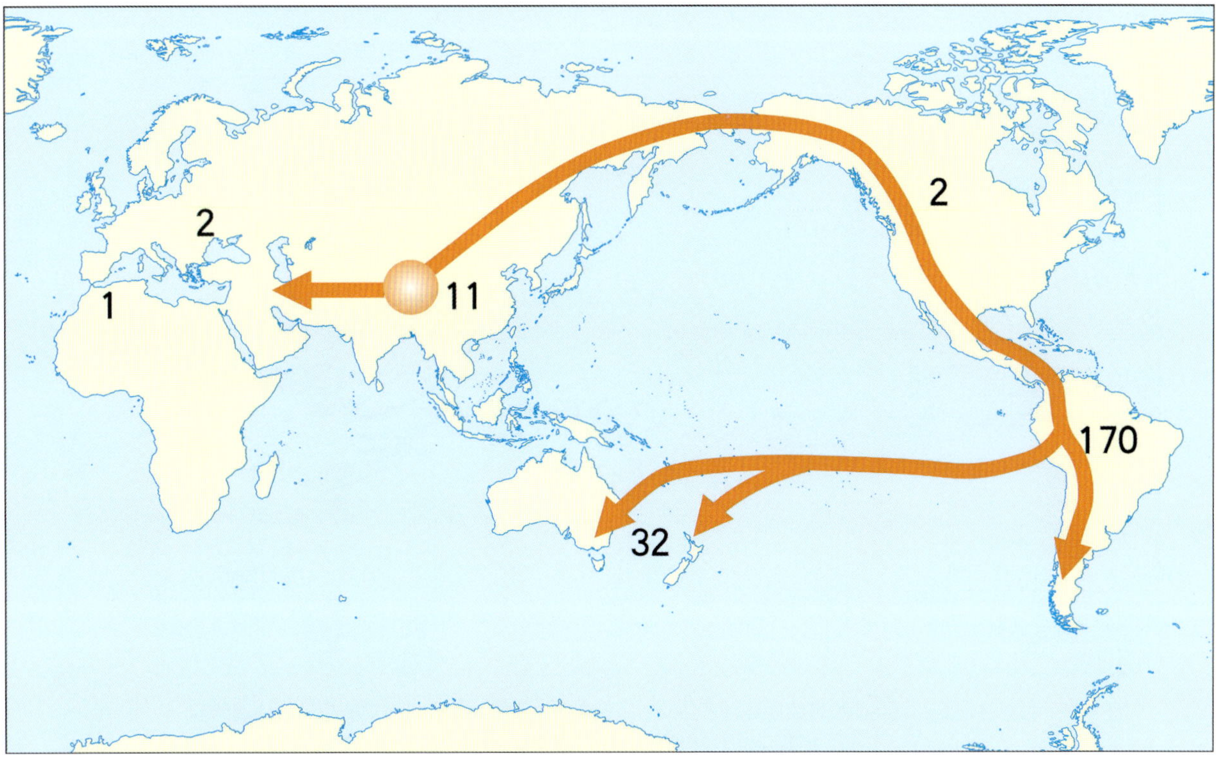

图 2-2　世界各地区假龙胆属花冠无流苏的种类数目及基于 DNA 分子序列的系统演化分析推测的起源与传播路径（根据 von Hagen & Kadereit, 2001 改编）

Fig. 2-2　Species numbers of different regions, origin center and dispersal routes deduced from phylogenetic analyses of DNA molecular data of the efimbrate species of the genus *Gentianella* (Modified after von Hagen & Kadereit, 2001)

欧洲和北美洲共有。基于 DNA 分子序列的系统演化分析揭示本属起源于东亚，通过北美洲传播到中美洲，然后通过 3 次独立的传播占据了南美洲（图 2-3）（von Hagen & Kadereit, 2003）。

獐牙菜属 (*Swertia*) 按照现在定义的范畴有约 137 种，分布于东亚、欧洲、北美洲和非洲，以东亚的中国 – 喜马拉雅地区最为丰富。然而，DNA 分子序列的系统演化分析初步表明，现在的獐牙菜属不是一个单系类群，如果接受獐牙菜亚族其他属地位的话，獐牙菜属就是一个高度并系或多系类群，它所包含的种类不是单一起源的，地理分布格局的形成历史也是十分复杂的，目前尚无确切结论。

无论如何，仅根据已确认的单系类群，龙胆属、假龙胆属和花锚属的系统演化分析已经可以确信，以中国西南部为中心的横断山 – 喜马拉雅山地区是龙胆族植物全球最大的起源、多样性分化和保存中心。龙胆族植物从这个中心起源，向亚洲其他地区、欧洲和北美洲辐射，再进一步向南美洲、澳大利亚和非洲散播（图 2-4）。这样的历史地理格局在不同属中重复出现，因此是比较可信的。这个过程在全球龙胆族植物演化、散播和今天分布格局的形成中可能起到了决定性作用。

下一节，我们来看看龙胆科植物的垂直分布式样，即龙胆科植物在地球表面的垂直方向即海拔高度上的自然分布式样，主要是考察温度梯度及其造就的植被类型对龙胆科物种分布的影响。与此同时，也来看看各种微环境，如湿地、流石滩等及土壤母质等要素对龙胆科植物分布的影响。

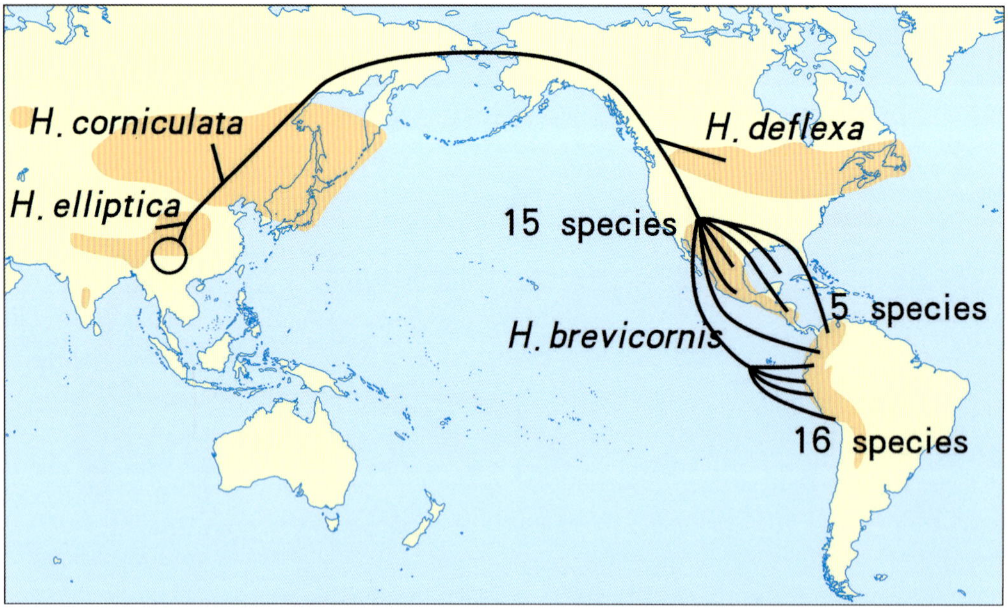

图 2-3　世界各地区花锚属的种类数目及基于 DNA 分子序列的系统演化分析推测的起源与传播路径（图片来源：von Hagen & Kadereit, 2003）

Fig. 2-3　Species numbers of different regions, origin center and dispersal routes deduced from phylogenetic analyses of DNA molecular data of the genus *Halenia*
(Source: von Hagen & Kadereit, 2003)

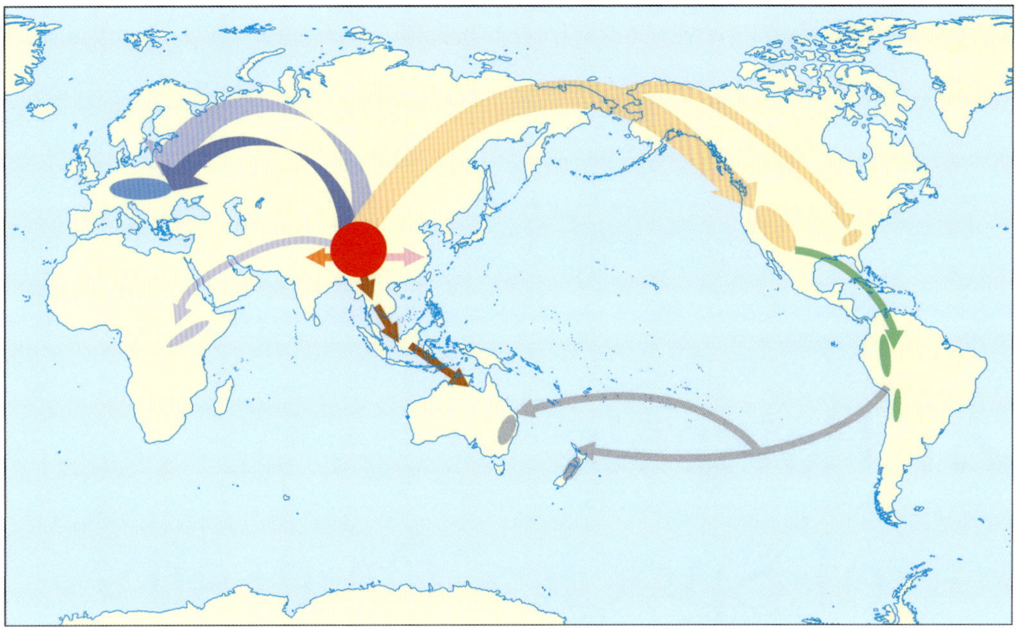

图 2-4　推测的龙胆族植物起源与散播过程的生物地理格局简图

Fig. 2-4　Hypothetic scheme of biogeographic pattern on the origin and dispersal routes of the tribe *Gentianeae*

Species of tribe *Gentianeae* are mainly concentrated in high mountain areas around the world.

2.1 Hotspots of Species Diversity

Generally speaking, the family Gentianaceae and even the tribe *Gentianeae* are cosmopolitan, with their presence on continents and islands in Asia, Europe, Africa, North and South Americas, and Oceania. However, regarding the composition of species and the number of species, the distribution patterns of different groups (tribes, or genera, or species) around the world are completely different. As shown in the main distribution areas listed in Table 1-1 above, roughly speaking, the species of tribe *Chironieae* are mostly distributed around the Atlantic Ocean and Mediterranean Sea. The tribe *Exaceae* is distributed around the Indian Ocean Basin as the center. The tribes *Helieae*, *Potalieae*, *Saccifolieae*, and *Voyrieae* are either disjunctively distributed in the pan tropical zone or centered around the new tropical zone dominated by Central and South America. Only the tribe *Gentianeae* is distributed in the temperate regions of the Northern Hemisphere, while also spreading to alpine and temperate areas in South America and Oceania.

Tribe *Gentianeae*, as a whole, appears subcosmopolitan, as its species grow on all continents of the world. From the richness of species diversity, it can be found that species of tribe *Gentianeae* are mainly concentrated in high mountain areas around the world, including the Hengduan Mountains and Himalayas in eastern Asia, the Kilimanjaro Mountains and Ethiopian Plateau in Africa, the Alps and Carpathians in Europe, the Rockies and Appalachians in North America, the Andes in South America, and the Great Divide in Australia. These mountainous areas constitute hot spots for the diversity of gentian species.

If we examine the distribution of the endemic taxa above species of the tribe *Gentianeae* in these hotspots, interesting patterns can be found. Among the 22 genera of the tribe *Gentianeae*, the East Asian hotspots centered around the Hengduan Mountains and Himalayas host the most endemic genera with 12 including *Crawfurdia, Kuepferia, Metagentiana, Sinogentiana, Tripteropermum, Jaeschkea, Latouchea, Lomatogoniopsis, Megacodon, Pterygocalyx, Sinoswertia* and *Veratrilla*. There are three endemic genera in North America, namely *Bartonia, Frasera,* and *Obolaria*. There is no endemic genus in any other hotspots. There are three genera widespread across Asia, Europe and North America, including *Comastoma*, *Gentianopsis*, and *Lomatogonium*. The remaining four genera, *Gentiana*, *Gentianella*, *Halenia*, and *Swertia*, are all widely distributed, with their distribution extending beyond Asia, Europe, and North America, to southern hemisphere in South America, Oceania and Africa. Very interesting biogeographic patterns can be found via close examination on distributions of species of those 4 widely spread genera, particularly when the phylogenetic history of the species is considered at the same time.

There are about 354 species in the genus *Gentiana*, which can be classified into 13 sections as shown in Table 1-2 above. The diversity of species is also the highest in the East Asian hotspot region centered around the Hengduan-Himalayas, with 5 sections limited to this region only, including the sect. *Kudoa*, sect. *Microsperma*, sect. *Monopodiae*, sect. *Phyllocalyx*, and sect. *Tetramerae*. The Alps and Carpathians hotspots in Europe take the second place, with two endemic sections, namely sect. *Ciminalis* and sect. *Gentiana*, and another sect. *Calathianae* with diversity center in Europe and few species further spreading to North America and the arctic area. There is no endemic section in any other regions of the world. One section, *Cruciata*, is limited to the Eurasian continent. Three sections are common across Asia, Europe, and North America, namely sect. *Frigida*, sect. *Isomeria*, and sect. *Pneumonanthe*, the former 2 sections with their most species limited to East Asia, and only a few spreading to Europe and North America, however, in the latter section, i.e. sect. *Pneumonanthe*, only a few species distributed in the Eurasian continent, and most species are limited to the Rocky Mountains or Appalachian Mountains in North America. Section *Chondrophyllae* s.l. is the only one with its species further spread to parts of South America, North Africa, and northern Australia. A simplified diagram of the diversity distribution and possible origin and distribution history of the species in the genus *Gentiana* is shown in Fig. 2-1. Favre and others have also conducted in-depth researches on the origin and distribution history of the genus *Gentiana*. Based on DNA molecular sequence data, it is speculated that the genus *Gentiana* may have originated from the Qinghai-Tibet Plateau around 34 million years ago, followed by different lineages spreading to Europe, North America, and other regions. The uplift of the Qinghai-Tibet Plateau caused by the collision between the Indian plate and the Eurasian plate was the major driving force for the differentiation and dispersals of gentians; Based on this result, a detailed scatter map of each lineage in the genus was also proposed (Favre et al., 2016).

Gentianella is the second most species-rich genus in the tribe, second only to *Gentiana*, with approximately 256 species, including three major types: fimbriate (approximately 27 species) and efimbriate (approximately 218 species), as well as binectariate (approximately 11 species). If looking at their distributions, the Andes region in South America has the most abundant species, with nearly 173 species, followed by Australia and New Zealand, with a total of about 32 species, about 26 species in Asia, about 15 species (including some so called microspecies) in Europe, about 12 species in North America, and only 1 species in western North Africa, with a few species distributed across different regions. From the perspective of species quantity alone, the Southern Hemisphere seems to be more important in the origin and evolution of this genus. However, phylogenetic analysis based on multiple DNA sequences showed that the primitive taxa in this genus are distributed only in East Asia and some parts of Europe. At least the majority of efimbriate species originated from East Asia and spread to North America at the end of the Tertiary, and further to South America, Australia, and New Zealand. Most species in South America are the results of local secondary diversification, while species in Australia and New Zealand are only the results of secondary differentiation caused by a single long distance dispersal event, as shown in Fig. 2-2 (von Hagen & Kadereit, 2001).

The situation of the genus *Halenia* is somewhat similar to that of the genus *Gentianella*. *Halenia* has about 40-80 species. The majority of species in this genus are distributed in the Andes Mountains in South America, and there are different opinions regarding species numbers. The splitters believes that there are more than 60 species, while the lumpers believes that there are only about 20 species. There are about 15 species in Central America, 1 species endemic to North America. There are 2 species in East Asia, with 1 species shared with Europe and North America. The phylogenetic analysis based on DNA sequences revealed that the genus originated in East Asia and spread to Central America through North America, and then colonized South America via three independent dispersal events, as shown in Fig. 2-3 (von Hagen & Kadereit, 2003).

According to the current definition, there are approximately 137 species in the genus *Swertia*, distributed in East Asia, Europe, North America, and Africa, with the Sino-Himalayan region in East Asia being the most abundant. However, the phylogenetic analysis of DNA sequences preliminarily indicated that the current genus *Swertia* is not a monophyletic group. If we accept the status of other genera in subtribe *Swertiinae*, the genus *Swertia* is a highly paraphyletic or polyphyletic group, and the species it included are not of a single origin. The formation history of the geographical distribution pattern of the species it included might be very complex, and there is currently no definite conclusion.

Nevertheless, based on phylogenetic analysis results of monophyletic taxa including genera *Gentiana*, *Gentianella* and *Halenia*, one can confidently conclude a pattern, i.e. origin in East Asia around the Henduan-Himalayas center followed by dispersals to other regions in Asia, to Europe and North America, and then from there further spreading to other continents including South America, Oceania, and Africa, as shown in a simplified scheme Fig. 2-4. Such a pattern, occurred repeatedly in different genera and thus highly confident, may have played critical roles in the evolution, dispersal and formation of distribution patterns in tribe *Gentianeae* of global wide.

Next, let us take a look at the vertical distribution pattern of Gentianaceae family, which refers to the natural distribution pattern of Gentianaceae plants on the surface of the Earth in the vertical direction at different altitudes. The main focus is to examine the influence of temperature gradients and the vegetation types they create on the distribution of Gentianaceae species. At the same time, let us also take a look at the impact of various microenvironments such as wetlands, screes and taluses, and mineral material compositions in soils on the distribution of species of family Gentianaceae.

具体到龙胆族，它的成员则比较集中地分布在不同海拔的山地生态环境中。

2.2 生态与生境

如果说水平分布更多地体现了一个类群宏观和漫长的历史性分布格局，垂直分布则更多地体现了物种分布的局部和微观属性。弄清楚物种分布的微观属性，有利于我们更好地理解不同物种的生存小环境，即生境。这不仅有利于我们在野外更容易发现具体物种，也有利于我们对物种的生物多样性实施更有效的保护。

从科的层面笼统来讲，可以说龙胆科植物出现在从荒漠到雨林、从岛屿到内陆、从海岸到地球上最高山峰的几乎所有类型的陆地生态环境中。但具体到龙胆族，它的成员则比较集中地分布在不同海拔的山地生态环境中。在地球表面的垂直方向上，按海拔从低到高，龙胆族植物分布在草原、山地阔叶林、山地针叶林、高山灌丛、高山草甸及冰缘流石滩的几乎所有地带性生态环境中，以及亚高山草甸和各类湿地等非地带性隐域生态环境中，只有极个别种类分布在平原农业带生态环境中，绝大多数种类比较集中地分布在高山灌丛和高山草甸地带。这 9 类生态环境或者说植被类型及其在海拔梯度上的位置关系见图 2-5。

在龙胆族内，不同属的生态偏好也有着比较明显的差异，比如缠绕藤本类群蔓龙胆属、双蝴蝶属和翼萼蔓属主要占据海拔较低的森林生态环境，而耳褶属、喉毛花属、假龙胆属、辐花属、肋柱花属等主要占据高海拔生境；龙胆属和獐牙菜属的生境跨度比较大，但前者主要偏好高海拔地区，后者分布的海拔相对更低一些。龙胆属（*Gentiana*）共包含约 354 种，在垂直分布方面，是高度差异化的。分布海拔最低的种类是北极附近的蓝绿龙胆（*Gentiana glauca*），它贴近海平面生长，最低海拔记录是海平面上 5 m，最高为 2300 m。分布海拔最高的龙胆是乌奴龙胆(*Gentiana urnula*)，它生长在海拔 3100 m 以上的地方，最高可达 6150 m，是有花植物中为数不多的可以分布在海拔 6000 m 以上的一种。通过对文献记载（主要是 Ho & Liu, 2001）的龙胆属植物的垂直分布做简单统计分析（图 2-6）可以看出，本属中多数种类分布在海拔 2000~5000 m 这个区间，对应的是亚高山草甸、高山灌丛和高山草甸等生态环境；海拔 2000 m 以下的种较少，因此龙胆属是典型的高山类群。

龙胆族不同属的生境偏好简要汇总见表 2-1。注意表 2-1 中海拔的低、中、高划分是人为的主观印象，并不是统计分析的结果。

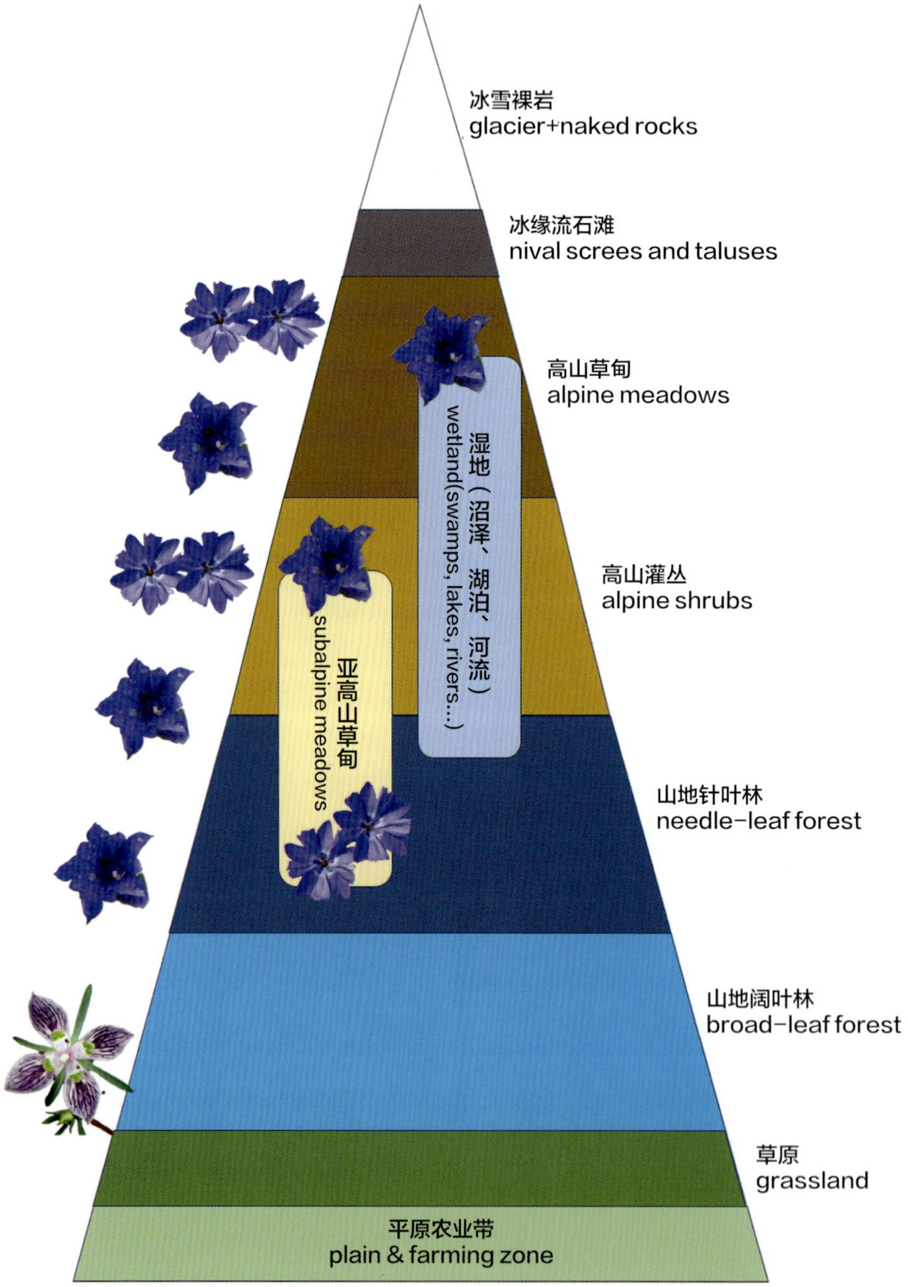

图 2-5　简化的龙胆科植物垂直分布生境类型示意
Fig. 2-5　Simplified schematic diagram of vertical distribution and habitats of gentians

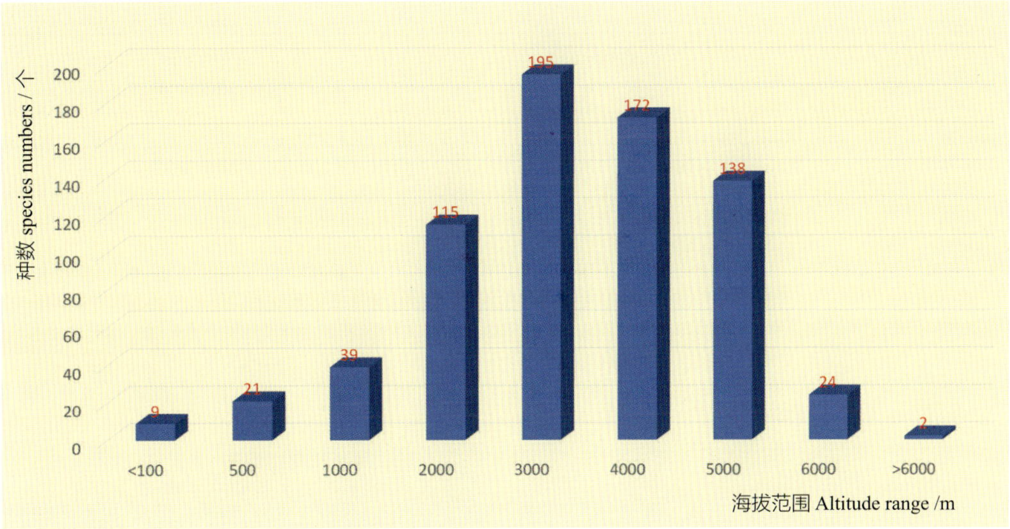

图 2-6　根据文献资料做出的龙胆属植物全部 354 种的垂直分布统计分析

Fig. 2-6　Distribution of altitude ranges of all 354 species of the genus *Gentiana* based on literature data

表 2-1　龙胆族不同属植物的典型生境

Table 2-1　Typical habitats of plants in different genera of tribe *Gentianeae*

属名 Genus	海拔* Altitude*	典型生境 Typical habitat
蔓龙胆属 *Crawfurdia*	中高海拔 medium to high	山地阔叶林到高山灌丛 broad-leaf forest to alpine meadows
龙胆属 *Gentiana*	高海拔为主，中海拔次之，少数低海拔 mostly high, some medium, occasionally low	草原到冰缘流石滩 Grassland to nival, screes and taluses
耳褶属 *Kuepferia*	高海拔 high	高山灌丛和高山草甸 alpine shrubs & alpine meadows
狭蕊属 *Metagentiana*	中海拔 medium	亚高山草甸、山地阔叶林和山地针叶林 Subalpine meadow, broad-leaf forest & needle-leaf forest
华龙胆属 *Sinogentiana*	中高海拔 medium to high	高山灌丛 alpine shrubs
双蝴蝶属 *Tripterospermum*	中海拔 medium	山地阔叶林 broad-leaf forest
巴脱泥属 *Bartonia*	中海拔 medium	山地阔叶林和山地针叶林 broad-leaf forest & needle-leaf forest
喉毛花属 *Comastoma*	高海拔 high	高山灌丛和高山草甸 alpine shrubs & alpine meadows
轮叶菜属 *Frasera*	中海拔 medium	山地阔叶林和山地针叶林 broad-leaf forest & needle-leaf forest
假龙胆属 *Gentianella*	中高海拔 medium to high	高山灌丛和高山草甸 alpine shrubs & alpine meadows

（续表 continued）

属名 Genus	海拔 * Altitude*	典型生境 Typical habitat
扁蕾属 *Gentianopsis*	中高海拔 medium to high	高山灌丛和高山草甸 alpine shrubs & alpine meadows
花锚属 *Halenia*	中高海拔 medium to high	山地阔叶林到高山灌丛 broad-leaf forest to alpine shrubs
口药花属 *Jaeschkea*	高海拔 high	高山灌丛和高山草甸 alpine shrubs & alpine meadows
匙叶草属 *Latouchea*	中海拔 medium	山地针叶林 needle-leaf forest
辐花属 *Lomatogoniopsis*	高海拔 high	高山灌丛和高山草甸 alpine shrubs & alpine meadows
肋柱花属 *Lomatogonium*	高海拔 high	高山灌丛和高山草甸 alpine shrubs & alpine meadows
大钟花属 *Megacodon*	中高海拔 medium to high	山地阔叶林和山地针叶林 broad-leaf forest & needle-leaf forest
铜币草属 *Obolaria*	中海拔 medium	山地阔叶林和山地针叶林 broad-leaf forest & needle-leaf forest
翼萼蔓属 *Pterygocalyx*	中海拔 medium	山地阔叶林 broad-leaf forest
异型花属 *Sinoswertia*	高海拔 high	高山灌丛和高山草甸 alpine shrubs & alpine meadows
獐牙菜属 *Swertia*	中高海拔，少数低海拔 medium to high, some low	山地阔叶林到高山灌丛 broad-leaf forest & needle-leaf forest
黄秦艽属 *Veratrilla*	高海拔 high	山地针叶林到高山灌丛 needle-leaf forest to alpine shrubs

* 低海拔，低于 1000 m; 中海拔，1000~3000 m; 高海拔，超过 3000 m。
low：<1000 m; medium：1000~3000 m; high：> 3000 m.

上述生境分析只是粗略概括，现实中的生态环境远比这些类别复杂多样。很多因素造就了生态环境的多样化，除了地带性因素和水体等大的环境因素外，很多相对微小的因素，比如山体的具体位置、地形、地貌、坡度、坡向、基岩等都能带来生态环境的差异。而且，不同生态环境还可以有大量的梯度变化和复杂的组合，进一步丰富了植物生长和分化的环境条件。以中国西南部为中心的横断山 – 喜马拉雅山地区，山大谷深，高山耸立，大江横流，地质、地理和生态环境极其复杂多样，造就了大量优越的适宜龙胆类植物生存、繁衍、变异和隔离的环境条件，使得这里成为龙胆族植物全球最大的起源、多样性分化和保存中心。

下面，简要介绍几种比较典型的龙胆族植物生境。

草原：草原是温带地区半干旱半湿润气候造就的以多年生草本植物为主的地带性植被，草原相对比较干燥，只有很少的龙胆科植物分布在草原生境，如欧洲的黄龙胆和亚洲北部地区的斜升秦艽等（图 2-7 A）。

森林生境：由阔叶林、混交林和针叶林构成的森林植被是地表最广泛和最重要的植被。然而，由于龙胆科植物多为比较矮小的草本，在树木比较茂盛的这类生境中它们并不占任何优势，只有在林窗和林缘等边缘性森林环境中才能见到一些龙胆的身影，比如缠绕藤本类的蔓龙胆属、双蝴蝶属、翼萼蔓属和有些龙胆属种类如十字秦艽等（图 2-7 B），只有少数比较特化的种类生长在林下，比如巴脱泥属、铜币草属、匙叶草属及龙胆属的个别种如笔龙胆等（图 2-7 C，图 2-8）。

亚高山草甸：亚高山草甸是非地带性的林间、溪流边或山间沟谷区域的草甸植被，常见于从阔叶林到高山灌丛间的不同地带，在针叶林带较为常见，往往是森林植被被人为或自然因素破坏后形成的以草本植物为主的碎片化植被。这里往往是报春花的天堂，龙胆科植物也相对比较丰富，如狭蕊属、龙胆属部分种、花锚属等（图 2-9）。

高山灌丛：高山灌丛就是林木线以上的灌木带，是杜鹃花的天堂，也是龙胆花和许多高山花卉的天堂，有大量的龙胆科植物分布在这个地带，如龙胆属部分种类，耳褶属、华龙胆属、肋柱花属、大钟花属、獐牙菜属和黄秦艽属等（图 2-10、图 2-11）。

高山草甸：高山草甸是高山地区没有任何木本植物生长的地带，是由很低矮的纯草本植物形成的植被。根据物种组成和水分条件不同可细分为高山冻原（苔藓和地衣类异常丰富）、湿生性草甸（季节性湿地、河边等小环境）、沼泽化草甸等类型。这里是龙胆花的另一个天堂和主场，有大量种类分布。另外，许多其他属的种类也常见，包括喉毛花属、假龙胆属、辐花属、肋柱花属等（图 2-12 至图 2-16）。

高山流石滩：高山流石滩是极高山区雪线附近的岩石碎屑带，是独特的小生境，只有一些适应极端严寒的植物才能生长。这里的龙胆种类不多，但却极具特色，最有名的就是乌奴龙胆，只有它紧密排列的莲座叶、发达的匍匐茎和强大的根系才能适应这里的极寒天气（图 2-17）。

特殊微生境：除了上述比较典型的生境外，有些龙胆科植物还选择了一些非常特化的小环境，如莲座獐牙菜（*Swertia rosularis*），只生长在中国四川峨眉山有限的几处滴水的岩壁上（图 2-7 D）。此外，有些龙胆属植物还对土壤的成土母质有较强的选择性，如欧洲常见的无茎龙胆（*Gentiana acaulis*）就是比较典型的嫌钙种（避钙种），只生长在硅质基岩的地方，而与它近缘的坷路龙胆（*Gentiana clusii*）却是典型的喜钙植物，只生长在高山以钙质岩为基质的石质山坡草地，二者从不会越界。

决定不同生境中龙胆族植物生存、生长与分布的重要因素包括温度、水分、光照、海拔和生物群落（微生物、植物群落组成、捕食者、传粉者等）。海拔是个综合因子，同时影响着其他物理因素（温度、光照、水分、风力）和生物因素（群落组成、传粉者等）。根据多年野外

图 2-7　龙胆族植物典型生境范例

Fig. 2-7　Examples of typical habitats of gentians

A. 低山草原，黄龙胆
B. 森林生境（林缘），十字秦艽
C. 森林生境（林下），笔龙胆
D. 特殊微生境：滴水岩壁，莲座獐牙菜

A. Low grassland, *Gentiana lutea*
B. Forest (forest margin), *Gentiana cruciata*
C. Forest (forest understory), *Gentiana zollingeii*
D. Special niche: dripping rocks, *Swertia rosularis*

的随机观察，龙胆属多生长在气候凉爽、土壤水分充足、地形开阔、阳光充沛的环境中，它们的生态适应性偏好可以简单地总结为耐寒怕热、喜水怕旱、喜光怕阴，只有高海拔地区才可以同时满足这些条件，图 2-18 展示的就是一个比较典型的场景。

图 2-8　龙胆族植物典型生境范例：森林生境（林窗），扭果柄龙胆 + 黄秦艽，中国云南省大理白族自治州

Fig. 2-8　Examples of typical habitats of gentians: Forest habit (gaps); *Gentiana harrowiana* + *Veratrilla baillonii*; Dali, Yunnan, China. 23.09.2023

图 2-9　龙胆族植物典型生境范例：亚高山草甸，椭圆叶花锚，中国云南省丽江市
注：这是 1990 年丽江甘海子的原初景观，如今这里已经修建成玉龙雪山旅游接待站和缆车站了。

Fig. 2-9　Examples of typical habitats of gentians: subalpine meadows; *Halenia elliptica*; Lijiang, Yunnan, China. 20.09.1990

图 2-10　龙胆族植物典型生境范例：高山灌丛，三歧龙胆，中国四川省康定市

Fig. 2-10　Examples of typical habitats of gentians: Alpine shrubs; *Gentiana trichotoma*; Kangding, Sichuan, China. 13.09.2023

图 2–11　龙胆族植物典型生境范例：高山灌丛，丝柱龙胆，中国西藏自治区错那市

Fig. 2-11　Examples of typical habitats of gentians: alpine shrubs; *Gentiana filistyla*; Cuona, Xizang, China. 14.08.2006

图 2–12　龙胆族植物典型生境范例：高山草甸，泽库秦艽（左）+ 蓝玉簪龙胆（右），中国青海省泽库县

Fig. 2-12　Examples of typical habitats of gentians: alpine meadows; *Gentiana zekuensis* (left) + *Gentiana veitchiorum* (right); Zeku, Qinghai, China. 09.09.2023

图 2-13　龙胆族植物典型生境范例：高山草甸，蓝玉簪龙胆（蓝色花）+ 大花龙胆（白色花），中国青海省泽库县城近郊

Fig. 2-13　Examples of typical habitats of gentians: alpine meadows; *Gentiana veitchiorum* (blue flowers) + *Gentiana szechenyii* (white flowers); Suburb of Zeku town, Qinghai, China. 09.09.2023

图 2-14　龙胆族植物典型生境范例：沼泽化草甸
注：这里有线叶龙胆、青藏龙胆、蓝白龙胆等数种龙胆生长，中国西藏自治区当雄县

Fig. 2-14　Examples of typical habitats of gentians: Swamp meadows
Note: several gentians including *Gentiana futtereri*, *Gentiana leucomelaena*, and *Gentiana lawrencei* var. *farreri* grow there; Dangxiong, Xizang, China. 19.09.2003

图 2-15　龙胆族植物典型生境范例：湿生草甸，线叶龙胆，中国青海省河南蒙古族自治县

Fig. 2-15　Examples of typical habitats of gentians: damp meadows; *Gentiana lawrencei* var. *farreri*; Zeku, Qinghai, China. 09.09.2023

图 2-16　龙胆族植物典型生境范例：高山冻原

Fig. 2-16　Examples of typical habitats of gentians: alpine tundra

列入联合国教科文组织世界遗产名录的瑞士著名冰川阿莱奇冰川，是阿尔卑斯山区最大的冰川。它拥有难以置信的广阔和无法言喻的美丽。冰川一直从艾格峰、僧侣峰和少女峰的北坡向瓦莱州延伸。从冰川下来的水流则通过原始的马萨峡谷（Massa Gorge）流入罗纳河。这个地区生长着许多珍稀植物，包括照片右下角显示的美丽的无茎巴伐利亚龙胆（*Gentiana bavarica* var. *subacaulis*）。

Aletsch - the Largest Alpine Glacier. The Aletsch region is part of a Unesco World Heritage Site. Incredibly immense, incredibly beautiful: the Large Aletsch Glacier is the Alps' mightiest ice flow. The ice extends from the northern slopes of Eiger, Mönch and Jungfrau down towards the Valais. The water flows through the untamed Massa Gorge into the Rhone. The entire region is a habitat for many rare plant species, including *Gentiana bavarica* var. *subacaulis* shown on the lower right corner of the picture. 17.07.2005.

图 2-17　龙胆族植物典型生境范例：高山流石滩，乌奴龙胆，中国西藏自治区浪卡子县

Fig. 2-17　Examples of typical habitats of gentians: alpine screes and taluses; *Gentiana urnula*; Langkazi, Xizang, China. 12.09.1992

图 2-18　龙胆属植物偏好的小生境范例：开阔向阳、水分充足的高山灌丛边缘，钟花龙胆

Fig. 2-18　Example of favorite microhabitat of gentians: open, sunny, and moist margins of bushes, *Gentiana nanobella*

As far as the tribe *Gentianeae* is concerned, its members are highly concentrated in mountainous environments of different elevations.

2.2 Ecology & Habitat

If horizontal distribution reflects more of the macro and long-term historical distribution pattern of a group, vertical distribution reflects more of the local and micro attributes of species distribution. Understanding the micro-properties of species distribution is beneficial for us to better understand the living environment, i.e. habitat, of different species, which could enable us to find a species more easily in the wild nature, and help us to implement more effective protection of the biodiversity of a specific species.

Looking from the family level, the plants of Gentianaceae family appear in all kinds of ecological habitats on lands of the earth, from desert to rainforest, from island to continent, and from seashores to the highest mountains. As far as the tribe *Gentianeae* is concerned, its members are highly concentrated in mountainous environments of different elevations. In the vertical direction of the surface of the Earth, from low to high altitude, the plants of gentian family are distributed in almost all zonal ecological environments such as grasslands, mountain broad-leaved forests, mountain coniferous (needle-leaf) forests, alpine shrubs, alpine meadows, and screes and taluses at nival zone, as well as in non-zonal ecological environments such as subalpine meadows and various wet lands. Only a very small number of species are distributed in low altitude plain and farmland ecological environments. The vast majority of species are concentrated in alpine shrubs and alpine meadows. These 9 types of ecological environments or vegetation types and their relative locations along different altitudes are illustrated in Fig. 2-5.

Within tribe *Gentianeae*, there are also significant differences in ecological preferences among different genera. For example, those genera with twisting vine stems, *Crawfurdia*, *Tripterospermum* and *Pterygocalyx*, mainly occupy forest ecological environments at lower altitudes, while the genera *Kuepferia*, *Comastoma*, *Gentianella*, *Lomatogoniopsis* and *Lomatogonium* mainly occupy high-altitude habitats. The habitat span of *Gentiana* and *Swertia* is relatively large, but the former mainly prefers high-altitude areas, while the latter is distributed at relatively lower altitudes. *Gentiana* contains approximately 354 species and is highly differentiated in vertical distribution. The species with the lowest altitude distribution is *Gentiana glauca* growing in the Arctic region. It grows near the sea level, with the lowest altitude recorded at 5 meters above sea level and the highest at about 2300 meters. The gentian with the highest growth altitude is *Gentiana urnula*, which grows above an altitude of 3100 meters and can reach up to 6150 meters. It is one of the few flowering plants with a distribution of over 6000 meters. A simple statistical analysis of the vertical distribution of the species in the genus *Gentiana* recorded in literature (mainly Ho & Liu, 2001) is shown in Fig. 2-6. It can be seen from this diagram that most species in this genus are distributed in the altitude range between 2000 and 5000 meters above the sea, corresponding mostly to ecological environments such as subalpine meadows, alpine shrubs, and alpine meadows. There are fewer species below an altitude of 2000 meters, so *Gentiana* is a typical alpine group.

A brief summary of the typical habitats of plants in different genera of tribe *Gentianeae* is shown in Table 2-1. Note that the division of low, medium, and high altitudes in the table is a subjective impression, rather than a result of statistical analysis.

In fact, the table above provides only a rough summary of typical habitats. In reality, habitats are more complex and diverse, such as the differences caused by the specific location, terrain, topography, slope, bedrock, etc. of mountains. Secondly, each type of habitat also has a wide range of gradient changes and complex combinations of multiple factors, diversified the conditions beneficial to plant growth and differentiation. The Hengduan-Himalayan region, centered in southwestern China, has large mountains, deep valleys, towering mountains, and rivers flowing across it. The geological, geographical, and ecological environment is extremely complex and diverse, creating a large number of rich and superior environmental conditions suitable for the survival, preservation, variation, and isolation of gentian plants, making it the largest center for the origin, diversification, and preservation of biodiversity of gentians in the world.

Below, we will briefly introduce several typical habitats of plants in the tribe *Gentianeae*.

Grassland: Grassland is a zonal vegetation mainly composed of perennial herbaceous plants in the semi-arid and semi-humid climate in temperate regions. Grassland is relatively dry, and a few species of Gentianaceae family can grow in grassland habitats, such as *Gentiana lutea* in Europe and *Gentiana decumbens* in northern Asia (Fig. 2-7 A).

Forest habitat: The forest vegetation composed of broad-leaved forests, mixed forests, and coniferous (needle-leaf) forests is the most extensive and important vegetation on the surface of the Earth. However, due to the fact that most plants in the Gentianaceae family are relatively short herbaceous, they do not have any advantage in forest habitats dominated by lush trees. Only in marginal forest environments such as gaps and edges can some gentians be seen, such as *Crawfurdia, Tripterospermum, Pterygocalyx,* and some species of the genus *Gentiana*, such as *Gentiana cruciata* (Fig. 2-7 B). Only a few specialized species grow under the forest. For example, the genera *Bartonia, Obolaria, Latouchea*, and a few species of *Gentiana* like *Gentiana zollingeri*, and so on (Fig. 2-7 C, Fig. 2-8).

Subalpine Meadows: Subalpine meadows are non-zonal meadows vegetation located in forest, along stream edge, or in mountain valleys. Subalpine meadows are commonly found in different areas from broad-leaved forests to alpine shrubs, and are more common in coniferous forest belts, often formed after deforestation due to human or natural factors, and thus often fragmented vegetation mainly composed of herbaceous species. This is often a paradise for primroses, and gentian species are also relatively rich in this habitat. Species of *Metagentiana, Halenia* and quite some species of the genus *Gentiana* are seen there (Fig. 2-9).

Alpine shrubs: Alpine shrubs are the shrub belt above the tree line, which is a paradise for azaleas, gentians and many other alpine flowers. There are a large number of gentian species distributed in this habitat, such as species of *Gentiana*, *Kuepferia*, *Sinogentiana*, *Lomatogonium*, *Megacodon*, *Swertia*, and *Veratrilla* (Fig. 2-10, Fig. 2-11).

Alpine Meadows: Alpine meadows are areas in high mountain areas that do not have any woody plant. The vegetation is formed by very low and herbaceous species only. According to species composition and water conditions, alpine meadows can be further subdivided into some subtypes, such as alpine tundra (extremely rich in moss and lichens), moist meadows (seasonal wetlands, riverside), swamp meadows, and other subtypes. This is another paradise and home ground for many *Gentiana* species, with a large number of species distributed, in addition, species of *Comastoma, Gentianella, Lomatogoniopsis, Lomatogonium* etc. are also common (Fig. 2-12 to Fig. 2-16).

Nival screes and taluses: Nival screes and taluses are a special microenviroment to the nival line, the upper limit plant can grow in extremely high mountain areas. This zone is covered by lots of debris of decayed rocks, and only some plants that adapt to extreme cold climate can grow. There are not many species of gentian here, but they are very unique ones. The most well known one is *Gentiana urnula*, which can only adapt to the extremely cold weather due to its tightly arranged rosette leaves, well developed stolon, and strong roots (Fig. 2-17).

Special Microhabitats: In addition to the typical habitats mentioned above, some Gentianaceae plants also choose some very specialized small environments, such as *Swertia rosularis*, which only grows on a few dripping rock walls in the Emei Mountains of Sichuan, China (Fig. 2-7: D). Some species of the genus *Gentiana* also have strong selectivity for soil forming rocks. For example, the commonly seen *Gentiana acaulis* in Europe is a typical calcifuge (calcium avoiding) species, which only grows in areas with siliceous bedrocks, while the closely related *Gentiana clusii* is a typical calcicole (calcium loving) species, which only grows in rocky mountain slopes and grasslands with calcareous bedrocks, and the two species never cross their boundaries.

The most important factors that affect the survival, growth, and distribution of gentians in different habitats include temperature, water, light, altitude, and biological communities (microorganisms, community composition, predators, pollinators, etc.). Altitude is a comprehensive factor that simultaneously affects other physical factors (temperature, light, water, wind power) and biological factors (community composition, pollinators etc.).

Based on our random observations during field expeditions, the favorite habitats of gentians can be summarized as cool weather, abundant soil water, open field with plentiful sunshine. The general ecological preference of gentians can be simplified as cold resistant, heat sensitive, strong hydrophilous and drought sensitive (between hygrophyte to mesophyte), strong photophilous and shady sensitive. Alpine environment is then a good fit. A typical scenario is shown in Fig. 2-18.

线叶龙胆 *Gentiana lawrencei* var. *farreri*; 中国甘肃省卓尼县
Zhuoni, Gansu, China. 07.09.2023

川东龙胆 *Gentiana arethusae*

中国陕西省太白山 Taibaishan Mt., Shaanxi, China. 02.09.2023

3

龙胆之美

Beauty of Gentians

作为三大高原名花之一，龙胆花的美丽在世界各地都是毫无争议的，在亚洲、欧洲、美洲皆是如此。无论是与龙胆花初次相遇，还是老友重逢，一个人看见龙胆花的瞬间，总会唤起他最原始的审美本能，激发对它的挚爱，从而发出由衷的赞叹。龙胆花开时，远观灿烂若云霞，华丽壮阔，近察鲜丽如孩提，珠圆玉润，玲珑剔透，让人欲罢不能。

那么，龙胆花到底美在哪里呢？概括起来就五个字：形、色、势、饰、质。

形：花样繁多

形者，花之外形。龙胆花的美首先在于它们的形多。龙胆族植物分布广、差异大、种类繁多。它们的花形多变，婀娜多姿，有的似繁星，在瀚海里闪烁；有的如号角，对着蓝天高歌；有的像少女，雀斑未褪；有的像新娘，仪态万方。信手拈来，皆是佳丽，目不暇接，好似每一朵龙胆花都在努力地让自己与众不同。从图 3-1 和图 3-2 的个例可以看出，龙胆花的差异既体现在大的花冠形态不同，比如不同属的花冠，也体现在细微之处，如反折花龙胆与刺芒龙胆的不同就主要体现在褶的形态上。

色：绚丽多彩

色者，花冠之颜色。龙胆花的美也在于它们的色艳。龙胆花的颜色几乎囊括了可见光谱的所有波段，有红、黄、绿、白、紫，黄者黄到耀眼，白者白得透彻，红者红至发紫，蓝者蓝到极致。但它们最为光彩夺目、让所有人着迷的颜色还是龙胆那特有的、鲜亮的、蓝宝石一样的蓝，只能称之为“龙胆蓝”。它是一种湛蓝，蓝到只有湛蓝才可以到达的境界，深邃、广阔，只有蓝天和大海才配与它相媲美。它是全世界认可的“国际克莱因蓝 IKB”，是绝对之蓝，是青金石的颜色；正如法国艺术家伊夫克莱因指出的，它是天空，是水，是空气，是深度和无限，是自由和生命，是宇宙最本质的颜色。图 3-1 和图 3-2 展示了不同的花形，也可以看出龙胆花丰富的花色。管中窥豹，可见一斑。

此外，不同的龙胆种类对色彩的诠释和表达也是不完全相同的，有些种总是保持着自己的本色，不论地域和季节的变换，比如蓝玉簪龙胆，无论你在什么时候、什么地方见到它，它都保持着那份特有的湛蓝、那份纯洁和浓郁的湛蓝（图 3-3：E~H）；而相反，有些色彩差异还体现在种内，随时随地都有变化，比如反折花龙胆，就像一群喜欢化妆的少女，自带眼影，有时化浓妆，有时化淡妆，有些化重彩，有些则化烟熏（图 3-3：A~D），始终令观者赏心悦目，大饱眼福。只想说，无论是纯色的玉簪，还是善变的少女，蓝天大海无限好，浓妆淡抹总相宜。

势：整齐辉煌

势者，花之阵仗。龙胆花开之时，不像其他植物，开完一朵或几朵再开新花、次第开放；龙胆花开时，一个个体、一个群体的全部花朵都像是听从着统一的号令，统一步调，同时开放。如图 3-4，不管是同一个个体的不同花朵、同一个居群中的不同个体还是同一个群落里的不同种，所有花朵都聚集好了自己的能量，只等号令一来，就倾尽全力，瞬间释放，任性地、尽情地开放，像爆炸一样，在草甸上炸出一片花的海洋。

图 3-1　龙胆族植物多姿多彩的花

Fig. 3-1　Examples of gentians with diverse shapes and gorgeous colors

A. 反折花龙胆
B. 刺芒龙胆
C. 无茎龙胆
D. 无茎巴伐利亚龙胆
E. 开张龙胆
F. 云南獐牙菜
G. 大花蔓龙胆
H. 开展獐牙菜

A.*Gentiana choanantha*
B. *Gentiana aristata*
C. *Gentiana acaulis*
D. *Gentiana bavarica var. subacaulis*
E. *Gentiana aperta*
F. *Swertia yunnanensis*
G. *Crawfurdia angustata*
H. *Swertia patula*

图 3-2 龙胆族植物多姿多彩的花

Fig. 3-2 Examples of gentians with diverse shapes and gorgeous colors

A. 卵萼花锚（白色型）
B. 高贵狭蕊（摄影 /Photo：Adrien Favre)
C. 川东龙胆
D. 大花肋柱花

A. *Halenia elliptica* (white type)
B. *Metagentiana gentilis*
C. *Gentiana arethusae*
D. *Lomatogonium macranthum*

图 3-3　反折花龙胆（A-D）和蓝玉簪龙胆（E-H）的花色变异

Fig. 3-3　Color variation of *Gentiana choanatha*(A-D)and *Gentiana veitchiorum* (E-H)

图 3-4　龙胆花个体内不同花和群体内不同个体间，甚至同一群落中的不同种间高度整齐统一的开花步调

Fig. 3-4　Highly synchronized blooming momentums of gentians, among different flowers of a single individual, among different individuals within a population of a single species, or even among different species within a community

龙胆花整齐开花的习性是它们所处环境造成的。高原和高山地区海拔高，气候严酷，秋天很短，秋高气爽、适合开花的好日子总是非常短暂，龙胆花儿们都必须抓住难得的时机，在适宜开花的日子里，集中精力，尽快地开放，尽情地开放、开放。

此外，龙胆花们也展现了高度的团结和和谐之美。它们之间尽管有竞争，竞争空间、阳光，却依然展现出了团结和和谐。面对严寒和风霜，一起共享短暂的温暖和舒适，一起共命运，一起同美丽。这无疑是人类纷争世界中最好的榜样。

饰：标新立异

饰者，即龙胆花冠上的纹饰，位于花冠筒、花冠裂片或褶等不同位置，龙胆花们别出心裁、匠心独运地用各种琳琅满目的斑点和纹路来装饰和打扮自己，使得本就色彩绚丽、形态各异的龙胆花更加丰富多彩、精彩纷呈。这些斑点和纹路其实就是色彩的式样和形状，是形与色的完美结合。不同龙胆花的纹饰都极具创意，仿佛是才艺的比拼，推陈出新，在诠释什么是真正“花样百出”。不信就仔细看图 3-5 和图 3-6。龙胆花的纹饰有点、有线，也有圈、有斑，有浓、有淡，也有曲、有直，有的在花冠裂片上，有的在褶上，也有的在花冠筒上，不同的种一般各不相同，但也有同一物种拥有多变而不同的饰纹。这些纹饰就像纹章和印信，每一种图案和式样可能都有着特定含义和用途。有意思的是，我注意到欧洲龙胆花的纹饰大多比较简单粗旷，如无茎龙胆、匈牙利龙胆等的斑点都简单粗大、对比强烈，而亚洲，尤其是中国龙胆花的纹饰大多比较细腻曲折，复杂灵巧，如蓝白龙胆、阿坝龙胆、反折花龙胆等，它们的斑点和条纹都是那样地细腻精巧和委婉复杂。难道龙胆花也跟人一样，有着类似的文化传承？

花的纹饰一般是花留给它传粉者的信标和导引，告诉它中意的昆虫哪里有花蜜、哪里可着陆、哪里是跑道。高原上的花蜜资源和传粉者资源都比较稀缺，不管是龙胆花，还是它们的主要传粉者——熊蜂和独居蜂，都必须相互依赖，都必须精准高效，所以必然建立了对应的识别和适应体系。我相信它们的故事一定异常复杂精彩，希望有人能早日来揭秘。

图 3-5 龙胆花美丽多样的纹饰

Fig. 3-5 The beautiful and diversified ornamentations of gentian flowers

图 3-6　更多龙胆花美丽多样的纹饰

Fig. 3-6　Some more beautiful and diversified ornamentations of gentian flowers

质：出类拔萃

质者，即气质，是姿态，是特色，是精气神，是内在美，也是反差，是龙胆花与背景和周边植物的反差，是鹤立鸡群，是出类拔萃，是与众不同。另外，龙胆花也有着内在的英雄气质，秋风萧瑟之时，高原上草木开始枯黄，只有龙胆花不畏严寒，霜愈重色愈浓，甚至雪压花更艳（图 3-7）。

图 3-7 “与众不同”的龙胆
Fig. 3-7 “Outstanding” gentians

A. 高草丛中醒目的线叶龙胆
B. 顶雪绽放的滇西龙胆
C. 枯草丛中绽放的线叶龙胆
D. 蓝天与白云掩映下的亮黄色东俄洛龙胆花

A. Eye-catching *Gentiana lawrencei* var. *farreri*
B. *Gentiana georgei* blooming in snows
C. *Gentiana lawrencei* var. *farreri* blooming in dying grasses
D. Showy bright yellow flowers of *Gentiana togolensis* under blue sky and white clouds

当然，龙胆花的美不仅仅是上述五个方面。龙胆花的美是一个整体，是全方位的。正如俗话说的，这个世界并不缺少美，缺少的只是发现美的眼睛，就让我们一起去探索、去发现，去看到龙胆花和大自然更多的美吧!

花为啥美？龙胆花又为啥美？这是个好问题。花美可能有一千个理由，但龙胆花美不需要理由。

花为谁美？龙胆花又为谁美？这也是个好问题。也许是为传粉的蜂蝶，也许是为给它们拍照的拍摄者，也许是为路过的游客，也许只是为它们自己。云想衣裳花想容……

As one of the three well known flowers from the high mountains and plateaus, the beauty of gentians is undisputed around the world, in Asia, Europe, or the Americas. Whether it is the first encounter with a gentian or a reunion with it as an old friend, the moment a person sees a gentian always evokes his/her most primitive instinctual aesthetic, stimulates his/her love for beauty, and thus expresses sincere admiration. When gentian flowers bloom, they are as brilliant as clouds when viewed from a distance, beautiful and magnificent, whereas, when looking at close up, they are as fresh and lovely as a new born baby, smooth as pears and jade, exquisite and translucent, making them irresistible.

So, here comes a question. Where is the beauty of the beautiful gentian flower and why? In fact, the beauty of gentian flowers can be summarized in five words: shape, color, array, design, and temperament.

Shape: Diverse and Lovely

The shape of a flower is the outline of a flower. The beauty of gentian flowers lies first and foremost in their numerous shapes. The numerous species of the tribe *Gentianeae* has a wide distribution and significant differences. Their flower shapes are highly varied, some like stars shining in the vast sea, some like horns, singing to the blue sky; Some are like little girls, with their freckles still on face, while others are like new brides, with charming demeanor. They are all beautiful and dazzling, as if every gentian flower is striving to make itself unique, different from others. From the examples shown in Fig. 3-1 and Fig. 3-2, it can be seen that the differences in the morphology of gentian flowers include significant differences in corolla morphology, such as those of different genera, as well as subtle differences, such as the difference between *Gentiana choanantha* and *Gentiana aristata,* mainly reflected in the form of their plicae.

Color: Gorgeous and Splendid

Color is the color of flower corollas. The beauty of gentian flowers also lies in their vibrant colors. The color of gentian flowers covers almost all bands of the visible spectrum, including red, yellow, green, white, and purple. Yellow flowers are yellow to dazzling, white to translucent, red to purple, and blue to the extreme blue as blue can be. But their most dazzling and enigmatic color must be the unique bright sapphire like blue of gentians, reaching a state where only azure blue can reach deep and vast, and only the blue sky and sea can match it. It is the globally recognized "International Klein Blue (IKB)", the most abstract blue, the color of lasurite. As the French artist Yves Klein had pointed out, it is the sky, water, air, the depth and infinity, the freedom and life, and the most essential color of the universe. Fig. 3-1 and Fig. 3-2 show different flower shapes and the rich colors of gentian flowers.

In addition, different species of gentians have their different expressions and interpretations of colors. Some species always maintain their true colors, regardless of the changing regions and seasons, such as the so called jade blue *Gentiana veitchiorum*; Whenever and wherever you see it, it maintains its unique jade blue, pure, smooth and dense (Fig. 3-3: E-H); On the contrary, some gorgeous color differences are also reflected among different individuals of the same species whenever and wherever they are, for examples, individuals of *Gentiana choanantha*, just like a group of girls who love to make up with their eye shadows, sometimes with heavy makeup, sometimes with light makeup, some using heavy colors, some paint smoked (Fig. 3-3: A-D). As viewers, we are pleased to see them all eye-catching. We just want to say that whether it is a constant jade blue or a changing dressed girl, they all are just right and suitable.

Array: Synchronized and Tremendous

Array is the action of many flowers performing together. When gentian flowers bloom, unlike other plants that bloom one flower or a few each time before opening new ones, all the flowers of gentians of an individual or a group of individuals seem to follow a unified command, with a unified pace, and bloom at the same time. As shown in Fig. 3-4, regardless of whether it is different flowers from the same individual, different individuals from the same population, or different species within the same community, all flowers gather their energy, exert all their strength, and instantly release to open when the single sign is issued, they freely open, and fully open, like an explosion, blasting out a sea of flowers on the grassland.

The habit of gentian flowers blooming neatly is caused by their environment. In high-altitude plateau and high mountains, the climate is harsh, and autumn is very short. The good days of autumn are very short and limited. Gentian flowers must seize the rare opportunity, concentrate on blooming as soon as possible, fully open and bloom on the day that is suitable for flowering.

The gentian flowers display a high degree of solidarity and harmonization. Despite competition on space and sunshine, they highly demonstrate their solidarity and harmonization, to face the harsh climate, cold, wind and frost together, to share the short summer warmth and autumn sunshine together, to bloom together, i.e. to share their destiny peacefully together. What a wonderful example for humans in such a current world full of conflicts.

Design: Innovative and Artistic

Design is the ornamentation of gentian flowers, either on the corolla lobes, plicae, or surfaces of corolla throat. Gentians use their unique, ingenious and diverse spots and patterns to decorate and dress their flowers, making the already colorful and diverse forms of gentian flowers much further diversified and fascinating. These spots and patterns are actually the patterns and shapes of colors, a perfect combination of form and color. The patterns of different gentian flowers are highly creative, as if they are a competition of talents, innovating and truly interpreting what "diverse patterns" mean. If you do not believe it, take a closer look at the Fig. 3-5 and Fig. 3-6 I have pieced together. The patterns of sculpture of gentian flowers are dots, lines, circles, spots, dense and light, curved and straight, on corolla lobes, plicae and surface of throat. Different species generally have different patterns, but there are also different and varied patterns of the same species. These decorations are like badges and seals, each pattern and style may have a specific meaning and purpose. It is interesting that I have noticed that the patterns of European gentian flowers are often simple and rough, such as the spots on *Gentiana acaulis* and *Gentiana pannonica* etc., which are simple and thick, with strong contrast. In contrast, the patterns of Asian gentian flowers, especially Chinese gentian flowers, are more delicate and intricate, complex and agile, such as *Gentiana leucomelaena*, *Gentiana abaensis*, and *Gentiana choanantha*. Their spots and stripes are so delicate, circuitous, tactful and complex. Do gentians have a similar cultural heritage as humans?

The floral decorations are usually beacons and guides left by the flowers to their pollinators, telling their favorite insects where the nectaries are, where they can land, and where the runways are. The nectar resources and pollinator resources on the high mountains and plateaus are rather scarce. Whether gentian flowers or their main pollinators, bumblebees and solitary bees, they must rely on each other and be precise and efficient. Therefore, corresponding identification and adaptation systems must be established. I believe their stories must be exceptionally complex and exciting, and I hope someone can come and reveal them as soon as possible.

Temperament: Outstanding and Contrasting

Temperament is flower’s spirit, uniqueness, and posture, the inner beauty. It is also all about contrast, contrast with the background environment and surrounding plants, standing out from the crowd and being unique. In addition, the gentian flower also has an inherent heroic temperament. When the autumn wind blows, the grass and bushes on the plateau begin to wither and dry out. Only the gentian flowers are not afraid of severe cold, becoming more colorful and more showy with frost and even snowflakes concentrate on top of them (Fig. 3-7).

Of course, the beauty of gentian flowers is not limited to the five separate aspects mentioned above. The beauty of gentian flowers is a whole and comprehensive. Of course, as the common saying goes, this world is not lacking in beauty, what is lacking is only discovery, and all beauty needs to be discovered. Let us explore and discover together, to see more beauty in gentian flowers and nature!

Why are flowers beautiful? Why are gentian flowers beautiful? This is a good question. There may be a thousand reasons for the beauty of gentian flowers, but gentian flowers do not require a reason for their beauty.

For whom are flowers beautiful? For whom are gentian flowers beautiful? This is also a good question. Perhaps it's the bees and butterflies that pollinate them, maybe it is the photographers who take pictures of them, maybe it is the tourists passing by, maybe it is just themselves. Flowers are beautiful for whom they feel pleased with ...

线叶龙胆 *Gentiana lawrencei* var. *farreri*

中国甘肃省合作市 Hezuo, Gansu, China. 19.10.2018

线叶龙胆 *Gentiana lawrencei* var. *farreri* 长梗喉毛花 *Comastoma penduncululatum*
中国青海省河南蒙古族自治县 Henan, Qinghai, China. 09.09.2023

4

龙胆谱录

Galaxy of Gentians

4.1 龙胆亚族

Subtribe *Gentianinae*

线叶龙胆 *Gentiana lawrencei* var. *farreri*

中国青海省河南蒙古族自治县 Henan, Qinghai, China. 09.09.2023

4.1.1 蔓龙胆属 *Crawfurdia*

大花蔓龙胆的生态环境——中国西藏自治区墨脱县
Habitat of *Crawfurdia angustata*, Motuo, Xizang, China. 26.09.2003

多年生草本，茎缠绕藤本；花萼筒具 10 条维管束；花冠管状、漏斗状或钟状，有褶；雄蕊 5 枚，同型，直伸；蜜腺 5 个，生于子房基部；种子具宽翅，盘状。

本属约 16 种，主要分布于喜马拉雅山东部地区及中国西南部横断山地区。

Perennial herb, stems twisting; calyx tube with 10 vascular bundles; nectaries 5, on gynophore; stamens equal, straight; seeds broadly winged, discoid.

About 16 species, mainly in the mountains in eastern Himalayas and the Hengduan Mountains in southwestern China.

大花蔓龙胆

Crawfurdia angustata

多年生草质藤本，茎缠绕，光滑；叶有短柄，叶片卵形到椭圆形；花萼筒管状，整齐或一侧开裂，花萼裂片平展到弯曲，三角形；花冠粉色到亮紫色，漏斗状，花冠裂片卵状三角形，褶三角形，偏斜，边缘不整齐齿状。

散布于喜马拉雅山东部地区的印度、缅甸和中国西藏、云南等地。生长在山谷疏林中及边缘和灌丛中。海拔 1500~2000 m。花果期 9~10 月。

Perennial, stems twisting, glabrous; leaves with short petioles, blades ovate to elliptic; calyx tubes tubular, entire or split by one side, calyx lobes triangular, spreading to recurved; corolla pink to bright purple, funnelform, lobes ovate-triangular, plicae triangular, oblique, margins irregularly crenulate.

Scattered in the eastern Himalayas, in India, Myanmar, and Yunnan and Xizang provinces of China. In sparse forest, forest margin and bushes. Altitude 1500-2000 m. Flowering & fruiting: September-October.

A-B. 中国西藏自治区墨脱县
Motuo, Xizang, China. 26.09.2003

C. 中国西藏自治区墨脱县 摄影 /Photo: 葛学军
Motuo, Xizang, China. 12.11.2019

C

云南蔓龙胆

Crawfurdia campanulacea

多年生草质藤本，茎缠绕，光滑；叶有短柄，叶片阔卵形到椭圆形，革质；花萼筒管状，整齐，花萼裂片三角形，基部合生成杯状的檐；花冠淡紫色，钟状，具深色条纹，花冠裂片卵状三角形，褶半圆形，偏斜，边缘啮齿状。

特产于中国云南省西北部。生长在山谷疏林及边缘和灌丛中。海拔 1800~2800 m。花果期 9~11 月。

Perennial, stems twisting, glabrous; leaves with short petioles, blades broad ovate to elliptic, leathery; calyx tubes tubular, entire, calyx lobes triangular, base connate into a cupular limb; corolla pale purple with dark stripes, campanulate, lobes ovate-triangular, plicae semiorbicular, oblique, margins erose.
Endemic to northwestern Yunnan of China. In sparse forest, forest margin and bushes. Altitude 1800-2800 m. Flowering & fruiting: September-November.

A

B

A-B. 中国云南省丽江市 摄影 /Photo: Adrien Favre
Lijiang, Yunnan, China. 03.10.2004

裂萼蔓龙胆

Crawfurdia crawfurdioides

多年生草质藤本，茎缠绕，光滑；叶无柄或有短柄，叶片阔卵形，叶脉上具刚毛；花单生，花萼筒管状，一侧开裂，花萼裂片披针形；花冠较大，长达 5 cm，钟状，紫色，具深色条纹，花冠裂片卵状三角形，褶平截，边缘齿状。

特产于中国西藏东南部到云南西北部。生长在山谷疏林中及边缘和灌丛中。海拔 1800~3200 m。花果期 9~10 月。

Perennial, stems twisting, glabrous; leaves sessile or with short petioles, blades broad ovate, hirsute along veins; flowers solitary, calyx tubes tubular, split on one side, calyx lobes lanceolate; corolla big, up to 5 cm long, purple with dark stripes, campanulate, lobes ovate-triangular, plicae truncate, margins crenulate.

Endemic to southeastern Xizang and northwestern Yunnan of China. In sparse forest, forest margin and bushes. Altitude 1800-3200 m. Flowering & fruiting: September-October.

A–C. 中国云南省大理白族自治州

Dali, Yunnan, China. 07.10.1999

披针叶蔓龙胆

Crawfurdia delavayi

多年生草质藤本，茎缠绕，光滑；叶无柄或有短柄，基部叶片卵形，上部叶披针形，宽不及 1 cm；花单生；花萼筒管状到狭钟形，多整齐，花萼裂片披针形；花冠中小型，长不及 4 cm，钟状，粉色到紫红色，具深色条纹，花冠裂片卵状三角形，褶平截，边缘齿状。

特产于中国云南西北部。生长在高山疏林中及箭竹丛中。海拔 2800~3500 m。花果期 9~10 月。

Perennial, stems twisting, glabrous; leaves sessile or with short petioles, lower blades ovate, upper blades lanceolate, width less than 1 cm; flowers solitary; calyx tubes tubular to narrow campanulate, mostly entire, calyx lobes lanceolate; corolla medium-small, less than 4 cm long, campanulate, pink to reddish purple with dark stripes, lobes ovate-triangular, plicae truncate, margins crenulate.

Endemic to northwestern Yunnan of China. In sparse alpine forest and bamboo thickets. Altitude 2800-3500 m. Flowering & fruiting: September-October.

A

B

A-B. 中国云南省大理白族自治州 摄影 /Photo: Adrien Favre
Dali, Yunnan, China. 12.10.2004

半侧蔓龙胆

Crawfurdia dimidiata

多年生草质藤本，茎缠绕，光滑；叶有柄，叶片阔卵形，基部圆形，光滑；聚伞花序，有数朵花；花萼筒管状到狭钟形，整齐或一侧开裂，花萼裂片线状披针形，开展；花冠较大，长 4 cm 以上，钟状，蓝紫色，具深色条纹，花冠裂片卵状三角形，褶平截，边缘齿状。

分布于中国西藏东南部到云南西北部。生长在高山疏林中及箭竹和灌丛中。海拔 3000~3500 m。花果期 9~10 月。

Perennial, stems twisting, glabrous; leaves with petioles, blades broad ovate, base rounded, glabrous; flowers several in cymes; calyx tubes narrow campanulate, entire or split by one side, calyx lobes linear-lanceolate, spreading; corolla medium-large, more than 4 cm long, campanulate, blue-purple with dark stripes, lobes ovate-triangular, plicae truncate, margins crenulate.

In southeastern Xizang and northwestern Yunnan of China. In sparse alpine forest, bamboo and bush thickets. Altitude 3000-3500 m. Flowering & fruiting: September-October.

A–B. 中国西藏自治区墨脱县
Motuo, Xizang, China.25.09.2003

细柄蔓龙胆

Crawfurdia gracilipes

多年生草质藤本，茎缠绕，光滑；叶有短柄，叶片卵形到卵状披针形，基部圆形，光滑；聚伞花序，有 2 至数朵花；花萼筒狭钟形，一侧开裂，花萼裂片线状披针形，开展到稍反折；花冠中小型，长不及 3 cm，钟状，蓝紫色，具深色条纹，花冠裂片卵状三角形，褶半圆到平截，偏斜，边缘齿状。

分布于中国西藏东南部到云南西北部。生长在高山疏林和灌丛中。海拔 2800~3200 m。花果期 9~10 月。

Perennial, stems twisting, glabrous; leaves with petioles, blades broad ovate, base rounded, glabrous; flowers 2 to several in cymes; calyx tubes narrow campanulate, split by one side, calyx lobes linear-lanceolate, spreading to recurved; corolla small, less than 3 cm long, campanulate, blue-purple with dark stripes, lobes ovate-triangular, plicae semiorbicular to truncate, oblique, margins crenulate.

In southeastern Xizang and northwestern Yunnan of China. In sparse alpine forest and bush thickets. Altitude 2800-3200 m. Flowering & fruiting: September-October.

A

B

A–C. 中国西藏自治区波密县
Bomi, Xizang, China. 04.10.2003

C

中国西藏自治区林芝市鲁朗镇附近，雪峰与针阔叶混交林交相辉映。

Snow covered peaks in touch with mixed forest of broad leaf and needle leaf trees, vicinity of Lulang, Linzhi, Xizang, China. 06.10.2003.

老街蔓龙胆

Crawfurdia poilanei

多年生草质藤本，茎缠绕，光滑；叶有短柄，叶片卵形到卵状披针形，基部圆形，光滑；聚伞花序，有 1 至数朵花；花萼筒狭钟形，一侧开裂，萼筒先端有明显的膜质萼内膜，花萼裂片线状披针形，开展到稍反折；花冠中大型，长约 4 cm，钟状，紫红色，具深色条纹，花冠裂片卵形，褶平截，边缘啮齿状。

特产于越南北部。生长在高山疏林中及灌丛中。海拔约 2000 m。花果期 9~10 月。

Perennial, stems twisting, glabrous; leaves with short petioles, blades ovate, base rounded, glabrous; flowers 1 to several in cymes; calyx tubes narrow campanulate, split by one side, tube apex with intracalycular membrane, calyx lobes linear-lanceolate, spreading to recurved; corolla medium, about 4 cm long, campanulate, reddish purple with dark stripes, lobes ovate, plicae truncate, margins erose.

Endemic to northern Vietnam. In sparse alpine forest and bush thickets. Altitude 2000 m. Flowering & fruiting: September - October.

A

B

A–C. 越南老街 Lào Cai, Viêtnam, 14.09.2003.
摄影 /Photo: Adrien Favre

C

福建蔓龙胆

Crawfurdia pricei

多年生草质藤本，茎缠绕，光滑；叶有短柄，叶片卵形到卵状披针形，基部圆形，光滑；聚伞花序，有 1 至数朵花；花萼筒狭钟形，整齐或偶尔一侧开裂，萼筒先端有叶质萼内膜，花萼裂片钻形，反折；花冠中大型，钟状，白色、粉色到红紫色，具深色条纹，花冠裂片卵形，褶平截，边缘啮齿状。

散布于中国南部湖南、福建、广东和广西等省份。生长在高山疏林和灌丛中。海拔 500~2000 m。花果期 9~11 月。

Perennial, stems twisting, glabrous; leaves with short petioles, blades ovate, base rounded, glabrous; flowers 1-several in cymes; calyx tubes narrow campanulate, entire or occasionally split by one side, tube apex with intracalycular membrane, calyx lobes subulate, spreading and recurved; corolla medium, campanulate, white, pink, to reddish purple with stripes, lobes ovate, plicae truncate, margins erose.
Scattered in Hunan, Fujian, Guangdong and Guangxi provinces in southern China. In sparse alpine forest and bush thickets. Altitude 500-2000 m. Flowering & fruiting: September - November.

A

B

C

A-D. 中国广西壮族自治区南宁市
Nanning, Guangxi, China. 11.09.2001

D

无柄蔓龙胆

Crawfurdia sessiliflora

多年生草质藤本，茎缠绕，光滑；叶几无柄，叶片卵状披针形，基部圆形，光滑；聚伞花序，有 2 至数朵花；花萼筒管状狭钟形，基本不开裂，先端有窄萼内膜，花萼裂片三角形，基部稍联合成檐状；花冠筒细长型，长约 5 cm，钟状，蓝紫色，具深色条纹，花冠裂片卵状三角形，褶平截，边缘齿状。

特产于中国四川省西部。生长在高山针叶林林缘。海拔 2800~3200 m。花果期 9~10 月。

Perennial, stems twisting, glabrous; leaves nearly without petioles, blades ovate-lanceolate, base rounded, glabrous; flowers 2-several in cymes; calyx tubes tubular campanulate, almost entire, apex with narrow intracalycular membrane, calyx lobes triangular, slightly connate into a narrow limb; corolla narrow and long, ~5 cm long, campanulate, blue-purple with dark stripes, lobes ovate-triangular, plicae truncate, margins crenulate.

Endemic to western Sichuan of China. In forest margins of needle-leaf trees. Altitude 2800-3200 m. Flowering & fruiting: September - October.

A-B. 中国四川省海螺沟
Hailuogou, Sichuan, China. 04.10.1993

穗序蔓龙胆

Crawfurdia speciosa

多年生草质藤本，茎缠绕，光滑；叶有短柄或无柄，叶片椭圆形、卵形或卵状披针形，基部圆形，光滑；聚伞花序，有 1 至数朵花；花萼筒钟形，整齐，偶尔一侧开裂，先端有明显萼内膜，花萼裂片小，三角形，平展；花冠中小型，长 3~4 cm，钟状，蓝紫色，具深色条纹，花冠裂片卵状三角形，褶平截，边缘齿状。

特产于中国西藏东南部。生长在高山疏林中及灌丛中。海拔 3000~3800 m。花果期 9~10 月。

Perennial, stems twisting, glabrous; leaves sessile or with short petioles, blades elliptic, ovate, or ovate-lanceolate, base rounded, glabrous; flowers 1-several in cymes; calyx tubes campanulate, entire, occasionally split by one side, apex with intracalycular membrane, calyx lobes small, triangular, spreading; corolla medium-small, 3-4 cm long, campanulate, blue-purple with dark stripes, lobes ovate-triangular, plicae truncate, margins crenulate.

Endemic to southeastern Xizang of China. In sparse alpine forest and bush thickets. Altitude 3000-3800 m. Flowering & fruiting: September - October.

A-E. 中国西藏自治区墨脱县

Motuo, Xizang, China. 24.09.2003

C

D

F. 中国西藏自治区墨脱县拉格附近景观；附近有穗序蔓龙胆 *Crawfurdia speciosa* 和半侧蔓龙胆 *Crawfurdia dimidiata*
Motuo, Xizang, China. 24.09.2003

4.1.2
龙胆属
Gentiana

一年生或多年生草本；茎直立、斜升或铺散；叶对生或轮生；花冠有褶，蜜腺位于子房基部；种子球形、卵形或椭球形，无棱。

本属约 354 种，是龙胆科最大的属。主要分布于亚欧大陆及美洲的高山地区，以喜马拉雅山、横断山、阿尔卑斯山、落基山、阿帕拉契亚山和安第斯山等山系种类最为丰富，个别种扩展到澳洲和北非。

关于属下的分类研究较多。本书采用最新的 DNA 分子系统学研究结论（Favre et al., 2020），属下划分为 13 个组，如右图所示。有关本属分类学，还请参见何廷农和刘尚武的英文专著（Ho & Liu, 2001）。

Annual or perennial herb; stem erect, ascending or prostrate; leaves opposite or whorled; corolla with plicae, nectary at base of ovary; seeds ellipsoid, ovoid to global without ridge.

About 354 species, the largest genus in the family; highly concentrated in the alpine areas in Asia, Europe and Americas, mostly in the Himalayas, Hengduan Mountains, Alps, Rockies, Appalachians, and Andes; few species extended to Australia and northern Africa.

There are lots of studies regarding infrageneric taxonomy of this genus. In this book, we adopted the classification of the most recent conclusions generated from DNA molecular phylogenetic studies (Favre et al., 2020) and divided the species into 13 sections, as shown in the figures on the right. For more taxonomic details, please refer to the monograph of the genus published by Ho & Liu (2001).

sect. *Calathiane*【*Gentiana bavarica* var. *subacaulis*】

sect. *Ciminalis*【*Gentiana acaulis*】

sect. *Chondrophyllae*【*Gentiana choanantha*】

sect. *Frigida*【*Gentiana algida*】

sect. *Cruciata*【*Gentiana dahurica*】

sect. *Monopodiae*【*Gentiana rigiscens*】

sect. *Gentiana*【*Gentiana lutea*】

sect. *Kudoa*【*Gentiana veitchiorum*】

sect. *Tetramerae*【*Gentiana lineolata*】

sect. *Isomeria*【*Gentiana georgei*】

sect. *Phyllocalyx*
【*Gentiana phyllocalyx*】

sect. *Microsperma*【*Gentiana togolensis*】

sect. *Pneumonanthe*【*Gentiana septemfida*】

春龙胆 *Gentiana verna*

瑞士伯尔尼州尼德宏 Niderhorn, Bern, Switzerland. 19.05.2007

盘柱组

sect. *Calathianae*

一年生或多年生小草本；两片柱头裂片靠合，形成膨大扁平的、圆环形、微微凹陷的圆盘。

本组约15种（含亚种、变种），主要分布于欧洲阿尔卑斯山及周边的高山地区，个别种扩展到北非（摩洛哥）、西亚、中亚和北美高纬度地区。

Annual or perennial diminutive herb; stigmata lobes contiguous forming an enlarged, circular, slightly concave disc.

About 15 species (including subspecies & varieties); highly concentrated in the alpine areas in Europe around the Alps; some species extended to northern Africa (Morocco), western and central Asia, and high latitude areas in North America.

巴伐利亚龙胆

Gentiana bavarica

多年生草本，高 5~20 cm；叶多基生，叶片较小，倒卵形到匙形，宽度大于长度，边缘软骨质；萼齿披针形，萼筒具极窄翅，花冠深蓝色。

主要分布于阿尔卑斯山及周边的高山地区，包括奥地利、德国、瑞士、法国及前南斯拉夫等国。一般生长在较潮湿的草地，海拔 1500~3500 m。花果期 6~9 月。

Perennial herb, 5-20 cm; leaves crowded towards the base of stem, blade small, obovate to spathulate, width slightly larger than length; margin cartilaginous; calyx-teeth lanceolate, calyx-tube narrowly winged; corolla deep blue. Widely spread in the Alps and adjacent mountains, including Austria, Germany, Switzerland, France and former Yugoslavia. Usually growing in damp places of alpine meadows. Altitude 1500-3500 m. Flowering & fruiting: June to September.

A

B

C

A. 瑞士瓦莱州
Oberwald to Gletsch, Valais, Switzerland. 16.07.2005

B. 德国巴伐利亚鹰巢峰
Eagle's Nest, Berchtesgaden, Germany. 19.08.2005

C-D. 瑞士瓦莱州
Oberwald to Gletsch, Valais, Switzerland. 16.07.2005

D

无茎巴伐利亚龙胆

Gentiana bavarica var. *subacaulis*

多年生草本，高 3~5 cm；叶片与巴伐利亚龙胆相似，但更小更密集，叶片几近圆形，贴生地面；茎极短；花单生，贴近地面。

仅分布于阿尔卑斯山区的奥地利、瑞士和法国等地。一般生长在较干燥、土质瘠薄的高山分化岩石边缘。海拔 2500~3500 m。花果期 6~9 月。

Diminutive perennial, 3-5 cm high; leaves similar to that of Bavarian Gentian, but smaller and denser, leaf blades nearly round crowded to ground; stems very short; flowers solitary, close to ground.

Spread in the Alps in Austria, Switzerland and France. Growing in dry and sterile soils around decayed rocks in alpine areas. Altitude 2500-3500 m. Flowering & fruiting: June - September.

A–C. 瑞士瓦莱州爱阁宏
Eggishorn, Valais, Switzerland. 17.07.2005

D. 瑞士阿莱奇冰川附近高山上生长的无茎巴伐利亚龙胆 *Gentiana bavarica* var. *subacaulis*
Gentiana bavarica var. *subacaulis* growing in the alpine meadows nearby the Aletsch Glacier in Switzerland. 17.07.2005

D

短叶龙胆

Gentiana brachyphylla

多年生草本，高 3~10 cm；莲座叶叶片菱形到近圆形，长度不超过宽度的两倍，边缘软骨质；茎生叶 1~2 对，有时无；花单生茎顶，花萼筒无翅或仅具狭翅，萼齿狭三角形；花冠深蓝色。

分布于欧洲的喀尔巴阡山、阿尔卑斯山、比利牛斯山及西班牙南部的内华达山。生长在中高山较潮湿的草甸中，海拔 1500~3500 m。花果期 7~9 月。

本种与春龙胆容易混淆，区别是本种的莲座叶叶片菱形或近圆形，长不及宽的两倍。

Perennial 3-10 cm high; rosette leaf blades rhombic to suborbicular, length usually less than twice as width, margin cartilaginous; cauline leave 1-2 pairs, occasionally none; flowers solitary and terminal, calyx tubes angled without wines or very narrowly winged, calyx-teeth narrowly triangular; corolla deep blue.

In the Carpathians, Alps, Pyrenees, and Sierra Nevada in Southern Spain. Wet alpine meadows. Altitude: 1500-3500 m. Flowering & fruiting: July - September.

A. 瑞士伯尔尼州施尼格普拉特
Schynige Platte, Bern, Switzerland. 25.06.2006

B-C. 瑞士瓦莱州大沙瓦拉
Grand Chavalard, Valais, Switzerland. 27.05.2005

C

雪地龙胆

Gentiana nivalis

一年生草本，高 3~30 cm，莲座叶密集成小丛；茎常多分枝，偶单一；茎生叶无柄，叶片卵形到椭圆形；花萼筒具棱、无翅，萼齿狭三角形，急尖；花冠深蓝色。

广泛分布于欧洲大陆及英伦三岛的山地地区，扩展到西亚、高加索及北欧高纬度格陵兰和加拿大等地。生长在山坡、溪边、冻原、草地及高山草场。海拔 100~2500 m。花果期 6~9 月。

Annual, 3-30 cm high, rosette leaves crowed; stem much divided, occasionally simple; cauline leaves sessile, blades ovate to elliptic; calyx tube angled without wing, calyx teeth narrowly triangular, acute; corolla intense blue. Widely spread in most mountains in continental Europe and British Islands, extending to western Asia, the Caucasus, and high latitude areas in northern Europe, Greenland, and Canada. In wide habitats such as hill slopes, stream banks, tundra, meadows and alpine pastures. Altitude: 100-2500 m. Flowering & fruiting: June-September.

A

B

A-B. 瑞士瓦莱州爱阁宏
Eggishorn, Valais, Switzerland. 17.07.2005

C. 雪地龙胆 *Gentiana nivalis* 花的访客；瑞士瓦莱州爱阁宏
Visitor of Snow Gentian flowers, Eggishorn, Valais, Switzerland. 17.07.2005

D. 雪地龙胆 *Gentiana nivalis* 的生境；瑞士瓦莱州爱阁宏
Habitat of Snow Gentian flowers, Eggishorn, Valais, Switzerland. 17.07.2005

C

D

本都山龙胆

Gentiana pontica

多年生草本，高 2~20 cm；莲座叶发达，阔卵形，长超过宽两倍；茎单一，在花后显著伸长；茎生叶无柄，1~3 对，叶片卵形；花萼筒具棱，棱上有 2~3 mm 的宽翅，萼齿三角形；花冠深蓝色。

分散分布于东亚西部（中国新疆、蒙古西北部、俄罗斯阿尔泰地区）、中亚到西亚及东欧和巴尔干地区。生长在石灰岩高山与亚高山草地。海拔 1500~3500 m。花果期 5~9 月。

本种常被作为亚种并入春龙胆，但它分布区相对独立，萼筒具较宽的翅而容易区分。

Perennial, 2-20 cm high; rosette leaves well developed, broad ovate, twice as long as wide; stem solitary, significantly elongating after flowering; cauline leaves 1-3 pairs, ovate; calyx tubes angled, with wings 2-3 mm wide, calyx teeth triangular; corolla deep blue.

Sporadically scattered from western part of Eastern Asia (Xinjiang of China, northwestern Mongolia, Russian Altai), Central and western Asian countries to Eastern Europe and Balkan peninsula. In meadows of alpine and subalpine lime stone slopes. Altitude 1500-3500 m. Flowering & fruiting: May - September.

Often included in *Gentiana verna* as a subspecies. It can be easily distinguished due to its wider calyx wings and distinct distribution.

A. 中国新疆维吾尔自治区赛里木湖
摄影 /Photo: 詹雪菁
Sailimu Lake, Xinjiang, China. 06. 2011

B. 中国新疆维吾尔自治区喀纳斯湖
Hanasi Lake, Xinjiang, China. 09.08.1995

施耐氏龙胆

Gentiana schleicheri

多年生矮小草本，高 2~5 cm；莲座叶显著、密集，呈覆瓦状排列，卵状披针形，质地较厚；茎单一，矮小；花萼筒具棱，棱上几无翅，萼齿狭三角形；花冠深蓝色。

仅分布于瑞士西部及法国和意大利间的阿尔卑斯山西南部。生于石质干山坡。海拔 2000~3000 m。花果期 6~9 月。

本种时常被作为亚种并入覆瓦叶龙胆 *Gentiana terglouensis* 中。

Perennial diminutive herb, 2-5 cm high; rosette leaves patent, densely crowed, often imbricate, blades ovate-lanceolate, thick; stem solitary, inconspicuous; calyx tube angles without wings, calyx teeth narrowly triangular; corolla deep blue.

Seen only in western Switzerland and the Maritime Alps. In dry stony slopes. Altitude 2000-3000 m. Flowering & fruiting: June-September.

Often included in *Gentiana terglouensis* as a subspecies.

A-B. 瑞士瓦莱州大沙瓦拉
Grand Chavalard, Valais, Switzerland. 07.1992

C. 瑞士瓦莱州策尔马特
Zermatt, Valais, Switzerland. 27.06.1996

覆瓦叶龙胆

Gentiana terglouensis

多年生矮小草本，高 2~5 cm；基生叶直立，覆瓦状排列，形成莲座；卵形到卵状披针形，质地较厚，边缘软骨质。其他与施耐氏龙胆接近。

仅分布于阿尔卑斯山东南部高山区。生于钙质岩石质干山坡。海拔 2000~3000 m。花果期 6~9 月。

Perennial diminutive herb, 2-5 cm high; base leaves erect forming a rosette, often imbricate, blades ovate to ovate-lanceolate, thick; otherwise similar to Gentiana schleischeri. Distributed in Southeastern Alps. In calcicole stony slopes. Altitude 2000-3000 m. Flowering & fruiting: June-September.

A. 斯洛文尼亚卡拉万肯山 摄影 /Photo: Adrien Favre
Karavanken Mt., Slovenia. 30.06.2018

B. 意大利波得龙省切莫莱斯
Cimolais, Friuli Venezia Giulia, Pordenone, Italy. 07.07.1996

胞果龙胆

Gentiana utriculosa

一年生草本，高 5~30 cm；无明显莲座叶；茎纤细，直立；茎生叶无柄，倒卵形到披针形；萼筒较宽阔，多少膨胀，直径 3~7 mm，具棱，棱上有明显的翅，翅宽 2~3 mm；萼齿披针形，急尖；花冠深蓝色。

仅分布于中欧到意大利及巴尔干半岛的山地。多生长在石灰岩石质干山坡和林缘。海拔 500~2000 m。花果期 6~8 月。

Annual, 5-30 cm high; no rosette leaves; stem slender, erect; cauline leaves sessile, obovate to lanceolate; calyx tubes wide, somewhat inflated, 3-7 mm wide, angled, with conspicuous wings 2-3 mm wide; calyx teeth lanceolate, acute; corolla deep blue.

Spread in the mountainous areas from central Europe to Italy and the Balkan Peninsula. In lime stone rocky slopes and forest margins. Altitude: 500-2000 m. Flowering & fruiting: June-August.

A-C. 意大利科莫格锐尼亚山

Mt. Grigna, Como, Italy. 03.07.1996

春龙胆

Gentiana verna

矮小的常绿多年生草本，会缓慢长成小垫状；莲座叶狭卵圆形到狭条形；茎生叶短小，长不及莲座叶的一半；花单生，直立，花枝在花后伸长；花冠深天蓝色，喉部白色，晚春到初夏开花；花萼具狭翅。

广泛分布在西欧、中欧和南欧的山区，俄罗斯北极地区和高加索山区也有。生长在岩石及山顶附近的潮湿高山草甸。海拔 300~3500 m。花果期 5~9 月。

A diminutive evergreen perennial slowly forming a mat; rosette leaves narrowly ovate to oblong; cauline leave 1/2 shorter than rosette leaves; flowers solitary, erect, elongating after flowering; corolla deep sky-blue with white throats, opening in late spring and early summer; calyx narrowly winged.

Wide spread in mountains of western, central and southern Europe, Arctic Russia, the Caucasus. On rocks, in damp alpine pastures near mountain summits. Altitude 300-3500 m. Flowering & fruiting: May - September.

A. 瑞士伯尔尼莎斯哈尔山
Mt. Chasseral, Bern, Switzerland. 06.06.2004

B. 瑞士圣莫里茨伯尔尼娜山口
Pso del Bernina, St.-Moritz, Graubünden, Switzerland. 22.05.2004

C. 瑞士伯尔尼格利得瓦尔德
First, Grindelwald, Bern, Switzerland. 26.05.2006

D. 瑞士纳沙泰尔鼠头山
Tete de Ran, Neuchatel, Switzerland. 15.05.2005

E. 瑞士伯尔尼格利得瓦尔德
First, Grindelwald, Bern, Switzerland. 26.05.2006

F. 瑞士伯尔尼大舍得格山口
Grosse Scheidegg, Bern, Switzerland. 26.05.2007

G. 瑞士伯尔尼州的尼得宏山
Niederhorn, Bern, Switzerland. 19.05.2007.

在瑞士伯尔尼州的尼得宏山上漫步，既能赏脚下的春龙胆（*Gentiana verna*）和高山蒲公英（*Taraxacum alpinum*）花，又可以观远处的阿尔卑斯山美景，两全其美。
You can enjoy the beautiful flowers of the Spring Gentian *Gentiana verna* and the alpine Dandelion *Taraxacum alpinum* near your feet, and view the spectacular scenery of the Alps when you hike on the meadows of Niederhorn, Bern, Switzerland. 19.05.2007.

反折花龙胆 *Gentiana choanantha*
中国四川省康定市折多山 Zheduoshan Mt., Kangding, Sichuan, China. 04.07.2021

小龙胆组（广义）

sect. *Chondrophyllae* s. l.

一年生小草本，个别多年生；花单一，顶生；花萼内膜退化；褶对称；种皮具细网纹。本组约 182 种，广布于北半球，个别种类散布到南美及大洋洲，以喜马拉雅山和横断山地区种类最为丰富。

广义小龙胆族包括以前单独成组的柱果组和喉毛组。DNA 分子系统学研究证明二者均非单系类群，而应该是小龙胆组的一部分。

Minute annual, occasionally perennial; flowers solitary and terminal; intracalycular membrane absent; plicae symmetrical; seedcoat minutely reticulate. About 182 species, widely spread over northern hemisphere with a few extended to South America and Oceania. The Himalayas and Hengduanshan mountains harbor the most species.

Section *Chongdrophyllae* s. l. includes previously separated sect. *Dolichocarpa* and sect. *Fimbricorona*. Phylogenetic studies based on DNA sequences revealed both sect. *Dolichocarpa* and sect. *Fimbricorona* were not monophyletic and should be parts of sect. *Chondrophyllae*.

阿坝龙胆

Gentiana abaensis

一年生草本，茎直立或斜升，高5~10 cm，基部多分枝，全株密被紫色乳突；叶片卵形到匙形，边缘具睫毛；花单生枝端；花冠筒漏斗状，紫红色，具深紫色斑点。

仅见于中国四川省北部、甘肃省南部和青海省东南部三省相邻的高原地区。生长于高山草甸。海拔2800~3500 m。花果期7~9月。

Annual, stems erect or ascending, 5-10 cm, much branched from base, densely purple papillate; leaf blades ovate to spathulate, margins ciliate; flowers solitary and terminal; corolla funnelform, reddish purple, with dark purple spots.

In the plateau areas connecting northern Sichuan, southern Gansu and southeastern Qinghai provinces of China. In alpine meadows. Altitude 2800-3500 m. Flowering & fruiting: July - September.

A

B

A-C. 中国甘肃省卓尼
Zhuoni, Gansu, China. 23.07.2011

D. 中国四川省若尔盖
Ruoergai, Sichuan, China. 27.07.2004

请注意，不同花中雄蕊的不同生长姿态可能表明该花功能性的性别状态：左侧的花为雌性期，右侧的花为雄性期。

Please note that different positions of anthers in different flowers may suggest different stage of sexual function of the flower: left flower in female stage while the right flower in male stage.

异药龙胆

Gentiana anisostemon

二年生草本，高 3~5 cm，坚硬，近光滑；茎生叶密集，叶片长圆状披针形，先端有小尖头；花数枚簇生枝端；花萼筒倒锥形，花萼裂片披针形，先端尖细；花冠筒漏斗状，黄白色到淡蓝色，喉部具黄绿色斑点和条纹；花冠裂片卵形；褶卵形，较花冠裂片稍短，全缘或齿状。

仅见于中国云南省西北部。生长于较干燥的山坡、森林、灌丛和草地边缘，海拔 2000~3500 m。花果期 4~7 月。

Biannual, 3-5 cm, erect, rigid and glabrous; stem leaves overlapping, densely spaced, leaf blades oblong-lanceolate, apex mucronate; flowers multiple clustered at terminal; calyx tube obconic, calyx lobes lanceolate, apiculate; corolla tube funnelform, yellowish white to pale blue, with dark greenish yellow spots and stripes; corolla lobes ovate; plicae ovate, shorter than lobes, entire or denticulate.

Endemic to northwestern Yunnan of China. In dry slope and margins of forest, bushes and grassland. Altitude 2000-3500 m. Flowering and fruiting: April - July.

A-C. 中国云南省昆明市
Kunming, Yunnan, China. 18.04.1998

D. 中国云南省丽江市
Seedling, Lijiang, Yunnan, China. 04.10.1992

E. 中国云南省昆明市
Kunming, Yunnan, China. 18.04.1998

D

E

开张龙胆

Gentiana aperta

一年生草本，茎直立或斜升，高 3~10 cm，基部多分枝，全株光滑；叶片卵形到椭圆形，较节间短许多；花单生枝端；花冠筒漏斗状，黄白色到淡蓝色，具黄绿色斑点和条纹；花冠裂片卵状椭圆形，褶长圆形，2 裂齿。

仅见于中国青海东北部及相邻近的甘肃中西部高原地区。生长于高山草甸，海拔 2500~4000 m。花果期 7~8 月。

Annual, stems erect or ascending, 3-10 cm, much branched from base, glabrous; leaf blades ovate to elliptic, much shorter than internodes; flowers solitary and terminal; corolla funnelform, yellowish white to pale blue, with dark greenish yellow spots and stripes; corolla lobes ovate-elliptic; plicae oblong, 2-clift.

Endemic to the high plateau in northeastern Qinghai and adjacent central to western Gansu provinces of China. In alpine meadows. Altitude 2500-4000 m. Flowering & fruiting: July - August.

A–G. 中国青海省门源回族自治县
Menyuan, Qinghai, China. 18.07.2011

D

E

F

G

阿里山龙胆

Gentiana arisanensis

多年生草本，有根状茎，茎直立或斜升，高 5~10 cm，基部多分枝，全株光滑；叶密集，叶片卵形到卵状椭圆形，外弯，近革质；花单生枝端；花萼筒漏斗状，萼裂片卵状，革质；花冠筒蓝色到蓝紫色，喉部淡黄色，有时有褐色斑点；花冠裂片卵状三角形，褶长圆形，边缘整齐或啮齿状。

特产于中国台湾。生长于高山草甸和灌丛边缘向阳处，海拔 2500~3500 m。花果期 7~8 月。

Perennial, with short stolon; stems erect or ascending, 5-10 cm, much branched from base, glabrous; leaves crowded, leaf blades ovate-elliptic, recurved, leathery; flowers solitary and terminal; corolla funnelform, blue to blue-purple, yellowish and sometimes with brown spots in throat; corolla lobes ovate-triangular; plicae oblong, margins entire or erose.

Endemic to Taiwan province of China. In sunny places of alpine meadows and bush margins. Altitude 2500-3500 m. Flowering & fruiting: July - August.

A

B

C

A–C. 中国台湾省太鲁阁国家公园
摄影 /Photo: Cheng-Tao Lin
Taroko National Park, Taiwan, China. 07.2009

D–E. 中国台湾省花莲市 摄影 /Photo: 付鹏程
Hualian, Taiwan, China. 09.07.2022

D

E

刺芒龙胆

Gentiana aristata

一年生草本，茎直立或斜升，高 3~10 cm，基部多分枝，全株光滑；叶片线形到线状椭圆形，对折，先端具芒尖；花单生枝端；花冠筒漏斗状，淡蓝紫色到淡粉红色，具蓝灰色条纹；花冠裂片卵形，褶长圆形，先端平截，边缘呈细锯齿状。

仅见于中国西藏、四川、青海及甘肃相邻接的高原地区。生长于高山草甸和灌丛边缘，海拔 2000~4000 m。花果期 6~9 月。

本种与反折花龙胆相似，但褶的形状和边缘明显不同。

Annual, stems erect or ascending, 3-10 cm, much branched from base, glabrous; leaf blades linear to linear-elliptic, folded, apex awned; flowers solitary and terminal; corolla funnelform, pale blue-purple to pale pink, with dark blue-grey stripes; corolla lobes ovate; plicae oblong, apex truncate, margin laciniate. Endemic to the high plateau areas connecting Xizang, Sichuan, Qinghai and Gansu provinces of China. In alpine meadows and bush margins. Altitude 2000-4000 m. Flowering & fruiting: June - September.

A

B

C

A-B. 中国四川省红原县
Hongyuan, Sichuan, China. 28.07.2004

C-E. 中国青海省门源回族自治县
Menyuan, Qinghai, China. 17.07.2011

F-G. 中国四川省红原县
Hongyuan, Sichuan, China. 28.07.2004

H. 中国青海省河南蒙古族自治县
Henan, Qinghai, China. 09.09.2023

D
E
F
G
H

星萼龙胆

Gentiana asterocalyx

一年生草本，茎直立或铺散，高 5~10 cm，基部以上多分枝，全株具乳突；叶较节间更长，越向顶端叶越长，叶片钻形到线状披针形，硬质，先端渐尖，尖刺状；花单生枝端；花萼裂片开散，针状；花冠筒漏斗状，淡蓝色，具蓝灰色条纹和斑点；花冠裂片卵形到卵状椭圆形，褶长圆形，先端圆钝。

仅见于中国云南省西北部。生长于高山草甸和灌丛边缘，海拔 3000~3500 m。花果期 7~10 月。

Annual, stems erect or spreading, 5-10 cm, much branched from base and upper part, densely papillate; leave longer than internodes, becoming longer toward apex; leaves subulate to lanceolate, rigid, apex acuminate and apiculate; flowers solitary and terminal; calyx lobes spreading, needle-like; corolla funnelform, pale blue, with blue-grey stripes and spots; corolla lobes ovate to ovate-elliptic; plicae oblong, apex obtuse.

Endemic to northwestern Yunnan province of China. In alpine meadows and bush margins. Altitude 3000-3500 m. Flowering & fruiting: July - October.

A-B. 中国云南省丽江市
Lijiang, Yunnan, China. 02.10.1992

卵萼龙胆

Gentiana bryoides

一年生草本，茎直立或斜升，高 2~5 cm，基部多分枝，全株疏被乳突；叶与节间等长或稍短，叶片倒卵形，边缘软骨质；花单生枝端；花萼裂片开展外弯，卵形，先端具尾尖；花冠筒管状，淡蓝色，具蓝灰色条纹；花冠裂片卵形，褶阔卵形，先端圆钝。

分布于喜马拉雅地区及中国西藏东南部和云南西北部。生长于高山林下、林缘及灌丛边缘，海拔3500~4000 m。花果期6~8月。

Annual, stems erect or ascending, 2-5 cm, much branched from base, sparsely papillate; leave equal to or slightly shorter than internodes; leaf blades obovate, margin cartilaginous; flowers solitary and terminal; calyx lobes spreading or recurved, ovate; corolla tubular, pale blue, with blue-grey stripes; corolla lobes ovate; plicae broadly ovate, apex obtuse.

Spread in the Himalayas, including southern eastern Xizang and northwestern Yunnan provinces of China. In alpine forest, forest margin and bush margins. Altitude 3500-4000 m. Flowering and fruiting: June - August.

A-B. 中国云南省贡山独龙族怒族自治县
Gongshan, Yunnan, China. 04.08.2000

C. 中国西藏自治区错那县
Cuona, Xizang, China. 14.08.2006

蓝灰龙胆

Gentiana caeruleogrisea

一年生草本，基部多分枝，茎斜升或直立，蕾期花枝下弯，高 5~10 cm，光滑；茎生叶倒卵形到匙形；花单生分枝顶端；花萼筒倒锥形，花萼裂片锥形；花冠筒漏斗状，外面灰蓝色，内面白色，具深绿色条纹；花冠裂片卵状三角形，褶卵圆形，较花冠裂片稍短，先端平截，边缘啮齿状；花冠极易受遮光和触碰而闭合。蒴果狭柱状。

特产于中国甘肃、四川和青海邻接地区。生长于沼泽草甸，海拔 3500~4500 m。花果期 6~9 月。

Annual, stems much branched from base, ascending or erect, branches recurved to ground before flowering, 5-10 cm, glabrous; stem leaves obovate to spathulate; flowers solitary; calyx tube obconic, calyx lobes subulate; corolla funnelform, grey to greenish blue outside, white inside, with dark green stripes; corolla lobes ovate-triangular; plicae ovate-orbicular, shorter than corolla lobes, apex truncate, margins erose; corolla easily closed upon shadow & touching. Capsules narrowly cylindric.

Endemic to the high plateau connecting Sichuan, Gansu and Qinghai Provinces of China. In bog meadows. Altitude 3500-4500 m. Flowering & fruiting: June - September.

A

B

C

A–C. 中国青海省河南蒙古族自治县
Henan, Qinghai, China. 10.09.2023

D. 蓝灰龙胆 *Gentiana caeruleogrisea* + 线叶龙胆 *Gentiana lawrencei* var. *Farreri*
中国青海省河南蒙古族自治县
Henan, Qinghai, China. 10.09.2023.

D

反折花龙胆

Gentiana choanantha

一年生草本，茎直立或斜升，高 3~8 cm，基部多分枝，全株光滑；叶片线形到披针形，对折，具芒尖；花单生枝端；花萼裂片三角形，直立；花冠筒漏斗状，淡蓝色到淡粉红色，具蓝灰色条纹；花冠裂片卵形到卵状长圆形，有时反折，褶长圆形，先端圆钝，有时 2 裂，边缘整齐。

特产于中国四川西部高山地区。生长于高山草甸、沼泽草甸和山坡，海拔 3000~4000 m。花果期 6~8 月。

本种与刺芒龙胆很相似，但褶的形状和边缘明显不同。

Annual, stems erect or ascending, 3-8 cm, much branched from base, glabrous; leaf blades linear to lanceolate, folded, apex awned; flowers solitary and terminal; calyx lobes triangular, erect; corolla funnelform, pale blue-purple to pale pink, with dark blue-grey stripes; corolla lobes ovate to ovate-oblong sometimes recurved; plicae oblong, apex obtuse, sometimes 2-cleft, margin entire.

Endemic to the mountains in western Sichuan of China. In alpine meadows, bog meadows and slopes. Altitude 3000-4000 m. Flowering & fruiting: June - August.

A-B. 中国四川省丹巴县
Danba, Sichuan, China. 02.07.2021

C. 中国四川省康定市
Kangding, Sichuan, China. 04.07.2021

D. 中国四川省丹巴县
Danba, Sichuan, China. 02.07.2021

E-F. 中国四川省康定市
Kangding, Sichuan, China. 04.07.2021

D

E

F

西域龙胆

Gentiana clarkei

一年生草本，茎直立或斜升，高 2~5 cm，基部多分枝，全株光滑；叶片匙形到倒卵形，先端圆钝；花单生枝端；花萼裂片三角形，直立；花冠筒漏斗状，蓝色到蓝紫色，具深蓝紫色条纹；花冠裂片卵形到卵状三角形，褶卵形，先端圆钝，边缘整齐或锯齿状。

特产于青藏高原的中国西藏和青海。生长于高山草甸干燥向阳处，海拔 4000~5000 m。花果期 8~9 月。

Annual, stems erect or ascending, 2-5 cm, much branched from base, glabrous; leaf blades spathulate to obovate, apex rounded; flowers solitary and terminal; calyx lobes triangular, erect; corolla funnelform, blue to blue-purple, with dark blue-purple stripes; corolla lobes ovate to ovate-triangular; plicae ovate, apex obtuse, margin entire to serrulate. Endemic to the Qinghai-Xizang Plateau, Xizang and Qinghai provinces of China. In dry and sunny places of alpine meadows. Altitude 4000-5000 m. Flowering and fruiting: August - September.

A-B. 中国西藏自治区聂拉木县
Nielamu, Xizang, China. 16.09.1992
C. 中国西藏自治区定日县
Dingri, Xizang, China. 15.09.1992

景天叶龙胆

Gentiana crassula

一年生草本，茎斜升，高 5~8 cm，基部及以上多分枝，紫红色，全株密被乳突；茎生叶肉质，叶片匙形到倒卵形，先端急尖，外弯，边缘软骨质；花单生分枝端；花萼裂片卵状圆形，先端具外弯的尾尖；花冠筒管状，蓝到蓝紫色；花冠裂片卵形到卵状长圆形，褶阔卵形，先端圆钝，边缘整齐或不规则啮蚀状。

散布于中国西藏、云南和四川等省区。生长于高草甸及灌丛边缘，海拔 3500~4200 m。花果期 7~8 月。

Annual, stems ascending, 5-8 cm, much branched from base and above, reddish purple, densely papillate; stem leaves fleshy, leaf blades spathulate to obovate, apex acute, cuspidate, margin cartilaginous; flowers solitary; calyx lobes ovate-orbicular, apex recurved and cuspidate; corolla tubular, blue to blue-purple; corolla lobes ovate to ovate-oblong; plicae broadly ovate, apex obtuse, margin entire or erose.

Scattered in Xizang, Yunnan and Sichuan provinces of China. In alpine meadows and bush margins. Altitude 3500-4200 m. Flowering and fruiting: July - August.

A–C. 中国西藏自治区亚东县
Yadong, Xizang, China. 11.08.2006

对折龙胆

Gentiana conduplicata

一年生草本，茎斜升或铺散，高 2~5 cm，基部多分枝，近光滑；茎生叶较密集，较节间长，叶片狭椭圆形到狭披针形；花单生枝端；花萼筒狭管状，花萼裂片狭三角形；花冠筒狭管状，蓝色；花冠裂片卵形，褶卵形，与花冠裂片近等长，整齐或偶有细裂齿；花冠极易受遮光和触碰而闭合。

仅见于中国四川和西藏。生长于高山草原和高山草甸，海拔 3500~4500 m。花果期 5~7 月。本种居群中常有蓝花和黄白色花个体混生，是否属同种，需进一步研究。

Annual, stems ascending or prostrate, 2-5 cm, much branched from base, nearly glabrous; stem leaves crowded, longer than internodes, leaf blades narrowly elliptic to narrowly lanceolate; flowers solitary and terminal; calyx tube narrowly tubular, calyx lobes narrowly triangular; corolla tube narrowly tubular, blue; corolla lobes ovate; plicae ovate, nearly as long as corolla lobes, entire or rarely denticulate; corolla easily closed upon shadow or touching. Endemic to the regions Sichuan and Xizang provinces of China. In alpine grassland and alpine meadows. Altitude 3500-4500 m. Flowering and fruiting: May - July.

A

B

C

A–G. 中国四川省康定市
Kangding, Sichuan, China. 04.07.2021

D

黄白色花个体 Individual with yellowish white flowers

E

F

黄白色花和蓝个体混生在一起 Individuals with yellowish white flowers and blue flowers growing together

G

肾叶龙胆

Gentiana crassuloides

一年生草本，茎斜升，高 3~5 cm，基部多分枝，全株密被乳突；叶片卵形到卵状圆形，先端急尖，具尾，边缘软骨质；花单生枝端；花萼裂片外弯，倒卵形到肾形，先端具外弯的尾尖；花冠筒管状，淡蓝到蓝紫色；花冠裂片卵形到卵状长圆形，褶阔卵形，先端圆钝，边缘不规则啮蚀状。

分布于喜马拉雅到中国中西部地区，较常见。生长于高山草甸、林缘及灌丛边缘，海拔 2500~4000 m。花果期 6~9 月。

Annual, stems ascending, 3-5 cm, much branched from base, densely papillate; stem leaf blades obovate to ovate-circular, apex acute, cuspidate, margin cartilaginous; flowers solitary; calyx lobes recurved, ovate to reniform; corolla tubular, pale blue, to blue-purple; corolla lobes ovate-oblong; plicae broadly ovate, apex obtuse, margin erose.

From the Himalayas to central and western provinces of China, common. In alpine meadows, forest margin and bush margins. Altitude 2500-4000 m. Flowering and fruiting: June - September.

A

B

A–D. 中国四川省康定市
Kangding, Sichuan, China. 05.07.2021

E. 中国四川省康定市
Kangding, Sichuan, China. 04.07.2021

F. 中国青海省久治县
Jiuzhi, Qinghai, China. 11.09.2023

C
D
E
F

圆齿褶龙胆

Gentiana crenulatotruncata

一年生草本，茎直立或斜升，高 2~3 cm，基部分枝，全株疏被乳突；叶片匙形到倒卵形，先端圆钝，边缘软骨质；花单生枝端；花萼裂片卵状三角形，直立；花冠筒高脚杯状，灰蓝色到蓝紫色，具深紫色条纹；花冠裂片卵形到卵状三角形，褶卵状长圆形，先端平截，边缘啮齿状。

特产于青藏高原的中国西藏、青海和甘肃。生长于高山草甸干燥向阳处，海拔 4000~5000 m。花果期 8~9 月。

Annual, stems erect or ascending, 2-3 cm, branched from base, sparsely papillate; leaf blades spathulate to obovate, apex rounded; flowers solitary; calyx lobes ovate-triangular, erect; corolla salverform, grey-blue to blue-purple, with dark purple stripes; corolla lobes ovate to ovate-triangular; plicae ovate-oblong, apex truncate, margin erose.

Endemic to the Qinghai-Xizang Plateau, Xizang, Qinghai and Gansu provinces of China. In dry and sunny places of alpine meadows. Altitude 4000-5000 m. Flowering and fruiting: August - September.

A

B

C

A-C. 中国青海省格尔木市
Geermu, Qinghai, China. 04.08.2006
D. 中国甘肃省玛曲县
Maqu, Gansu, China. 10.08.1992
E. 中国青海省沱沱河
Tuotuohe, Qinghai, China. 05.08.2006
F. 中国青海可可西里国家级自然保护区
Kekexili, Qinghai, China. 04.08.2006
G. 中国青海省格尔木市
Geermu, Qinghai, China. 04.08.2006
H. 中国青海可可西里国家级自然保护区
Kekexili, Qinghai, China. 04.08.2006

D
E
F
G
H

弯叶龙胆

Gentiana curviphylla

一年生草本，茎斜升，高 3~7 cm，基部多分枝，全株密被乳突，开花后茎节伸长；叶片卵形，先端急尖，外弯，具尾，边缘硬质软骨质；花单生枝端；花萼裂片外弯，卵形到卵状三角形，先端具外弯的尾尖；花冠筒管状，蓝色到蓝紫色，具密集的深蓝紫色斑点；花冠裂片卵形，褶阔卵形，先端圆钝，边缘近整齐。

特产于中国四川西部。生长于向阳高山草甸、灌丛边缘及空旷山坡，海拔 2500~4000 m。花果期 6~9 月。

Annual, stems ascending, 3-7 cm, much branched from base, densely papillate; internodes elongated after flowering; leaf blades ovate, apex acute, recurved and cuspidate, margin thick cartilaginous; flowers solitary; calyx lobes recurved, ovate to ovate-triangular; corolla tubular, blue to blue-purple, with dense dark purple spots; corolla lobes ovate; plicae broadly ovate, apex obtuse, margin nearly entire.

Endemic to western Sichuan of China. In sunny alpine meadows, bush margins, and open places. Altitude 2500-4000 m. Flowering and fruiting: June - September.

A. 中国四川省康定市
Kangding, Sichuan, China. 03.07.2021

B. 中国四川省丹巴县
Danba, Sichuan, China. 02.07.2021

C-E. 中国四川省康定市
Kangding, Sichuan, China. 02.07.2021

F-G. 中国四川省道孚县
Daofu, Sichuan, China. 02.07.2021

D

E

F

G

黄山龙胆

Gentiana delicata

一年生草本，茎直立，高 5~15 cm，基部以上多分枝，茎紫色，被乳突；花期莲座叶宿存，阔卵形；茎生叶显著较小，披针形，较节间短；花 3 到多数，排列成松散的总状花序；花冠筒漏斗状，黄白色到淡蓝色，具黄绿色条纹；花冠裂片卵状三角形，褶阔卵形，长不及花冠裂片的一半，边缘较整齐。

特产于中国中东部地区的安徽、福建及浙江。生长于山地林缘、灌丛边缘等向阳处，海拔 800~1500 m。花果期 5~7 月。

Annual, stems erect, 5-15 cm, branched from base and above, papillate; rosette leaves persistent at anthesis, broad ovate; stem leaves much smaller, lanceolate, shorter than internodes; flowers 3 to several in raceme; corolla funnelform, yellowish white to pale blue, with greenish yellow stripes; corolla lobes ovate-triangular; plicae broad ovate, less than ½ of the corolla lobes, margin nearly entire.

Endemic to central to eastern China, in Anhui, Fujian and Zhejiang provinces. In open and sunny places of mountain forest and bush margins. Altitude 800-1500 m. Flowering and fruiting: May - July.

A–D. 中国浙江省百丈岭

Baizhangling Mt., Zhejiang, China. 09.05.2023

中国安徽省黄山 Huangshan Mt., Anhui, China. 08.10.2010

沼泽龙胆

Swamp Gentian

Gentiana douglasiana

一年生草本，茎直立，高 5~20 cm，基部以上多分枝，光滑；花期莲座叶宿存，卵形到椭圆形；茎生叶较小，卵形到椭圆形，较节间短；花多数，排列成松散的总状花序；花冠筒管状，黄白色到淡蓝色，具黄绿色或紫色斑点；花冠裂片卵状长圆形，褶卵形，长约花冠裂片的一半，2 裂。

特产于北美西海岸，从美国华盛顿州、加拿大不列颠哥伦比亚省直到阿拉斯加的太平洋沿岸低山到中山区。生长于潮湿向阳的沼泽及沼泽化草甸处，海拔 50~1000 m。花果期 6~8 月。

Annual, stems erect, 5-20 cm, branched from base and above, glabrous; rosette leaves persistent at anthesis, ovate to elliptic; stem leaves smaller, ovate to elliptic, shorter than internodes; flowers several in raceme; Corolla tubular, yellowish white to pale blue, with greenish yellow or purple spots; corolla lobes oblong; plicae ovate ~½ of the corolla lobes, 2-cleft.

Endemic Pacific coast of North America, from Washington state, to Alaska through British Colombia. In wet and sunny places of low to subalpine bogs and wet meadows. Altitude 50-1000 m. Flowering and fruiting: June - August.

美国阿拉斯加 摄影 /Photo: Erica Simonson

Hoonah-Angoon County, Alaska, USA. 06.07.2021

丝瓣龙胆

Gentiana exquisita

多年生草本，具纤细较短的匍匐茎，茎纤细直立，高 10~15 cm，光滑；茎生叶卵形到卵状披针形，较节间短许多；花单生枝顶端；花萼筒倒锥形，花萼裂片丝状直立；花冠筒钟形到倒锥形，淡蓝色到蓝灰色，外侧基部具深蓝紫色条纹，内面具深紫色斑点；花冠裂片卵状三角形，褶阔卵形，与花冠裂片等长，边缘具丝状流苏。

特产于中国云南省西北部及邻近的缅甸山区。生长于山地林缘及灌丛中，海拔 3000~4000 m。花果期 8~9 月。

Perennial, with short slender stolon; stems erect, slender, 10-15 cm, glabrous; stem leaves ovate to ovate-lanceolate, much shorter than internodes; flowers solitary and terminal; calyx tube obconic, calyx lobes filiform, erect; corolla broad campanulate to obconic, pale blue to grey-blue, with dark blue-purple stripes outside at base and dark purple spots inside at base; corolla lobes ovate-triangular; plicae broad ovate, equal to corolla lobes, margin filiform fringed.

Endemic to mountains in northwestern Yunnan province of China and adjacent Myanmar. In forest margins and bushes. Altitude 3000-4000 m. Flowering and fruiting: August - September.

A

B

A–E. 中国云南省贡山独龙族怒族自治县
Gongshan, Yunnan, China. 03.08.2000

C
D
E

毛喉龙胆

Gentiana faucipilosa

一年生草本，茎斜升或铺散，高 5~7 cm，基部多分枝，有乳突；茎生叶较小，线形到线状披针形，先端渐尖，具尾尖；花单生分枝顶端；花萼筒管状到倒锥形，花萼裂片线状披针形；花冠筒管状，蓝色，喉部黄色，具深蓝色条纹和分散的喉毛；花冠裂片卵形，褶卵形，与花冠裂片近等长，边缘近整齐或齿状。

产于中国云南西北部和西藏东南部及不丹。生长于高山草地及森林和灌丛边缘，海拔 2500~3500 m。花果期 6~9 月。

Annual, stems ascending to prostrate, 5-7 cm, much branched from base, papillate; stem leaves small, linear to linear-lanceolate, apex acuminate, cuspidate; flowers solitary at branch terminals; calyx tube tubular to obconic, calyx lobes linear-lanceolate; corolla tubular, blue, throat yellow with dark blue stripes and scattered hairs; corolla lobes ovate; plicae ovate, almost as long as the corolla lobes, margin nearly entire or denticulate.

Scattered in northwestern Yunnan and southeastern Xizang provinces of China and Bhutan. In open and sunny alpine slopes and margins of forest bushes. Altitude 2500-3500 m. Flowering and fruiting: June - September.

A. 中国云南省贡山独龙族怒族自治县
摄影 /Photo: Adrien Favre
Gongshan, Yunnan, China. 02.08.2011.

B-C. 中国云南省贡山独龙族怒族自治县
摄影 /Photo: 付鹏程
Gongshan, Yunnan, China. 26.08.2018

弯茎龙胆

Gentiana flexicaulis

一年生草本，茎斜升或铺散，高 3~7 cm，基部以上多分枝，光滑；花期莲座叶宿存，卵形到卵圆形；茎生叶较小，卵形到匙形，明显较节间短，先端外弯；花单生分枝顶端；花萼筒倒锥形，花萼裂片锥形；花冠筒漏斗状，蓝色，喉部深蓝色条纹；花冠裂片卵形，褶卵形，与花冠裂片近等长，边缘较整齐。

特产于中国四川北部。生长于高山草甸和灌丛边缘等向阳处，海拔 2500~3500 m。花果期 6~8 月。

Annual, stems ascending to prostrate, 3-7 cm, much branched from base and above, glabrous; rosette leaves persistent at anthesis, ovate to ovate-orbicular; stem leaves smaller, ovate to lanceolate, much shorter than internodes, apex recurved; flowers solitary at branch terminals; calyx tube obconic, calyx lobes subulate; corolla funnelform, blue, with dark blue stripes; corolla lobes ovate; plicae ovate, almost as long as the corolla lobes, margin nearly entire.
Endemic to northern Sichuan province of China. In open and sunny places of bush margins and alpine meadows. Altitude 2500-3500 m. Flowering and fruiting: June - August.

A–B. 中国四川省红原县
Hongyuan, Sichuan, China. 28.07.2004

圆球龙胆

Gentiana globosa

一年生草本，茎矮小，高 2~3 cm，茎中部以上多分枝，花和叶密集成球状；叶片倒卵形，先端外弯，具尖头，边缘明显的软骨质；花 3~5 朵，密集成头状；花萼筒筒状，花萼裂片近心形，先端具尖头，边缘明显软骨质；花冠筒状，淡蓝色，喉部深蓝色和黄绿色条纹及斑点；花冠裂片卵状三角形，褶卵圆形，长仅约花冠裂片的一半，边缘较整齐。

特产于中国四川西部。生长于高山草甸和灌丛边缘等向阳处，海拔 3000~4000 m。花果期 7~10 月。

Annual, stems very short, 2-3 cm, much branched from middle, leaves and flowers crowded to capitate; blade obovate, apex recurved and cuspidate, margins distinctly cartilaginous; flowers 3-5 crowded; calyx tube tubular, calyx lobes suborbicular; corolla tubular, pale blue, with dark blue and green stripes and dots; corolla lobes ovate-triangular; plicae ovate-rounded, almost ½ as long as the corolla lobes, margin nearly entire.

Endemic to western Sichuan of China. In open and sunny places of bush margins and alpine meadows. Altitude 3000-4000 m. Flowering and fruiting: July - October.

A–C. 中国四川省德格县雀儿山

Queershan Mt., Sichuan, China. 26.09.1993

阿尔泰龙胆

Gentiana grandiflora

多年生草本，具有纤细较短的匍匐茎和发达的莲座叶丛，茎极短，直立，高 2~3 cm，光滑；莲座叶钻形到披针形，茎生叶钻形，集中在花的基部；花大型，单生枝顶端；花萼筒倒锥形，花萼裂片披针形，直立；花冠筒管状钟形，深蓝色；花冠裂片阔卵状，褶阔卵形，与花冠裂片近等长，先端平截，边缘啮齿状。

特产于阿尔泰山及其周边的哈萨克斯坦、俄罗斯和中国新疆。生长于山地草甸及灌丛边缘，海拔 2000~3000 m。花果期 7~9 月。

Perennial, with short slender stolon, rosettes persistent; stems short, erect, 2-3 cm, glabrous; rosette leaves subulate to lanceolate; stem leaves subulate, crowded at the base of flowers; flowers big, solitary; calyx tube obconic, calyx lobes lanceolate, erect; corolla tubular-campanulate, dark blue; corolla lobes broad ovate; plicae broad ovate, nearly as long as corolla lobes, apex truncate, erose.

Endemic to the Altai Mountains and areas around it, in Kazakhstan, Russia, and Xinjiang of China. In meadows and bush margins. Altitude 2000-3000 m. Flowering and fruiting: July - September.

A-D. 中国新疆维吾尔自治区喀纳斯
Kanasi, Xinjiang, China. 09.08.1995

D

长流苏龙胆

Gentiana grata

多年生草本，有短根状茎，茎纤细，直立，高 8~15 cm，黄绿色，光滑；茎生叶卵形到卵状椭圆形，排列稀疏，较节间短许多；花单生枝顶端；花萼筒筒状，花萼裂片锥形，整齐，直立；花冠筒漏斗形，淡蓝色到蓝灰色，外侧基部深蓝紫色，内面具深紫色条纹；花冠裂片卵形，褶卵形，长约花冠裂片之半，边缘具丝状短流苏。

特产于中国西藏东南部、云南省西北部及邻近的缅甸山区。生长于山地草地，海拔 3000~4000 m。花果期 8~9 月。

Perennial, with short rhizome; stems erect, slender, 8-15 cm, greenish yellow, glabrous; stem leaves ovate to elliptic, sparsely arranged, much shorter than internodes; flowers solitary and terminal; calyx tube tubular, calyx lobes subulate, erect; corolla funnelform, pale blue to grey-blue, dark blue-purple outside at base and with dark purple stripes inside; corolla lobes ovate; plicae ovate, ~ ½ corolla lobes, margin shortly filiform fringed.

Endemic to mountains in southeastern Xizang and northwestern Yunnan provinces of China and adjacent Myanmar. In forest slopes. Altitude 3000-4000 m. Flowering and fruiting: August - September.

A-B. 中国云南省贡山独龙族怒族自治县
摄影 /Photo: 付鹏程
Gongshan, Yunnan, China. 27.08.2018

A

B

C

南山龙胆

Gentiana grumii

一年生草本，基部多分枝，茎斜升或铺散，高 2~4 cm，光滑；花期莲座叶凋萎，茎生叶线形到披针形，较节间稍长，对折；花单生分枝顶端；花萼筒倒锥形，花萼裂片锥形，先端尾尖；花冠筒漏斗状，淡蓝色，内面具深蓝色斑点；花冠裂片卵状三角形，褶卵形，较花冠裂片稍短，边缘较整齐或具短齿。

特产于中国青海和甘肃相邻的祁连山地区。生长于高山草甸和灌丛边缘等向阳处，海拔 2500~3500 m。花果期 6~7 月。

Annual, stems branched from base, ascending to prostrate, 2-4 cm, glabrous; rosette leaves withered at anthesis; stem leaves linear to lanceolate, slightly longer than internodes, folded; flowers solitary at branch terminals; calyx tube obconic, calyx lobes subulate; corolla funnelform, pale blue, with dark blue spots; corolla lobes ovate-triangular; plicae ovate, slightly shorter than corolla lobes, margin entire or denticulate.

Endemic to the Qilian Mountains between Qinghai and Gansu provinces of China. In open and sunny places of bush margins and alpine meadows. Altitude 2500-3500 m. Flowering and fruiting: June - July.

A-B. 中国青海省门源回族自治县
Menyuan, Qinghai, China. 18.07.2011
C. 中国甘肃省肃南裕固族自治县
Sunan, Gansu, China. 21.07.1995

钻叶龙胆

Gentiana haynaldii

一年生草本，基部多分枝，茎斜升或直立，高 3~10 cm，光滑；茎生叶钻形，密集排列，较节间长，坚硬，近革质，有光泽；花单生分枝顶端；花萼筒倒锥形，花萼裂片锥形，先端尾尖，近革质；花冠筒漏斗状，淡蓝色，具深紫色斑点；花冠裂片卵形，褶卵圆形，较花冠裂片稍短，先端平截，边缘不规则啮齿状。

特产于中国青藏高原的西藏、云南、四川和青海等省区。生长于高山草甸和灌丛边缘等向阳处，海拔 2500~4500 m。花果期 7~10 月。

Annual, stems much branched from base, ascending to erect, 3-10 cm, glabrous; stem leaves subulate, crowded and longer than internodes, rigid, subleathery, glossy; flowers solitary; calyx tube obconic, calyx lobes subulate, caudate, subleathery; corolla funnelform, pale blue, with dark blue spots; corolla lobes ovate; plicae ovate-orbicular, shorter than corolla lobes, truncate and erose. Endemic to the Qinghai-Xizang Plateau, Xizang, Yunnan, Sichuan and Qinghai Provinces of China. In open and sunny places of bush margins and alpine meadows. Altitude 2500-4500 m. Flowering and fruiting: July - October.

A

B

C

A. 中国四川省道孚县
Daofu, Sichuan, China. 01.10.1993

B. 中国西藏自治区那曲县
Naqu, Xizang, China. 07.08.2006

C-D. 中国四川省康定市
Kangding, Sichuan, China. 14.09.2023

E. 中国西藏自治区丁青县
Dingqing, Xizang, China. 25.09.1992

F. 中国云南省香格里拉市
Xianggelila, Yunnan, China. 18.09.2023

D

E

F

针叶龙胆

Gentiana heleonastes

一年生草本，基部多分枝，茎斜升或直立，蕾期花枝下弯，高 5~10 cm，光滑；茎生叶钻形或针形；花单生分枝顶端；花萼筒倒锥形，花萼裂片锥形；花冠筒漏斗状，外面灰蓝色，内面白色，具深绿色条纹；花冠裂片卵状三角形，褶卵圆形，较花冠裂片稍短，先端平截，边缘啮齿状；花冠极易受遮光和触碰而闭合；蒴果倒卵状。

特产于中国甘肃、四川和青海邻接地区。生长于沼泽草甸，海拔 3500~4500 m。花果期 6~9 月。

Annual, stems much branched from base, ascending or erect, branches recurved to ground before flowering, 5-10 cm, glabrous; stem leaves subulate to needle-like; flowers solitary; calyx tube obconic, calyx lobes subulate; corolla funnelform, grey to greenish blue outside, white inside, with dark green stripes; corolla lobes ovate-triangular; plicae ovate-orbicular, shorter than corolla lobes, apex truncate, margins erose; corolla easily closed upon shadow & touching. Capsules obovoid. Endemic to the high plateau connecting Sichuan, Gansu and Qinghai Provinces of China. In bog meadows. Altitude 3500-4500 m. Flowering and fruiting: June - September.

A

B

C

A. 中国青海省河南蒙古族自治县
Henan, Qinghai, China. 30.08.1994

B. 中国甘肃省玛曲县 Maqu, Gansu, China. 27.07.1990

C. 中国青海省河南蒙古族自治县
Henan, Qinghai, China. 30.08.1994

D. 中国甘肃省玛曲县 Maqu, Gansu, China. 10.08.1992

E. 针叶龙胆 *Gentiana heleonastes*（右上 top-right）and
蓝白龙胆 *Gentiana leucomelaena* （左下 bottom-left）
中国青海省河南蒙古族自治县
Henan, Qinghai, China. 30.08.1994

D

E

膜果龙胆

Gentiana hyalina

一年生草本，高 2~5 cm，基部多分枝，全株光滑，花期前茎铺散，果期后茎直立伸长；叶片匙形到倒卵形，先端圆钝；花单生枝端；花萼裂片三角形，直立；花冠筒漏斗状，蓝色，偶尔白色，具深蓝紫色条纹；花冠裂片卵状三角形，先端渐尖，褶卵形，2 裂，边缘整齐；果膜质，半透明。

特产于青藏高原的中国西藏和青海。生长于高山草甸干燥向阳处，海拔 4000~5000 m。花果期 8~9 月。本种与西域龙胆和圆齿褶龙胆接近，区别在于本种花冠裂片先端尖，蒴果开裂时膜质，半透明。

Annual, 2-5 cm, much branched from base, glabrous, stems short and prostrate before anthesis, becoming erect and elongated after anthesis, leaf blades spathulate to obovate, apex rounded; flowers solitary and terminal; calyx lobes triangular, erect; corolla funnelform, blue, occasionally white, with dark blue-purple stripes; corolla lobes ovate-triangular, apex acuminate; plicae ovate, 2-cleft, margin entire; capsules membranous. semitransparent.

Endemic to the Qinghai-Xizang Plateau, Xizang and Qinghai provinces of China. In dry and sunny places of alpine meadows. Altitude 4000-5000 m. Flowering and fruiting: August - September.

A. 中国青海省格尔木市
Geermu, Qinghai, China. 04.08.2006

B. 中国青海省玛多县
Maduo, Qinghai, China. 23.09.1993

C-D. 中国青海省格尔木市
Geermu, Qinghai, China. 04.08.2006

E-F. 白花型膜果龙胆 *Gentiana hyalina*
中国青海省沱沱河
Tuotuohe, Qinghai, China. 05.08.2006

帚枝龙胆

Gentiana intricata

一年生草本，高 3~5 cm，茎基部单一，中部以上多分枝，全株密被紫红色乳突；莲座叶花期凋萎；茎生叶片匙形到倒卵形，先端圆钝，边缘软骨质；花多数，单生小枝顶端，密集成头状；花萼裂片卵形，外弯，具尾尖；花冠筒管状，淡蓝色，具紫色条纹；花冠裂片卵状，先端渐尖，褶卵形，边缘整齐或啮齿状。

特产于中国云南西北部。生长于高山草地及灌丛边缘向阳处，海拔 2200~3500 m。花果期 5~9 月。

Annual, 3-5 cm, stem simple at base, much branched from middle up, densely purple papillate; rosette leaves withered at anthesis; stem leaves spathulate to obovate, apex rounded; flowers many, solitary at branch terminals; crowded and capitate; calyx lobes ovate, recurved; corolla tubular, pale blue, with dark purple stripes; corolla lobes ovate, apex acuminate; plicae ovate, margin entire to erose. Endemic to southeastern Xizang and northwestern Yunnan of China. In dry and sunny places of mountain meadows and bush margins. Altitude 2200-3500 m. Flowering and fruiting: May - September.

A

B

A–B. 中国云南省丽江市
Lijiang, Yunnan, China. 20.09.1990

C–D. 中国云南省香格里拉市
Xianggelila, Yunnan, China. 30.09.1992

C

D

长白山龙胆

Gentiana jamesii

多年生草本，高 5~15 cm，茎基部多分枝，光滑；莲座叶花期凋萎；茎生叶卵状披针形到椭圆形，先端圆钝；花多数，单生或 2~3 朵生于小枝顶端；花萼裂片卵形，外弯；花冠筒管状，蓝色到蓝紫色；花冠裂片卵状，先端圆钝，褶卵形，边缘流苏状。

特产于东北亚地区俄罗斯、朝鲜、日本及中国的黑龙江、吉林和辽宁省。生长于高山草地及灌丛边缘向阳处，海拔 1200~2500 m。花果期 7~9 月。

Perennial, 5-15 cm, stem much branched from base, glabrous; rosette leaves withered at anthesis; stem leaves ovate-spathulate to elliptic, apex rounded; flowers many, solitary or 2-3 clustered at branch terminals; calyx lobes ovate, recurved; corolla tubular, blue to blue-purple; corolla lobes ovate, apex rounded; plicae ovate, margin fimbriate.

Endemic to northeastern Asia, in Russia, Korea peninsula, Japan and Heilongjiang, Jilin, and Liaoning provinces of China. In dry and sunny places of alpine meadows and bush margins. Altitude 1200-2500 m. Flowering and fruiting: July - September.

A. 中国吉林省长白山 摄影 /Photo: 张粤
Changbaishan Mt., Jilin, China. 22.07.2000

B. 中国吉林省长白山
Changbaishan Mt., Jilin, China. 19.08.2000

B

蓝白龙胆 A型

Gentiana leucomelaena Type A

一年生草本，茎铺散或斜升，高 2~3 cm，基部多分枝，全株光滑，略肉质；叶片披针形，较节间短，略肉质；花单生枝端；花冠筒漏斗状，外面蓝灰色，内面白色，喉部具黄紫色斑点；花冠裂片阔卵形，先端圆钝；褶轮廓阔卵形，与花冠裂片近等长，边缘啮齿状。

分散分布于锡金、青藏高原及周边地区，中国西藏、四川、青海和甘肃西南部等地区。生长于较干燥的高山草甸，海拔 3500~4500 m。花果期 7~9 月。蓝白龙胆分布广，变异较大。该类型体型矮小，花冠裂片与褶近等长，小生境干燥，为二倍体，2n=18。本类型新近被印度学者错误地作为耳褶属一个新种发表为 *Kuepferia pringlei*，并细分出了 2 个变种。

Annual, stems prostrate or ascending, 2-3 cm, much branched from base, glabrous, slightly fleshy; leaf blades lanceolate, shorter than internodes, fleshy; flowers solitary and terminal; corolla funnelform, blue-grey outside, yellowish white inside, with dark purple spots in throat; corolla lobes broad ovate, apex obtuse; plicae broad ovate, nearly as long as corolla lobes, margins erose.

Widespread in Sikkim & Qinghai-Xizang Plateau and adjacent areas, in Xizang, Sichuan, Qinghai and southern Gansu in China. In dry alpine meadows. Altitude 3500-4500 m. Flowering and fruiting: July - September. Diploid 2n=18.

A-B. 中国青海省格尔木市
Geermu, Qinghai, China. 04.08.2006

C-F. 中国青海省格尔木市可可西里
Kekexili, Geermu, Qinghai, China. 04.08.2006

G. 中国青海省格尔木市
Geermu, Qinghai, China. 04.08.2006

H. 中国西藏自治区康马县
Kangma, Xizang, China. 10.08.2006

D
E
F
G
H

蓝白龙胆 B型

Gentiana leucomelaena Type B

与A型较相似，但体型较大，高3~7 cm；花冠内面斑点细而密，花冠裂片卵状三角形，先端渐尖；褶轮廓狭卵形，长仅及花冠裂片的一半，边缘啮齿状。本类型为四倍体，2n=36。

本类型更为常见，分散分布于青藏高原及周边地区，中国西藏、四川、青海和甘肃西南部等地区常见。生长于较潮湿的溪边、高山草甸或沼泽草甸，海拔3500~4500 m。花果期7~10月。

Highly resemble Type A, but plants larger, 3-7 cm high; corolla with denser, and finer spots; corolla lobes narrowly ovate, apex acuminate; plicae narrowly ovate, about ½ as long as corolla lobes, margins erose. This type was shown to be tetraploid with 2n=36.

This type is more common, widespread in Qinghai-Xizang Plateau and adjacent areas, in Xizang, Sichuan, Qinghai and southern Gansu in China. Growing along stream-sides, in damp alpine meadows, or in bog meadows. Altitude 3500-4500 m. Flowering and fruiting: July - October.

Type A from Sikkim was wrongly recognized as a distinct species of the genus *Kuepferia*, as *K. pringlei*, and further split into 2 varieties, *K. pringlei* var. *lhonakensis* and *K. pringlei* var. *cavei*. [doi.org/10.1017/S0960428615000189]

A

B

C

A. 中国青海省河南蒙古族自治县
Henan, Qinghai, China. 30.08.1994

B. 中国四川省理塘县 Henan, Qinghai, China. 13.08.2002

C. 中国甘肃省碌曲县 Luqu, Gansu, China. 14.08.1992

D. 中国西藏自治区米拉山
Milashan Mt., Xizang, China. 16.08.2006

E. 中国西藏自治区那曲县 Naqu, Xizang, China. 07.08.2006

F. 中国四川省甘孜县 Ganzi, Sichuan, China. 29.09.1993

G. 蓝白龙胆（B型） Gentiana leucomelaena Type B 及
束花报春 Primula fasciculata
中国西藏自治区那曲县 Naqu, Xizang, China. 07.08.2006

H. 中国甘肃省夏河县 Xiahe, Gansu, China. 28.08.1994

I. 中国甘肃省碌曲县 Luqu, Gansu, China. 30.08.1994

D
E
F
G
H
I

蓝白龙胆　C 型

Gentiana leucomelaena Type C

与 A 型较相似，但体型较大，高 4~8 cm；花冠内面斑点疏而大，花冠裂片卵状，先端圆钝；褶轮廓阔卵形，长约花冠裂片的一半，边缘啮齿状。本类型染色体数目不详。

本类型较少见，仅在西藏自治区那曲县观察到。生长于沼泽草甸及其边缘，海拔约 4200 m。花果期 8~10 月。

Highly resemble Type A, but plants larger, 4-8 cm high; corolla with less but bigger spots; corolla lobes broadly ovate, apex rounded; plicae broad ovate, nearly as long as corolla lobes, margins erose. Chromosome number unknown.

This type was found only in Naqu, Xizang of China. Growing in bog meadows. Altitude 4200 m. Flowering and fruiting: August - October.

A-D. 中国西藏自治区那曲县
Naqu, Xizang, China. 07.08.2006

D

蓝白龙胆 D型

Gentiana leucomelaena Type D

与B型较相似，但体型较大，高4~8 cm；花冠淡蓝色到白色；褶轮廓阔卵形，长超过花冠裂片的一半，边缘啮齿状。本类型染色体数目不详。

本类型较少见，仅在西藏自治区隆子县观察到。生长于较潮湿的高山草甸及沟渠边缘，海拔约4000 m。花果期8~10月。

Highly resemble Type B, but plants larger, 4-8 cm high; corolla pale blue to white; plicae broad ovate, more than ½ as long as corolla lobes, margins erose. Chromosome number unknown. This type was found in Longzi, Xizang of China only. Growing in damp meadows or along tunnels. Altitude ~4000 m. Flowering and fruiting: August - October.

A

B

C

A-E. 中国西藏自治区隆子县
Longzi, Xizang, China. 13.08.2006

D

苞叶龙胆

Gentiana licentii

一年生草本，茎斜升或铺散，高 3~5 cm，基部多分枝，具乳突；花期莲座叶显著，卵形到卵圆形，边缘软骨质；茎生叶较小，较节间长，匙形，先端外弯；花单生分枝顶端；花萼筒倒锥形，花萼裂片三角形；花冠筒管状或漏斗状，淡蓝色，喉部黄白色，有绿褐色斑点；花冠裂片卵形，褶阔卵形，长约花冠裂片的一半，边缘啮齿状。

特产于中国甘肃和陕西的山区。生长于山地灌丛边缘等向阳处，海拔 1800~2500 m。花果期 3~5 月。

Annual, stems ascending to prostrate, 3-5 cm, much branched from base, papillate; rosette leaves conspicuous at anthesis, ovate to ovate-orbicular; stem leaves smaller, spathulate, shorter than internodes, apex recurved; flowers solitary at branch terminals; calyx tube obconic, calyx lobes triangular; corolla funnelform, pale blue, yellowish white in throat, with dark green spots; corolla lobes ovate; plicae broad ovate, about 1/2 as long as the corolla lobes, margin erose.

Endemic to mountains in Gansu and Shaanxi provinces of China. In sunny places of bush margins. Altitude 1800-2500 m. Flowering and fruiting: March - May.

A

B

A-C. 中国陕西省宁陕县

Ningshan, Shaanxi, China. 31.03.2016

c

亚麻状龙胆

Gentiana linoides

一年生草本，茎直立，高 5~15 cm，基部以上多分枝，茎紫色，被乳突；花期莲座叶凋萎；茎生叶较节间短，钻形到披针形，对折；花 3 到多数，排列成松散的总状花序；花萼筒倒锥形，花萼裂片钻形，内弯，中脉隆起；花冠筒管状，蓝色到蓝紫色，基部黄绿色；花冠裂片卵状三角形，褶阔卵形，长不及花冠裂片的一半，边缘较整齐。

特产于中国云南西北部。生长于山地林缘、灌丛边缘等向阳处，海拔 3000~3500 m。花果期 7~8 月。

Annual, stems erect, 5-15 cm, branched from base and above, purple, papillate; rosette leaves withered at anthesis; stem leaves shorter than internodes; subulate to lanceolate, folded, flowers 3 to several in raceme; calyx tube obconic, calyx lobes subulate, incurved, midveins prominent; corolla tubular, blue to blue-purple, base greenish; corolla lobes ovate-triangular; plicae broad ovate, less than ½ of the corolla lobes, margin nearly entire.
Endemic to northwestern Yunnan of China. In open and sunny places of forest and bush margins. Altitude 3000-3500 m. Flowering and fruiting: July - August.

A. 中国云南省香格里拉市
Xianggelila, Yunnan, China. 05.08.2002
B-D. 中国云南省香格里拉市翁水村
Xianggelila, Yunnan, China. 05.08.2002

C

D

华南龙胆

Gentiana loureiroi

多年生草本，根肉质；茎直立，高 5~8 cm，基部多分枝，紫色，密被乳突；有莲座叶；茎生叶显著较小，匙形到椭圆形，较节间短；花多单生分枝顶端；花萼筒倒锥形，萼齿披针形，外弯；花冠筒漏斗状，淡蓝色到蓝紫色；花冠裂片卵形，褶阔卵形，长约花冠裂片的一半，边缘较整齐。

散布于喜马拉雅低山区到东南亚及中国南部，中国广东、广西、江西、福建及浙江等省份有分布。生长于山地林缘、灌丛边缘等向阳处，海拔 500~2500 m。花果期 3~7 月。

Perennial, roots fleshy; stems erect, 5-8 cm, much branched from base, purple, densely papillate; rosette leaves persistent; stem leaves much smaller, spathulate to elliptic, shorter than internodes; flowers solitary, terminal; calyx tube obconic, calyx teeth lanceolate, recurved; corolla funnelform, pale blue to blue-purple; corolla lobes ovate; plicae broad ovate, about ½ as long as the corolla lobes, margin nearly entire.

Sparsely spread in low Himalayas to Southeast Asia. In open and sunny places of mountain forest and bush margins. Altitude 500-2500 m. Flowering and fruiting: March - July.

A

B

A-E. 中国广东省英德市 摄影 /Photo 曾佑派
Yingde, Guangdong, China. 26.03.2016
F. 中国福建省黄冈山
Huanggangshan Mt., Fujian, China. 21.07.2000

C

D

E

F

大颈龙胆

Gentiana macrauchena

一年生草本，茎直立或斜升，高 2~5 cm，基部多分枝，全株光滑；茎生叶密集，较节间长，叶片钻形到舟状，先端圆钝，边缘软骨质；花单生枝端；花萼裂片三角形，直立；花冠筒漏斗状，淡蓝色；花冠裂片卵形，褶卵形，较花冠裂片短，先端圆钝，边缘整齐或锯齿状。

特产于中国云南西北部。生长于高山山坡和灌丛边缘，海拔 3000~3600 m。花果期 5~8 月。

Annual, stems erect or ascending, 2-5 cm, much branched from base, glabrous; stem leaves densely spaced, longer than internodes; stem leaf blades subulate to navicular, apex rounded, margins cartilaginous; flowers solitary and terminal; calyx lobes triangular, erect; corolla funnelform, pale blue; corolla lobes ovate; plicae ovate, apex obtuse, margin entire to serrulate.

Endemic to northwestern Yunnan province of China. In alpine slopes and bush margins. Altitude 3000-3600 m. Flowering and fruiting: May - August.

A–C. 中国云南省香格里拉市大雪山

Xianggelila, Yunnan, China. 05.08.2002

C

马耳山龙胆

Gentiana maeulchanensis

一年生草本，茎直立，高 3~10 cm，基部以上多分枝，茎紫色，密被乳突；莲座叶阔卵形；茎生叶显著较小，卵形到披针形，较节间短；花 3 到多数，排列成较松散的总状花序；花冠筒管状，淡蓝色，具蓝灰色条纹；花冠裂片卵状三角形，褶阔卵形，长约花冠裂片的一半，边缘较整齐。

分布于喜马拉雅山中东部，包括不丹、锡金、缅甸及中国西藏和云南等地。生长于山地林缘、灌丛边缘等向阳处，海拔 2000~3500 m。花果期 5~7 月。

Annual, stems erect, 3-10 cm, branched from base and above, purple, densely papillate; rosette leaves broad ovate; stem leaves much smaller, ovate to lanceolate, shorter than internodes; flowers 3 to several in raceme; corolla tubular, pale blue, with blue-grey stripes; corolla lobes ovate-triangular; plicae broad ovate, about ½ as long as the corolla lobes, margin nearly entire.

In central to eastern Himalayan regions, including Bhutan, Sikkim, Myanmar, and Xizang and Yunnan provinces of China. In open and sunny places of mountain forest and bush margins. Altitude 2000-3500 m. Flowering and fruiting: May - July.

A-B. 中国云南省大理白族自治州苍山
Cangshan Mt., Dali, Yunnan, China. 13.06.1999

C-D. 中国云南省丽江市 摄影 /Photo: 贾留坤
Lijiang, Yunnan, China. 12.05.2020

C

D

寡流苏龙胆

Gentiana mairei

一年生草本，茎铺散或斜升，高 5~10 cm，基部以上二歧分枝，光滑；基部莲座叶在花期凋萎；茎生叶较节间短，叶片卵形到心形，膜质；花单生枝端；花萼裂片线状钻形，直立；花冠倒锥形，淡蓝色到白色，具蓝灰色短条纹；花冠裂片卵形，褶阔卵形，具 8~10 条先端卷曲的丝状流苏。

特产于中国四川省西南部和云南省西北部。生长于高山林下、林缘及草甸中，海拔 2500~4000 m。花果期 6~9 月。

Annual, stems prostrate or ascending, 5-10 cm, dichotomously branched from base, glabrous; rosette leaves withered at anthesis; stem leaves shorter than internodes; leaf blades ovate to cordate, membranous; flowers solitary and terminal; calyx lobes erect, filiform-subulate; corolla obconic, pale blue to white, with blue-grey short stripes; corolla lobes ovate; plicae broadly ovate, fringed with 8-10 curly filaments.

Endemic to the mountains in southwestern Sichuan and northwestern Yunnan provinces of China. In alpine forest, forest margin and meadows. Altitude 2500-4000 m. Flowering and fruiting: June - September.

A-D. 中国四川省泸定县

Luding, Sichuan, China. 11.08.2004

C

D

亮叶龙胆

Gentiana micans

一年生草本，高 3~5 cm，基部多分枝，分枝密集呈小垫状，茎光滑；茎生叶钻形，对折，密集排列，较节间长，坚硬，近革质，有光泽；花单生分枝顶端；花萼筒倒锥形，花萼裂片锥形，近革质，基部边缘有缘毛；花冠筒管状，淡蓝色，具深紫色短条纹；花冠裂片卵形，褶长圆形，整齐或 2 裂，裂片不规则齿状。

特产于喜马拉雅地区的尼泊尔、不丹、锡金和中国西藏南部。生长于高山草甸和灌丛边缘等向阳处，海拔 3500~4500 m。花果期 7~9 月。

Annual, 3-5 cm, stems much branched from base, short branches densely crowded as tuft, glabrous; stem leaves subulate, folded, crowded and longer than internodes, rigid, subleathery, glossy; flowers solitary; calyx lobes subulate, subleathery, margin base ciliolate; corolla tubular, pale blue, with blue-purple short stripes; corolla lobes ovate; plicae oblong, entire or 2-cleft segment denticulate.

Endemic to the Himalayas, including Nepal, Bhutan, Sikkim, and southern Xizang of China. In open and sunny places of bush margins and alpine meadows. Altitude 3500-4500 m. Flowering and fruiting: July - September.

A

B

A-F. 中国西藏自治区亚东县
Yadong, Xizang, China. 11.08.2006

C
D
E
F

类亮叶龙胆

Gentiana micantiformis

一年生草本，高 3~5 cm，基部多分枝，茎光滑；莲座叶花期凋萎；茎生叶线形到钻形，对折，密集排列，较节间长，坚硬，近革质，有光泽，先端有小尖头，中脉隆起；花单生分枝顶端；花萼筒倒锥形，花萼裂片锥形，近革质；花冠筒漏斗状，淡蓝色，具深紫色短条纹；花冠裂片阔卵形，褶长圆形，先端圆钝，边缘整齐。

特产于喜马拉雅地区的不丹、锡金和中国西藏自治区南部。生长于高山草甸，海拔 3800~4500 m。花果期 7~9 月。

Annual, 3-5 cm, stems much branched from base, glabrous; rosette leaves withered at anthesis; stem leaves linear to subulate, folded, crowded and longer than internodes, rigid, subleathery, glossy, apex cuspidate, midvein prominent; flowers solitary; calyx lobes subulate, subleathery; corolla funnelform, pale blue, with blue-purple short stripes; corolla lobes broad ovate; plicae oblong, apex obtuse, margins entire.

Endemic to the Himalayas, including Bhutan, Sikkim, and southern Xizang of China. In open and sunny alpine meadows. Altitude 3800-4500 m. Flowering and fruiting: July - September.

A-D. 中国西藏自治区亚东县

Yadong, Xizang, China. 11.08.2006

C

D

钟花龙胆

Gentiana nanobella

一年生草本，茎平铺或斜升，高 4~8 cm，基部多分枝，紫红色，被乳突；茎生叶较节间短，叶片匙形到倒卵形，先端急尖，具尖，边缘软骨质；花单生分枝端；花萼裂片三角形，先端尾尖，中脉隆起，下延到萼筒，呈翅状；花冠筒管状，淡蓝或蓝灰色到蓝紫色，具紫黑色斑点和斑块；花冠裂片卵形，褶卵形或长圆形，先端圆钝，边缘不规则啮蚀状。

分布于中国西藏、云南和四川等省区。生长于高山山坡及灌丛边缘，海拔 3000~4000 m。花果期 8~10 月。

Annual, stems prostrate or ascending, 4-8 cm, much branched from base, reddish purple, papillate; stem leaves shorter than internodes; leaf blades spathulate to obovate, apex acute, cuspidate, margin cartilaginous; flowers solitary; calyx lobes triangular, apex cuspidate, midveins keeled and decurrent into wings of calyx tube; corolla tubular, pale blue or blue-grey to blue-purple, with dark purple spots and stripes; corolla lobes ovate; plicae broadly ovate or oblong, apex obtuse, margins erose.

Scattered in Xizang, Yunnan and Sichuan provinces of China only. In alpine slopes and bush margins. Altitude 3000-4000 m. Flowering and fruiting: August - October.

A

B

C

A-G. 中国云南省白马雪山

Baimaxueshan Mountain, Yunnan, China. 19.09.2023

D

E

F

G

少叶龙胆

Gentiana oligophylla

一年生草本，茎铺散或斜升，高 8~15 cm，从基部二歧分枝，光滑；茎生叶较节间显著短，叶片卵状三角形；花单生枝端；花萼裂片线状钻形，直立；花冠倒锥形，淡蓝色到蓝灰色，具黄绿色条纹；花冠裂片卵状心形，褶长圆形，具 8~15 条先端膨大的棒状流苏。

特产于中国中部到西南部的湖南、湖北、四川和贵州等省。生长于高山林下、林缘及草丛中；海拔 1500~2800 m。花果期 5~8 月。

Annual, stems prostrate or ascending, 8-15 cm, dichotomously branched from base, glabrous; stem leaves conspicuously shorter than internodes; leaf blades ovate-triangular; flowers solitary and terminal; calyx lobes erect, subulate; corolla obconic, pale blue to blue-grey, with yellowish-green stripes; corolla lobes ovate-orbicular; plicae oblong, margin 8-15 clavate fringed with expanded tips.

Endemic to central to southwestern China including Hubei, Hunan, Sichuan and Guizhou provinces. In alpine forest, forest margin and grasses. Altitude 1500-2800 m. Flowering and fruiting: May - August.

A-B. 中国湖北省神农架林区
Shennongjia Mt., Hubei, China. 17.08.2004

B

流苏龙胆

Gentiana panthaica

一年生草本，茎铺散或斜升，高 5~20 cm，从基部二歧分枝，光滑；茎生叶较节间短，叶片卵状披针形；花单生枝端；花萼裂片线状钻形，直立；花冠倒锥形，淡蓝色到蓝灰色，具深蓝色条纹；花冠裂片卵形，褶长圆形，具 15~20 条先端卷曲的丝状流苏。

特产于中国中部到西南部的江西、湖北、湖南、广西、四川、贵州和云南等省。生长于高山林下、林缘及草丛中，海拔 1500~3500 m。花果期 5~8 月。

Annual, stems prostrate or ascending, 5-20 cm, dichotomously branched from base, glabrous; stem leaves shorter than internodes; leaf blades ovate-lanceolate; flowers solitary and terminal; calyx lobes erect, filiform-subulate; corolla obconic, pale blue to blue-grey, with dark blue stripes; corolla lobes ovate; plicae oblong, margin fringed with 15-20 curly filaments. Endemic to central to southwestern China including Jiangxi, Hubei, Hunan, Guangxi, Sichuan Guizhou and Yunnan provinces. In alpine forest, forest margin and grasses. Altitude 1500-3500 m. Flowering and fruiting: May - August.

A

B

A. C. D. 中国云南省大理白族自治州
Dali, Yunnan, China. 13.06.1999

B. E. 中国云南省云龙县
Yunlong, Yunnan, China. 22.09.2023

C

D

E

糙毛龙胆

Gentiana pedicellata

一年生草本，茎斜升或铺散，高 3~8 cm，基部以上多分枝，密被乳突；茎生叶卵形到披针形，先端有尖头；花单生分枝顶端；花萼筒管状，花萼裂片锥形到卵形，先端外弯，有小尖头；花冠筒管状到漏斗状，蓝色，喉部具蓝紫色条纹；花冠裂片卵形，褶卵形，与花冠裂片近等长，边缘较整齐，先端有小尖头。

散布于喜马拉雅地区。生长于高山草甸、灌丛和森林边缘等向阳处，海拔 2500~3500 m。花果期 4~8 月。

Annual, stems ascending to prostrate, 3-8 cm, much branched from base and above, densely papillate; stem leaves ovate to lanceolate, apex recurved, cuspidate; flowers solitary at branch terminals; calyx tube tubular, calyx lobes subulate to ovate, recurved, cuspidate; corolla tubular to funnelform, blue, with blue-purple stripes; corolla lobes ovate; plicae ovate, almost as long as the corolla lobes, margin nearly entire, cuspidate.

Widespread in the Himalayan regions. In open and sunny places of alpine forest and bush margins and meadows. Altitude 2500-3500 m. Flowering and fruiting: April - August.

A. 中国西藏自治区当雄县
Dangxiong, Xizang, China. 07.08.2006

B–C. 中国西藏自治区亚东县
Yadong, Xizang, China. 11.08.2006

D. 尼泊尔 摄影 /Photo: Christian Bravard
Nepal. 11.05.2007

E. 尼泊尔帕克丁昆琼 摄影 /Photo: Elizabeth Byers
Phakding, Khumjung, Nepal; 24.05.2014

D

E

陕南龙胆

Gentiana piasezkii

一年生草本，茎斜升或直立，高 5~10 cm，基部以上多分枝，密被乳突；茎生叶较节间短，卵状披针形；花单生分枝顶端；花萼筒倒锥形，花萼裂片钻形，中脉隆起下延到萼筒成翅；花冠筒高脚杯状，蓝紫色，喉部具深紫色斑点和条纹；花冠裂片卵形，褶卵形，边缘较整齐。

特产于中国陕西、甘肃和四川三省邻接的山区。生长于高山草坡、灌丛和森林边缘等向阳处。海拔 1500~3500 m。花果期 4~9 月。

Annual, stems ascending to erect, 5-10 cm, much branched from base and above, densely papillate; stem leaves shorter than internodes, ovate- lanceolate; flowers solitary at branch terminals; calyx tube obconic, calyx lobes subulate, midvein strongly keeled and decurrent into conspicuous wings of calyx tube; corolla tube salverform, blue-purple, with dark purple stripes; corolla lobes ovate; plicae ovate, margins nearly entire.

Endemic to the mountains connecting Shaanxi, Gansu, and Sichuan provinces of China. In open and sunny places of alpine forest and bush margins and grassland. Altitude: 1500-3500 m. Flowering and fruiting: April - September.

A

B

A. 中国甘肃省宕昌县
Tanchang, Gansu, China. 10.04.2023

B-C. 中国甘肃省宕昌县
Tanchang, Gansu, China. 08.08.2012

D-E. 中国甘肃省宕昌县
Tanchang, Gansu, China. 21.07.2011

F. 中国甘肃省宕昌县
Tanchang, Gansu, China. 02.10.2010

G. 中国甘肃省宕昌县；偶尔可以见到粉色花类型
Tanchang, Gansu, China. 08.08.2012

C
D
G
F

草甸龙胆

Gentiana praticola

多年生草本，根圆锥状，肉质；茎直立，高 3~8 cm，基部以上二歧分枝，密被乳突；莲座叶在花期宿存，椭圆形；茎生叶较节间短，椭圆形；花数朵，密集于分枝顶端叶腋；花萼裂片椭圆形，绿色，外弯；花冠筒管状，蓝紫色，具深紫色条纹；花冠裂片卵形，褶卵圆形，较花冠裂片短，边缘具细齿。

特产于中国贵州、云南和四川的山区。生长于开阔较干燥的高山草坡、灌丛和森林边缘等向阳处，海拔 1500~3000 m。花果期 6~9 月。

Perennial, roots cylindric, fleshy; stems erect, 3-8 cm, dichotomously branched from base and above, densely papillate; rosette leaves persistent at anthesis; stem leaves shorter than internodes, elliptic; flowers few clustered in axillary at branch terminals; calyx lobes elliptic, green, recurved; corolla tube tubular, blue-purple, with dark purple stripes; corolla lobes ovate; plicae ovate-orbicular, almost shorter than corolla lobes, margins denticulate. Endemic to the mountains in Guizhou, Yunnan and Sichuan provinces of China. In open and dry places of alpine grassland slopes, forest and bush margins. Altitude: 1500-3000 m. Flowering and fruiting: June-September.

A

B

A-B. 中国云南省丽江市
Lijiang, Yunnan, China. 20.09.1990

C-F. 中国云南省永平县
Yongping, Yunnan, China. 22.09.2023

C

D

E

F

黄白龙胆

Gentiana prattii

一年生草本，茎斜升或直立，高 2~5 cm，基部多分枝，密被乳突；茎生叶卵形到匙形，中脉隆起，边缘软骨质；花单生分枝顶端；花萼筒倒锥形，花萼裂片三角形，边缘软骨质，中脉隆起，先端有小尖头；花冠筒管状，黄白色，具深绿色条纹；花冠裂片卵形，褶长圆形，边缘啮齿状。

特产于中国西藏、青海、四川和云南。生长于高山草甸、灌丛和森林边缘等向阳处，海拔 3000~4000 m。花果期 6~9 月。

Annual, stems ascending to erect, 2-5 cm, much branched from base, densely papillate; stem leaves ovate to spathulate, midveins prominent, margins cartilaginous; flowers solitary at branch terminals; calyx tube obconic, calyx lobes triangular, margins cartilaginous, midveins prominent, cuspidate; corolla tubular, yellow-white, with dark green stripes; corolla lobes ovate; plicae oblong, margin erose.
Endemic to Xizang, Qinghai, Sichuan and Yunnan provinces of China. In open and sunny places of alpine meadows, forest and bush margins. Altitude: 3000-4000 m. Flowering and fruiting: June-September.

A

B

C

A-B. 中国青海省门源回族自治县
Menyuan, Qinghai, China. 17.07.2011
C. 中国青海省门源回族自治县
Menyuan, Qinghai, China. 18.07.2011
D-E. 中国西藏自治区那曲县
Naqu, Xizang, China. 07.08.2006

D

E

伸梗龙胆

Gentiana producta

一年生草本，茎斜升或直立，高 6~10 cm，基部多分枝，光滑；茎生叶显著较节间短，倒卵形，先端圆钝；花单生分枝顶端；花萼筒倒锥形，花萼裂片钻形，中脉隆起下延到萼筒成翅；花冠筒高脚杯状，蓝紫色，喉部具深紫色条纹；花冠裂片卵形，褶卵形，与花冠裂片近等长，边缘细锯齿状。

特产于中国四川西部。生长于高山草坡向阳处，海拔 3000~3500 m。花果期 8~10 月。

本种与偏翅龙胆 *G. pudica* 比较相似，区别是本种叶片先端钝，无尖头，褶边缘为细锯齿，不是粗锯齿。

Annual, stems ascending to erect, 6-10 cm, much branched from base, glabrous; stem leaves conspicuously shorter than internodes, obovate, apex obtuse; flowers solitary; calyx tube obconic, calyx lobes subulate, midvein keeled and decurrent into wings of calyx tube; corolla tube salverform, blue-purple, with dark purple stripes; corolla lobes ovate; plicae ovate, almost as long as the corolla lobes, margin laciniate.

Endemic to western Sichuan province of China. In alpine meadows. Altitude: 3000-3500 m. Flowering and fruiting: August-October.

Highly resemble *G. pudica*, but leaves not mucronate, plicae laciniate.

A

B

A–B. 中国四川省德格县 Gege, Sichuan, China. 28.09.1993

新疆龙胆（变种）

Gentiana prostrata var. *karelinii*

一年生草本，茎斜升或直立，高 3~5 cm，基部多分枝，光滑；茎生叶匙形到倒卵形，先端圆钝；花单生分枝顶端；花萼筒管状，花萼裂片三角形，中脉隆起下延到萼筒成翅；花冠筒管状，蓝色，喉部具紫色条纹；花冠裂片卵形，褶卵形，较花冠裂片稍短，边缘啮齿状。

分布于中亚到东北亚，包括中国新疆。生长于高山草甸、山坡、林缘等向阳处，海拔 2000~3000 m。花果期 6~9 月。

Gentiana prostrata 被划分为 3 个高度相似的变种。本变种植物体基本光滑无乳突，分布于中亚到东北亚，包括中国新疆。

Annual, stems ascending to erect, 3-5 cm, branched from base, glabrous; stem leaves spathulate to obovate, apex obtuse; flowers solitary; calyx tube tubular, calyx lobes triangular, midvein keeled and decurrent into wings of calyx tube; corolla tube tubular, blue, with purple stripes; corolla lobes ovate; plicae ovate, shorter than corolla lobes, margin erose. Distributed in central Asia to northeastern Asia including Xinjiang of China. In sunny places of alpine meadows, slopes and forest margins. Altitude: 2000-3000 m. Flowering and fruiting: June-September.

A-D. 中国新疆维吾尔自治区博格达山
Bogeda Mt., Xinjiang, China. 02.08.1995

C

D

短蕊龙胆（变种）

Gentiana prostrata var. *ludlowii*

短蕊龙胆（变种）与其他变种的区别在于仅茎及叶缘具乳突，花萼较短，长仅约花冠的 1/2；分布于喜马拉雅西部及邻近的帕米尔高原，包括尼泊尔及中国的西藏和青海等地区。

The difference of the variety *Gentiana prostrata* var. *ludlowii* includes plants papillate on stem and leaf margins only, calyx shorter, about ½ as long as corolla.

Distribution: the western Himalayas and adjacent Pamir plateau, including Nepal, Xizang and Qinghai provinces of China.

A-B. 中国新疆维吾尔自治区托木尔峰
Tuomuer Mt., Xinjiang, China. 22.08.1995

C. 中国青海省泽库县 Zeku, Qinghai, China. 31.08.1994

柔毛龙胆

Gentiana pubigera

一年生草本，茎斜升或铺散，高 2~4 cm，基部多分枝，光滑；茎生叶密集，幼时被柔毛，较节间长，匙形到近心形，先端外弯，有小尖头；花单生分枝顶端；花萼筒管状，花萼裂片三角形，有小尖头；花冠筒管状，淡蓝色；花冠裂片卵形，褶长圆形，与花冠裂片近等长，边缘整齐或啮齿状。

特产于中国四川和云南。生长于山坡草地和草甸向阳处。海拔 2000~3500 m。花果期 5~9 月。

Annual, stems ascending to prostrate, 2-4 cm, much branched from base, glabrous; stem leaves crowded, pubescent when young, longer than internodes, spathulate to suborbicular, apex recurved, mucronate; flowers solitary at branch terminals; calyx tube tubular, calyx lobes triangular, mucronate; corolla, pale blue; corolla lobes ovate; plicae oblong, almost as long as the corolla lobes, margins entire or erose.

Endemic to Sichuan and Yunnan provinces of China. In sunny places of slopes, grassland and meadows. Altitude: 2000-3500 m. Flowering and fruiting: May-September.

A–C. 中国四川省小金县
Xiaojin, Sichuan, China. 23.09.1994

假水生龙胆

Gentiana pseudoaquatica

一年生草本，茎斜升或铺散，高3~5 cm，基部多分枝，密被乳突；茎生叶匙形到倒卵形，先端圆钝，边缘软骨质；花单生分枝顶端；花萼筒倒锥形，花萼裂片三角形，中脉隆起；花冠筒管状到漏斗状，蓝色，喉部具紫色条纹；花冠裂片卵形，褶卵形，与花冠裂片近等长，边缘整齐或细齿状。

散布于亚洲东部到中部，包括中国北部及西部。生长于山坡草地、灌丛边缘等向阳处。海拔 1500~3500 m。花果期 4~8 月。

Annual, stems ascending to prostrate, 3-5 cm, much branched from base, densely papillate; stem leaves spathulate to obovate, apex obtuse, margins cartilaginous; flowers solitary at branch terminals; calyx tube obconic, calyx lobes triangular, midveins prominent; corolla tubular to funnelform, blue, with purple stripes; corolla lobes ovate; plicae ovate, almost as long as the corolla lobes, margins entire or denticulate.

Widespread in eastern to central Asia including northern and western China. In sunny places of slopes, grassland and bush margins. Altitude: 1500-3500 m. Flowering and fruiting: April-August.

A. 中国西藏自治区安多县
Anduo, Xizang, China. 06.08.2006

B-C. 中国青海省日月山
Riyueshan Mt., Qinghai, China. 02.08.2006

D. 中国西藏自治区安多县
Anduo, Xizang, China. 06.08.2006

E. 中国四川省石渠县 Shiqu, Sichuan, China. 25.09.1993

F. 中国青海省日月山
Riyueshan Mt., Qinghai, China. 02.08.2006

G. 中国甘肃省玛曲县 Maqu, Gansu, China. 28.07.1990

D
E
F
G

假鳞叶龙胆

Gentiana pseudosquarrosa

一年生草本，茎斜升或直立，高 3~6 cm，基部以上多分枝，密被乳突；茎生叶较节间短，匙形到近心形，先端外弯，有小尖头，边缘软骨质；花单生分枝顶端；花萼筒倒锥形，花萼裂片卵状心形，外弯，有小尖头；花冠筒管状到漏斗状，蓝色，喉部具紫色条纹；花冠裂片卵形，褶卵形，与花冠裂片近等长，边缘整齐或 2 裂齿。

特产于中国青海、四川、云南及西藏等省区。生长于山坡草地及草甸向阳处。海拔 2000~3500 m。花果期 4~8 月。

Annual, stems ascending to erect, 3-6 cm, much branched from base and above, densely papillate; stem leaves spathulate to suborbicular, apex recurved, mucronate, margins cartilaginous; flowers solitary at branch terminals; calyx tube obconic, calyx lobes ovate-orbicular, recurved, mucronate; corolla tubular to funnelform, blue, with purple stripes; corolla lobes ovate; plicae ovate, almost as long as the corolla lobes, margins entire or 2-cleft.

Endemic to Qinghai, Sichuan, Yunnan and Xizang provinces of China. In sunny places of slopes, grassland and meadows. Altitude: 2000-3500 m. Flowering and fruiting: April-August.

A

B

A–E. 中国四川省道孚县 Daofu, Sichuan, China. 03.07.2021

C
D
E

偏翅龙胆

Gentiana pudica

一年生草本，茎斜升或直立，高 3~8 cm，基部多分枝，光滑；茎生叶较节间短，倒卵形，先端圆钝，有小尖头；花单生分枝顶端；花萼筒管状，花萼裂片三角形，中脉隆起；花冠筒高脚杯状，淡蓝到蓝紫色，喉部对应每个花冠裂片有 2 条深紫色短条纹；花冠裂片卵形，褶卵形，较花冠裂片短，边缘啮齿状。

特产于中国青海、甘肃和四川相邻接地区。生长于高山草甸向阳处。海拔 2500~4000 m。花果期 7~9 月。

Annual, stems ascending to erect, 3-8 cm, much branched from base, glabrous; stem leaves shorter than internodes, obovate, apex obtuse, mucronate; flowers solitary; calyx tube tubular, calyx lobes triangular, midvein prominent; corolla tube salverform, pale blue to blue-purple, with 2 short dark purple stripes against each corolla lobes in the throat; corolla lobes ovate; plicae ovate, shorter than corolla lobes, margin erose.

Endemic to the regions connecting Qinghai, Gansu and Sichuan provinces of China. In alpine meadows. Altitude: 2500-4000 m. Flowering and fruiting: July-September.

A–C. 中国四川省甘孜县 Ganzi, Sichuan, China. 21.08.2006

D. 偏翅龙胆 *Gentiana pudica* + 线叶龙胆 *Gentiana lawrencei* var. *farreri*
中国青海省河南蒙古族自治县
Henan, Qinghai, China. 09.09.2023

D

比利牛斯龙胆

Gentiana pyrenaica

多年生草本，茎斜升，高 3~8 cm，基部多分枝，光滑；茎生叶线状披针形，先端有小尖头，边缘软骨质；花单生分枝顶端；花萼筒管状，花萼裂片狭三角形，中脉隆起；花冠筒管状到倒锥形，蓝色到蓝紫色，喉部黄绿色；花冠裂片卵形，褶卵形，较花冠裂片短，边缘整齐或啮齿状。

散布于欧洲比利牛斯山东部，喀尔巴阡山及保加利亚西南部山区。生长于山坡草地、灌丛边缘等向阳处。海拔 1200~2500 m。花果期 6~8 月。

Perennial, stems ascending, 3-8 cm, much branched from base, glabrous; stem leaves linear-lanceolate, apex mucronate, margins cartilaginous; flowers solitary at branch terminals; calyx tube tubular, calyx lobes narrowly triangular; corolla tubular to obconic, blue to blue-purple, yellowish green at throat; corolla lobes ovate; plicae ovate, shorter than corolla lobes, margins entire or erose.

Endemic to Europe in eastern Pyrenees, Carpathians, and the mountains in southwestern Bulgaria. In sunny places of slopes, grassland and bush margins. Altitude: 1200-2500 m. Flowering and fruiting: June-August.

A

B

A–C. 保加利亚穆沙拉山 Musala Mt., Bulgaria. 18.07.1993

辐射龙胆

Gentiana radiata

一年生草本，茎斜伸或直立，高 3~5 cm，基部多分枝，密被乳突；茎生叶较密集，与节间近等长或较短，叶片狭披针形到钻形，对折，近革质，先端尾尖；花单生枝端；花萼筒倒锥形，花裂片线状钻形或三角形，直立；花冠倒锥形，黄白色到淡蓝色，喉部具深色条纹；花冠裂片卵形，褶卵形，较花冠裂片短，啮齿状。

特产于中国四川西部。生长于高山草甸与山坡。海拔 3500~4500 m。花果期 5~8 月。

Annual, stems ascending to erect, 3-5 cm, branched from base, densely papillate; stem leaves somewhat densely spaced, equal to or shorter than internodes; leaf blades narrowly lanceolate to subulate, folded, subleathery, caudate; flowers solitary and terminal; calyx tube obconic, calyx lobes erect, filiform-subulate to triangular; corolla obconic, pale yellowish white to pale blue, with dark stripes in throat; corolla lobes ovate; plicae ovate, shorter than corolla lobes, erose.

Endemic to western Sichuan of China. In alpine meadows and slopes. Altitude: 3500-4500 m. Flowering and fruiting: May-August.

A

B

A–D. 中国四川省乡城县
Xiangcheng, Sichuan, China. 06.08.2002

E. 中国四川省稻城县
Daocheng, Sichuan, China. 09.07.2014

深红龙胆

Gentiana rubicunda

多年生草本，有时有丛生短根状茎；茎直立，高 8~15 cm，单一或中部以上分枝，光滑；基生营养叶排列密集，阔卵形到椭圆形；茎生叶稀疏，较节间短，叶片卵状披针形；花单生枝端；花萼裂片线状钻形，直立；花冠倒锥形，紫红色，喉部具深紫色斑点；花冠裂片卵形，褶卵形，整齐、啮齿状或 2 裂片。

产于中国中部到西南部的甘肃、四川、湖北、湖南、贵州和云南等省。生长于高山林下、林缘及草丛中。海拔 500~3500 m。花果期 5~10 月。

本种叶片和花冠的大小及形状变异较大，常被细分为几个变种。

Perennial, sometimes with short stolon; stems erect, 8-15 cm, simple or branched from middle, glabrous; stem leaves sparsely spaced, shorter than internodes; leaf blades ovate-lanceolate; flowers solitary and terminal; calyx lobes erect, filiform-subulate; corolla obconic, reddish purple, with dark purple spots; corolla lobes ovate; plicae ovate, entire, erose or 2-cleft. Widely spread in Gansu, Sichuan, Hubei, Hunan, Guizhou and Yunnan provinces of China. In alpine forest, forest margin and grasses. Altitude: 500-3500 m. Flowering and fruiting: May-October.

A

B

C

A–C. 中国四川省峨眉山
Emeishan Mt., Sichuan, China. 24.08.2004

D–E. 中国四川省海螺沟
Hailuogou, Sichuan, China. 29.09.1994

D

E

黑斑龙胆

Gentiana scabrida var. *punctata*

一年生草本，茎直立或斜升，高 3~15 cm，基部以上多分枝，密被乳突；茎生叶稀疏，较节间短，叶片卵形到卵状披针形，先端具尾尖；花单生枝端；花萼裂片匙形到卵形，开展到外弯，有小尖头；花冠倒锥形，淡黄色到亮黄色，喉部具深紫色斑点；花冠裂片卵形，有小尖头，褶卵形，整齐、啮齿状或有小尖头。

产于中国台湾。生长于较干燥的石质山坡草丛中。海拔 2500~3500 m。花果期 7~10 月。本种花色和花冠及其裂片和褶的大小及形状变异较大，常被细分为不同变种。

Annual, stems erect or ascending, 3-15 cm, simple or branched from base and above, densely papillate; stem leaves sparsely spaced, shorter than internodes; leaf blades ovate to ovate-lanceolate; flowers solitary and terminal; calyx lobes spathulate to ovate, spreading to recurved, mucronate; corolla obconic, pale to bright yellow, with dark purple spots in throat; corolla lobes ovate, mucronate; plicae ovate, entire, erose or mucronate.

Endemic to Taiwan province of China. In dry and sunny stony slopes and grassland. Altitude: 2500-3500 m. Flowering and fruiting: July-October.

A

B

C

A. 中国台湾太鲁阁国家公园
摄影 /Photo：Cheng-Tao Lin
Tailuge Park, Taiwan, China. 09.07.2011

B. 中国台湾太鲁阁国家公园
摄影 /Photo：Cheng-Tao Lin
Tailuge Park, Taiwan, China. 01.07.2009

C. 中国台湾太鲁阁国家公园
摄影 /Photo：Cheng-Tao Lin
Tailuge Park, Taiwan, China. 14.07.2012

D-E. 中国台湾省花莲县 摄影 /Photo：付鹏程
Hualian, Taiwan, China. 02.07.2022

尖叶龙胆

Gentiana sedifolia

多年生草本，茎斜升或直立，高 2~5 cm，基部多分枝，有时呈垫状；茎生叶密集，长于节间，线状披针形，中脉隆起，先端渐尖，有尖头；花单生分枝顶端；花萼筒管状，花萼裂片三角形，边缘软骨质；花冠筒漏斗状，苍白色、淡蓝色到蓝色，喉部黄色，有时有深色斑点；花冠裂片卵形，褶卵形，先端平截，边缘啮齿状。

广泛分布于中美洲到南美洲高山地区。生长于高山草甸及垫状植物中间。海拔 2200~3500 m。花果期 7~12 月。

Perennial, stems ascending to erect, 2-5 cm, branched from base, sometimes cushion-like; stem leaves crowded, longer than internodes, linear-lanceolate, apex acuminate and cuspidate; flowers solitary; calyx tube tubular, calyx lobes triangular, margins cartilaginous; corolla tube funnelform, pale white, pale blue to blue, throat yellow with dark spots; corolla lobes ovate; plicae ovate, apex truncate, margin erose.

Widely spread in alpine regions in the countries of Central and South Americas. In alpine meadows and cushion plants. Altitude: 2200~3500 m. Flowering and fruiting: July-December.

A

B

A-B. 厄瓜多尔 摄影 /Photo：葛学军
Pichincha Mt. Ecuator. 17.05.2023

厚边龙胆

Gentiana simulatrix

一年生草本，茎斜升或直立，高 2~6 cm，基部以上多分枝，紫红色，密被乳突；茎生叶卵形、卵状椭圆形到椭圆形，边缘增厚成软骨质；花单生分枝顶端；花萼筒倒锥形，花萼裂片卵形，直立或向外伸展，有小尖头，边缘增厚成软骨质；花冠筒管状，淡蓝色，喉部具紫色条纹；花冠裂片卵形，有小尖头，褶卵形，较花冠裂片短，先端钝，边缘整齐或齿状。

分布于不丹到中国西藏、四川和青海等地。生长于较干燥的山坡、草地及草甸向阳处。海拔 3000~4000 m。花果期 5~9 月。

Annual, stems ascending to erect, 2-6 cm, much branched from base and above, reddish purple, densely papillate; stem leaves ovate, ovate-elliptic to elliptic, margins thick cartilaginous; flowers solitary at branch terminals; calyx tube obconic, calyx lobes ovate, erect or spreading, mucronate, margins thick cartilaginous; corolla tubular, pale blue, with purple stripes in throat; corolla lobes ovate, mucronate; plicae ovate, shorter than corolla lobes, apex obtuse, margins entire or dentate.

Distributed in Bhutan and Xizang, Sichuan and Qinghai provinces of China. Growing in dry and sunny slopes, grassland, and meadows. Altitude: 3000-4000 m. Flowering and fruiting: May-September.

中国西藏自治区林芝市 Linzhi, Xizang, China. 22.09.2003

匙叶龙胆

Gentiana spathulifolia

一年生草本，茎直立或斜升，高 8~12 cm，基部多分枝，全株密被紫色乳突；莲座叶花期宿存；茎生叶稀疏，显著较节间短，叶片卵形到匙形，边缘具睫毛；花单生枝端；花萼筒倒锥形，花萼裂片三角形；花冠筒漏斗状，紫红色，具深紫色条纹；花冠裂片卵形，褶卵形，长约花冠裂片的一半，整齐或 2 裂。

仅见于中国青海、甘肃、四川和陕西。生长于高山草原和山坡草丛中。海拔 2800~3500 m。花果期 7~9 月。

Annual, stems erect or ascending, 8-12 cm, much branched from base, densely purple papillate; rosette leaves persistent at anthesis; stem leaves sparsely spaced, much shorter than internodes, leaf blades ovate to spathulate, margins ciliate; flowers solitary and terminal; calyx tube obconic, calyx lobes triangular; corolla tube funnelform, reddish purple, with dark purple lines; corolla lobes ovate; plicae ovate, about ½ as long as lobes, entire or 2-cleft. In the Qinghai, Gansu, Shaanxi, and Sichuan provinces of China. In alpine grassland and mountain slope grasses. Altitude: 2800-3500 m. Flowering and fruiting: July-September.

A. 中国甘肃省临潭县 Lintan, Gansu, China. 22.07.2011

B. E. 中国四川省松潘县
Songpan, Sichuan, China. 06.08.1992

C–D. 中国甘肃省临潭县 Lintan, Gansu, China. 22.07.2011

D

E

鳞叶龙胆

Gentiana squarrosa

一年生草本，茎斜升或直立，高 2~6 cm，基部以上多分枝，密被乳突；茎生叶匙形到近心形，先端外弯，边缘软骨质；花单生分枝顶端；花萼筒倒锥形，花萼裂片卵状心形，外弯，有小尖头；花冠筒管状到漏斗状，淡蓝色，喉部具紫色条纹；花冠裂片卵形，有小尖头，褶卵形，较花冠裂片短，边缘整齐或啮齿状。

较广泛分布于从喜马拉雅到中亚和东亚的广大地区。生长于较干燥的山坡、草地及草甸向阳处。海拔 500~4000 m。花果期 4~9 月。

Annual, stems ascending to erect, 2-6 cm, much branched from base and above, densely papillate; stem leaves spathulate to suborbicular, apex recurved, margins cartilaginous; flowers solitary at branch terminals; calyx tube obconic, calyx lobes ovate-orbicular, recurved, mucronate; corolla tubular to funnelform, pale blue, with purple stripes; corolla lobes ovate; plicae ovate, shorter than corolla lobes, margins entire or erose.

Widely spread from Himalaya to central and east Asia. Growing in dry and sunny slopes, grassland, and meadows. Altitude: 500-4000 m. Flowering and fruiting: April-September.

A–C. 中国陕西省太白县 Taibai, Shaanxi, China. 08.04.2023

D. 中国甘肃省夏河县 Xiahe, Gansu, China. 23.07.2011

D

珠峰龙胆

Gentiana stellata

一年生草本，茎斜升或直立，高 2~6 cm，茎基部多分枝，光滑；茎生叶密集，红紫色，卵形、披针形、匙形到倒卵形，先端外弯，有小尖头，边缘软骨质；花单生分枝顶端；花萼筒管状，花萼裂片钻形，中脉隆起，下延到萼筒呈翅状；花冠筒高脚杯状，深蓝色，喉部具蓝紫色斑点；花冠裂片卵形，褶倒卵形，较花冠裂片近等长，先端圆形或平截，边缘细齿状。

分布于喜马拉雅山区的印度、尼泊尔、不丹及中国西藏。生长于高山草甸及灌丛边缘。海拔 3500~6000 m。花果期 8~9 月。

Annual, stems ascending or erect, 2-6 cm, much branched from base; stem leaves densely spaced, reddish purple, ovate, lanceolate, spathulate to obovate, apex recurved, mucronate, margins cartilaginous; flowers solitary at branch terminals; calyx tube tubular, calyx lobes subulate, midveins strongly keeled and decurrent into conspicuous wings; corolla salverform, dark blue, with dark blue purple spots in throat; corolla lobes ovate; plicae obovate, nearly as long as corolla lobes, apex round or truncate, margins serrulate.

In the Himalayas in India, Nepal, Bhutan and Xizang of China. In sunny alpine meadows and bush margins. Altitude: 3500-6000 m. Flowering and fruiting: August-September.

A

B

C

A–E. 中国西藏自治区色季拉山
Sejilashan Mt., Xizang, China. 17.08.2006

地球屋脊上的“屋脊兽”：马卡鲁峰（8463 m）、珠穆朗玛峰（8848 m）和卓奥友峰（8201 m）；中国西藏自治区

The “ornament beasts” on the house-ridge of the Earth: the Makalu (8463 m), the Everest (8848 m), and the Cho Oyu (8201 m), Xizang, China. 17.09.1992

星状龙胆

Gentiana stellulata

一年生草本，茎直立到斜升，高 4~10 cm，基部及以上二歧分枝，光滑；茎生叶稀疏，较节间短，叶片卵状披针形；花单生枝端；花萼裂片线状钻形，直立；花冠倒锥形，灰紫色，喉部具深紫色斑块；花冠裂片狭卵形，褶狭卵形，与花冠裂片近等长，整齐或 2 深裂片。

特产于中国云南西北部。生长于高山林缘及灌丛边缘和草丛中。海拔 3200~3800 m。花果期 6~9 月。

Annual; stems erect to ascending, 4-10 cm, dichotomously branched from base and above, glabrous; stem leaves sparsely spaced, shorter than internodes; leaf blades ovate-lanceolate; flowers solitary and terminal; calyx lobes erect, filiform-subulate; corolla obconic, greyish purple, with dark purple spots in throat; corolla lobes narrowly ovate; plicae narrowly ovate, nearly as long as corolla lobes, deeply 2-cleft. Endemic to northwestern Yunnan of China. In alpine forest and bush margins and among grasses. Altitude: 3200-3800 m. Flowering and fruiting: June-September.

A–C. 中国云南省贡山县

Gongshan, Yunnan, China. 04.08.2000

单花龙胆

Gentiana subuniflora

一年生草本，茎直立到斜升，高 3~5 cm，基部分枝，近光滑；茎生叶较密集，叶片匙形到卵形，密生乳突，具小尖头，边缘软骨质；花单生枝端；花萼裂片三角状披针形，具小尖头；花冠漏斗形，淡蓝色，喉部有时具深紫色斑点；花冠裂片卵形，褶卵形，与花冠裂片近等长，整齐或圆齿状。

特产于中国四川省西部。生长于高山山坡和草地。海拔 3500~4000 m。花果期 4~7 月。

Annual; stems erect to ascending, 3-5 cm, branched from base, subglabrous; stem leaves densely spaced; leaf blades spathulate to ovate, densely papillate, margin cartilaginous, apex mucronate; flowers solitary and terminal; calyx lobes erect, triangular lanceolate, apex mucronate; corolla funnelform, pale blue, with dark purple dots around throat; corolla lobes ovate; plicae ovate, nearly as long as corolla lobes, entire to crenulate.

Endemic to western Sichuan of China. In alpine slopes and meadows. Altitude 3500-4000 m. Flowering and fruiting: April - July.

A. 中国四川省巴郎山 摄影 /Photo：孟德昌
Balang Mt., Sichuan, China. 20.05.2020

B. 中国四川省巴郎山 摄影 /Photo：孟德昌
Balang Mt., Sichuan, China. 30.04.2023

C. 中国四川省夹金山
Jiajin Mt., Sichuan, China. 01.07.2021

四川龙胆

Gentiana sutchuenensis

一年生草本，茎直立，高 3~8 cm，基部以上多分枝，近光滑；茎生叶较密集，较节间长，叶片狭椭圆形、披针形到卵形；花数枚簇生枝端；花萼筒倒锥形，花萼裂片狭披针形，外展，先端芒刺状；花冠筒漏斗状，淡蓝色到蓝紫色；花冠裂片卵形，褶卵形，较花冠裂片稍短，整齐或偶有细裂齿。

仅见于中国四川、云南和贵州等省。生长于山坡、森林及灌丛和草地边缘。海拔 1000~2500 m。花果期 4~8 月。

Annual, stems erect, 3-8 cm, much branched from base and above, nearly glabrous; stem leaves crowded, longer than internodes, leaf blades narrowly elliptic to lanceolate to ovate; flowers several clustered at terminal; calyx tube obconic, calyx lobes narrowly lanceolate, spreading, apex aristate; corolla tube funnelform, pale blue to blue-purple; corolla lobes ovate; plicae ovate, shorter than corolla lobes, entire or rarely denticulate.

Endemic to Sichuan, Yunnan and Guizhou provinces of China. In mountain slopes, the margins of forest, bushes, and grassland. Altitude: 1000-2500 m. Flowering and fruiting: April-August.

A

B

C

A. 中国四川省乡城县
Xiangcheng, Sichuan, China. 06.08.2002

B. 中国云南省昆明市
Kunming, Yunnan, China. 18.04.1999

C. 中国云南省昆明市
Kunming, Yunnan, China. 10.06.1999

紫花龙胆

Gentiana syringea

一年生草本，茎直立或斜升，高 3~8 cm，基部多分枝，近光滑；茎生叶较稀疏，较节间短，叶片狭长圆形到披针形，对折；花单生枝端；花萼筒倒锥形，花萼裂片狭三角形；花冠筒漏斗状，紫红色，具深紫色短条纹；花冠裂片卵形，褶卵形，较花冠裂片稍短，整齐或有细裂齿。

仅见于中国青海、甘肃和四川三省交会地区。生长于高山草原和高山草甸。海拔 2500~3500 m。花果期 6~8 月。

Annual, stems erect or ascending, 3-8 cm, much branched from base, nearly glabrous; stem leaves sparsely spaced, shorter than internodes, leaf blades narrowly oblong to lanceolate, folded; flowers solitary and terminal; calyx tube obconic, calyx lobes narrowly triangular; corolla tube funnelform, reddish purple, with dark purple short lines; corolla lobes ovate; plicae ovate, slightly shorter than corolla lobes, entire or denticulate.

Endemic to the regions connecting Qinghai, Gansu, and Sichuan provinces of China. In alpine grassland and alpine meadows. Altitude: 2500-3500 m. Flowering and fruiting: June-August.

A-B. 中国四川省若尔盖县
Ruoergai, Sichuan, China. 05.08.1992

大理龙胆

Gentiana taliensis

一年生草本，形态变异大；茎直立，紫色，坚硬，高 3~8 cm，基部多分枝，具乳突；茎生叶较稀疏，叶片披针形到卵状披针形；花多数，簇生枝端；花萼筒倒锥形，花萼裂片披针形，先端急尖成芒，直立不开展；花冠筒漏斗形，淡蓝到蓝紫色；花冠裂片卵形，褶卵形，较花冠裂片短，先端整齐或有缺刻。

仅见于中国四川、云南和贵州。生长于山坡、森林及灌丛和草地边缘。海拔 1000~2500 m。花果期 3~10 月。本种变异大，与四川龙胆不易区分，也许存在杂交或种质间渗，主要区别在于本种基部多分枝，萼裂片具芒不开展。

Annual, variable, stems erect, purple, rigid, 3-8 cm, much branched from base, papillate; stem leaves sparsely spaced, leaf blades lanceolate to ovate-lanceolate; flowers many, clustered at terminal; calyx tube obconic, calyx lobes lanceolate, acute, awned; corolla tube funnelform, pale blue to blue-purple; corolla lobes ovate; plicae ovate, shorter as corolla lobes, entire or emarginate.

Endemic to Sichuan, Yunnan and Guizhou of China. In mountain slopes, the margins of forest, bushes, and grassland. Altitude: 1000-2500 m. Flowering and fruiting: March-October.

A. 中国云南省昆明市
Kunming, Yunnan, China. 11.06.1999

B-D. 中国云南省楚雄市
Chuxiong, Yunnan, China. 04.13.2000

C

D

打箭炉龙胆

Gentiana tatsienensis

一年生草本，基部多分枝，茎斜升或铺散，高 2~5 cm，光滑；茎生叶密集，倒卵状匙形，具小尖头，边缘膜质或软骨质；花单生分枝顶端；花萼裂片三角状或卵状披针形，具小尖头；花冠筒筒状，蓝色，内面有时具紫色斑点和条纹；花冠裂片卵状三角形，具小尖头，褶卵形，较花冠裂片短，边缘整齐或具粗齿。

分布于中国西藏自治区中部、东部和四川西部地区。生长于高山草甸的开阔向阳处。海拔 3500~4500 m。花果期 4~8 月。

Annual, stems branched from base, ascending to prostrate, 2-5 cm, glabrous; stem leaves densely spaced, obovate-spathulate, apex mucronate, margin membranous to cartilaginous; flowers solitary at branch terminal; calyx lobes triangular- or ovate- lanceolate, apex mucronate; corolla tube tubular, blue, sometimes with dark blue spots and stripes; corolla lobes ovate-triangular, apex mucronate; plicae ovate, shorter than corolla lobes, margin entire or denticulate.

In central to eastern Xizang and western Sichuan provinces of China. In open and sunny alpine meadows. Altitude 3500-4500 m. Flowering and fruiting: April - August.

A. 中国四川省汶川县 摄影 /Photo：孟德昌
Wenchuan, Sichuan, China. 30.04.2023

B. 中国四川省九寨沟 摄影 /Photo：孟德昌
Jiuzhaigou, Sichuan, China. 15.04.2023

C. 中国西藏自治区错那县
Cuona, Xizang, China. 14.08.2006

D–E. 中国西藏自治区康马县
Kangma, Xizang, China. 10.08.2006

四列龙胆

Gentiana tetrasticha

一年生草本，高 2~5 cm，斜升，基部多分枝，茎光滑；茎生叶匙形到倒卵形，密集排列，较节间长，先端有小尖头，边缘软骨质；花单生分枝顶端；花萼筒管状，花萼裂片钻形，先端有软骨小尖头；花冠筒管状，淡蓝色，喉部具蓝紫色斑点；花冠裂片卵形，褶长圆形，整齐或不规则啮齿状。

特产于喜马拉雅地区的锡金和中国西藏。生长于高山草甸。海拔 4000~5000 m。花果期 7~9 月。

Annual, 2-5 cm, ascending, stems much branched from base, glabrous; stem leaves spathulate to obovate, densely spaced, longer than internodes, apex mucronate, margins cartilaginous; flowers solitary on branch terminals; calyx tube tubular, calyx lobes subulate, apex cartilaginous and mucronate; corolla tubular, pale blue, throat with blue-purple spots; corolla lobes ovate; plicae oblong, entire or irregular erose.

Endemic to the Himalayas, including Sikkim, and Xizang of China. In alpine meadows. Altitude: 4000-5000 m. Flowering and fruiting: July-September.

A–C. 中国西藏自治区那曲县
Naqu, Xizang, China. 07.08.2006

D–H. 中国西藏自治区错那市
Cuona, Xizang, China. 13.08.2006

母草叶龙胆

Gentiana vandellioides

一年生草本，茎纤细，铺散到斜升，高 5~10 cm，基部以上二歧分枝，光滑；茎生叶稀疏，叶片卵状到近心形；花单生枝端；花萼筒倒锥形，花萼裂片线形到线状钻形，直立；花冠倒锥形，淡蓝色，喉部黄绿色，檐部具深蓝紫色斑点；花冠裂片卵状心形，先端圆钝褶卵形，较花冠裂片短，整齐或 2 裂片。

产于中国中部的四川、陕西、湖北等省。生长于高山林下、林缘及草丛中向阳处。海拔 1500~3000 m。花果期 7~9 月。

Annual, stems slender, prostrate to ascending, 5-10 cm, dichotomously branched from base, glabrous; stem leaves sparsely spaced; leaf blades ovate to subcordate; flowers solitary and terminal; calyx tubes obconic, calyx lobes filiform to filiform-subulate, erect; corolla tube obconic, pale blue, greenish yellow in throat, with dark blue-purple spots and short lines; corolla lobes ovate-orbicular, apex obtuse; plicae ovate, shorter than corolla lobes entire or 2-cleft.

Sparsely spread in central China, Sichuan, Shaanxi and Hubei provinces. In alpine forest, forest margin and grasses. Altitude: 1500-3000 m. Flowering and fruiting: July-September.

A

B

C

A–C. 中国湖北省神农架
Shennongjia Mt., Hubei, China. 18.08.2004

D–E. 中国陕西省太白山
Taibaishan Mt., Shaanxi, China. 20.08.1990

F. 中国湖北省神农架
Shennongjia Mt., Hubei, China. 18.08.2004

D

E

F

灰绿龙胆

Gentiana yokusai

一年生草本，茎斜升或直立，高 3~8 cm，基部以上多分枝，密被乳突；茎生叶卵形到心形，先端外弯，有小尖头，边缘软骨质；花单生分枝顶端；花萼筒倒锥形，花萼裂片披针形到卵形，开展，外弯，有小尖头；花冠筒管状到漏斗状，淡蓝色，喉部具深色短条纹；花冠裂片卵形，褶卵形，较花冠裂片短，边缘啮齿状。

较广泛分布于东亚地区。生长于较潮湿的溪边、山坡、路边及田边。海拔 300~2500 m。花果期 3~9 月。

Annual, stems ascending to erect, 3-8 cm, much branched from base and above, densely papillate; stem leaves ovate to cordate, apex recurved, mucronate, margins cartilaginous; flowers solitary at branch terminals; calyx tube obconic, calyx lobes lanceolate to ovate, spreading, recurved, mucronate; corolla tubular to funnelform, pale blue, with dark stripes; corolla lobes ovate; plicae ovate, shorter than corolla lobes, margins erose.

Widely spread in eastern Asia. Growing in damp places such as slopes, stream side, roadside, and edges of cultivated fields. Altitude: 300-2500 m. Flowering and fruiting: March-September.

A–C. 中国贵州省石牌镇

Shipai, Guizhou, China. 27.04.2000

C

笔龙胆

Gentiana zollingeri

一年生草本，茎直立，高 3~5 cm，单一或基部以上分枝，光滑；茎生叶卵形，边缘软骨质；花单生或数朵簇生分枝顶端；花萼筒倒锥形，花萼裂片三角，直立，中脉隆起；花冠筒管状到漏斗状，淡蓝色，喉部黄绿色，具深色短条纹；花冠裂片卵形，褶卵形，较花冠裂片短，边缘啮齿状。

较广泛分布于东亚地区。生长于较潮湿的山坡、灌丛、林缘及林下。海拔 500~1500 m。花果期 3~6 月。

Annual, stems erect, 3-5 cm, simple or branched from base and above, glabrous; stem leaves ovate, margins cartilaginous; flowers solitary or a few clustered at terminals; calyx tube obconic, calyx lobes triangular, erect, midveins prominent; corolla tubular to funnelform, pale blue, yellowish white at throat, with dark stripes; corolla lobes ovate; plicae ovate, shorter than corolla lobes, margins erose.

Widely spread in eastern Asia. Growing in damp and shady places such as slopes, in bushes, forest margins and under forest. Altitude: 500-1500 m. Flowering and fruiting: March-June.

A

B

A. 中国陕西省佛坪县
Foping, Shaanxi, China. 30.03.2016

B-C. 中国浙江省天目山
Tianmushan Mt., Zhejiang, China. 17.04.2023

D. 中国陕西省佛坪县
Foping, Shaanxi, China. 30.03.2016

C

D

道孚小龙胆（新种）

Gentiana daofuensis sp. nov.

一年生草本，茎斜升，高 5~10 cm，基部多分枝，茎紫色，纤细，被乳突；茎生叶密集，无柄，叶片心形到阔卵形，先端急尖，具小尖头；花单生枝端；花萼裂片外弯，卵形到心形，先端具外弯的尖头；花冠筒管状，蓝色到蓝紫色，具密集深蓝紫色斑点；花冠裂片卵形，褶长圆形，边缘整齐或有不规则圆齿。

特产于中国四川西部的道孚。生长于向阳潮湿的山坡地。海拔 3500~3800 m。花果期 7~9 月。

本新种有待合规发表，与假鳞叶龙胆和弯叶龙胆较相似，但茎叶形态和质地及花纹饰明显不同，亦可能是二者之杂种。

Annual, stems ascending, 5-10 cm, much branched from base, stem purple, slender, papillate; stem leaves densely spaced; leaf blades cordate to broad ovate, apex acute, recurved and mucronate; flowers solitary; calyx lobes recurved, ovate to cordate, mucronate; corolla tubular, blue to blue-purple, with dark purple spots; corolla lobes ovate; plicae oblong, margin entire or irregularly crenulate.

Endemic to Daofu in western Sichuan province of China. In sunny and moist open slopes. Altitude: 3500-3800 m. Flowering and fruiting: July-September.

A

B

C

A–D. 中国四川省道孚县 Daofu, Sichuan, China. 22.08.2006

D

黄龙龙胆（新种）

Gentiana huanglongensis sp. nov.

一年生草本，茎紫色，纤细，铺散到斜升，高 5~10 cm，基部以上二歧分枝，光滑；茎生叶稀疏，叶片披针形，有刚毛；花单生枝端；花萼筒倒锥形，花萼裂片线形到线状钻形，直立；花冠倒锥形，淡粉色，喉部黄绿色，檐部及褶具深蓝紫色斑点；花冠裂片卵状心形，先端圆钝，褶三角形，较花冠裂片短，整齐或 2 裂。

特产于中国四川北部的松潘黄龙。生长于山地峡谷的林缘及草丛中向阳处。海拔 1900~2500 m。花果期 5~7 月。

本新种有待合规发表，与母草叶龙胆较相似，但叶片和花的形态明显不同。

Annual, stems purple, slender, prostrate to ascending, 5-10 cm, dichotomously branched, glabrous; stem leaves sparsely spaced; leaf blades lanceolate, hirsute; flowers solitary and terminal; calyx tubes obconic, calyx lobes filiform, erect; corolla tube obconic, pale pink, greenish yellow in throat, with dark purple spots; corolla lobes ovate-orbicular, apex obtuse; plicae triangular, shorter than corolla lobes entire or 2-cleft.

Endemic to northern Sichuan of China. In forest margin and grasses of alpine valley. Altitude: 1900-2500 m. Flowering and fruiting: May-July.

A

B

C

A-E. 中国四川省松潘县
Songpan, Sichuan, China. 21.05.2013

D

E

无茎龙胆 *Gentiana acaulis*

瑞士伯尔尼州施尼格普拉特 Schynige Platte, Bern, Switzerland. 29.05.2010

大花组
sect. *Ciminalis*

多年生草本；叶聚集基生；花单生茎顶端，花冠筒大，倒圆锥形，长达 3~7 cm，艳丽。

本组约 7 种，非常相似，常被作为不同的种下单元；基生叶叶片和花萼裂片的形态是区分它们的主要依据。主要分布于欧洲中部和南部的阿尔卑斯山、比利牛斯山、喀尔巴阡山及狄那里克山等高山地区。

Perennial herb; leaves crowded towards the base of stem; flowers solitary, terminal; corolla tubes large and conspicuous, obconical, 3-7cm long, showy.

About 7 species, highly resemble to each other, often treated as intraspecific taxa. Morphology of leaves and calyx teeth are key to distinguish them. Distributed in the high mountains in central and southern Europe, including the Alps, the Pyrenees, the Carpathian, and the Dinarides.

无茎龙胆

Gentiana acaulis

基生叶披针形、椭圆形或倒卵形；花单生茎顶端，花萼裂片较萼筒短，卵形，萼齿基部收缩变狭；花冠筒倒圆锥形，喉部具明显的绿色斑点，花冠裂片急尖或有小尖头。

广泛分布于中欧到南欧的主要高山地区，包括俄罗斯西南部、保加利亚、罗马尼亚、乌克兰、波兰、捷克、斯洛伐克、前南斯拉夫、奥地利、德国、瑞士、法国、西班牙和意大利。

生长在高山和亚高山地区的硅质岩石质山坡、草地和林缘。海拔 500~3000m。花果期 6~9 月。本种是嫌钙植物，其萼齿基部收缩变狭呈卵形，可与相近的坷路龙胆区别。

Basal leaves lanceolate, elliptic to obovate; flowers solitary, terminal; calyx lobes shorter than tube, ovate, narrowed at base; corolla tube obconical, with green spots in the throat, lobes acute or cuspidate.

Calcifuge, widely spread in the major mountains in central and southern Europe. In silicious soils of alpine and subalpine slopes, meadows and forest margin. Altitude: 500-3000 m. Flowering & fruiting: June-September.

A. 瑞士纳沙泰尔州
Fleurier, Neuchatel, Switzerland. 12.06.2006

B. 瑞士伯尔尼州格林德瓦尔得
Grindelwald, Bern, Switzerland. 26.05.2006

C-E. 瑞士伯尔尼州格里姆塞尔山口
Grimsel Pass, Bern, Switzerland. 01.07.2006

D

在瑞士伯尔尼州尼德宏山顶可以眺望远处白雪覆盖的阿尔卑斯山艾格峰、僧侣峰和少女峰及相邻诸峰的美景，也可以端详脚下的无茎龙胆（*Gentiana acaulis*）。

You can view the snow covered Eiger, Mönch, Jungfrau and other peaks of the Alps, and examine the beautiful flowers of the Stemless Gentian near your feet; Niderhorn, Bern, Switzerland. 19.05.2007.

阿尔卑斯龙胆

Gentiana alpina

基生叶近圆形，有时卵形或近心形；花单生茎顶端，萼齿（花萼裂片）较萼筒短，卵形，萼齿基部稍收缩；花冠筒钟状，长3~4 cm，喉部有绿色斑点，花冠裂片圆钝，近圆形。

分布于阿尔卑斯山西南到中西部及比利牛斯山中部的瑞士、法国、西班牙和意大利。嫌钙植物，生长在高山硅质岩石质山坡草地。海拔 2000~3000 m。花果期 6~9 月。本种基部叶片近圆形，花冠筒钟形，容易与其他种区分。

Basal leaves usually suborbicular, occasionally ovate or cordiform; flowers solitary, terminal; calyx teeth shorter than tube, ovate, narrowed at base; corolla tube campanulate, 3-4 cm long, with green spots in the throat, corolla lobes obtuse, usually rounded.

Distributed in southwestern and central to western Alps, and central Pyrenees, in Switzerland, France, Spain and Italy. Calcifuge, in silicious soil of alpine rocky slopes and meadows. Altitude: 2000-3000 m. Flowering & fruiting: June-September.

A–C. 瑞士瓦莱州大莎瓦拉
Grand Chavalard, Valais, Switzerland. 30.06.1996

D–E. 瑞士阿尔卑斯山 摄影 /Photo：Adrien Favre
Col de La Balme, Alps, Switzerland

D

E

狭叶龙胆

Gentiana angustifolia

基生莲座叶长条形、狭倒披针形；萼齿不及萼筒长度的一半，披针形，基部稍收缩；花冠筒倒锥形，长 5~6 cm，喉部有绿色斑点，花冠裂片急尖。

分布于阿尔卑斯山西南部、汝拉山及比利牛斯山。喜钙植物，生长在高山钙质岩石质山坡草地。海拔 1000~2800 m。花果期 6~9 月。可能由于本种相对比较耐热，在欧洲的植物园和家庭花园常有栽培。

Rosette leaves linear-lanceolate to oblanceolate; calyx teeth less that half as long as the tube, lanceolate, narrowed at the base; corolla obconical, 5-6 cm long, with green spots in the throat, lobes acute.

Wild in southwestern Alps, Jura and Pyrenees. Calcicole, in calcium soil alpine rocky slopes and meadows. Altitude: 1000-2800 m. Flowering & fruiting: June-September. Perhaps due to its higher tolerance to heat, this species is often cultivated in botanical gardens and private yards in Europe.

A

B

C

A–C. 瑞士伯尔尼格林德瓦尔得栽培

Cultivated in Grindelwald, Bern, Switzerland. 25.06.2006

D. 瑞士苏黎世大学植物园

Botanical Garden, Zurich University. 16.08.2006

D

E–G. 施尼格普拉特阿尔卑斯山花园和园中栽培的狭叶龙胆（*Gentiana angustifolia*）
The Alpengarten at Schynige Platte and the Narrow-Leaf Gentian *Gentiana angustifolia* cultivated in the garden. 25.06.2006

H. 施尼格普拉特阿尔卑斯山花园和历史悠久的齿轨火车火车站
The Alpengarten and the Schynige Platte station of the nostalgic cog railway. 25.06.2006

I. 瑞士伯尔尼格林德瓦尔得私家庭院的狭叶龙胆（*Gentiana angustifolia*）
Narrow-leaf Gentian cultivated in private yard, Grindelwald, Bern, Switzerland. 25.05.2006

H

I

在每年的 6~7 月间，乘坐产自 19 世纪充满历史感的齿轨铁路列车，行车 50 多分钟，就来到海拔 1967 m 的施尼格普拉特高山站。这里只有两个关键词：鲜花和美景。就在齿轨火车站的上方有一个袖珍植物园——阿尔卑斯山花园。这个袖珍植物园一半人工，一半天成，比一个人工假山大不了多少，但却生长着 650 多种高山植物，包括 3 种龙胆花，即狭叶龙胆（*Gentiana angustifolia*）、坷路龙胆（*Gentiana clusii*）和春龙胆（*Gentiana verna*），还有大片的欧洲金莲花（*Trollius europaeus*）和水仙银莲花（*Anemone narcissiflora*）。鲜花之外便是毫无遮拦的全景，一边是雪山，另一边是湖泊。阿尔卑斯山艾格峰、僧侣峰和少女峰等雪峰就眼前，仿佛触手可及；身后便是因特拉肯（朱自清先生称之为“交湖”）及它两侧的图恩湖和布里恩茨湖。

Spectacular views of the snow-covered mountains and beautiful flowers including 3 gentian species are the only 2 things that you can have in the Alpengarten at Schynige Platte, Bern, Switzerland. 25.06.2006.

坷路龙胆

Gentiana clusii

基生莲座叶椭圆形至长披针形；萼齿长约萼筒长度的一半，三角形，基部最宽、不收缩；花冠筒倒锥形，长 4~6 cm，喉部无斑点或有较少的绿色斑点，花冠裂片圆钝，有小尖头。

广泛分布于欧洲中部到南部的高山地区，从德国西南和喀尔巴阡山到巴尔干北部、奥地利、瑞士到法国南部及意大利北部。为喜钙植物，生长在高山钙质岩石质山坡草地。海拔：800~2800 m。花果期 6~9 月。本种花萼裂片三角形，基部不收缩与其他种不同。

Rosette leaves elliptic to oblong-lanceolate; calyx teeth about half as long as the tube, triangular, widest at the base; corolla obconical, 4-6 cm long, without or with few green spots in the throat, lobes obtuse, apiculate.

Widely spread in mountains of central to southern Europe, from southern Germany and the Carpathians, northern Balkans, Austria, Switzerland, to southern France and northern Italy. Calcicole, in calcium soil alpine rocky slopes and meadows. Altitude: 800-2800 m. Flowering & fruiting: June-September.

A

B

C

A. D. E. H. 瑞士伯尔尼州莎斯哈尔山 Chasseral, Bern, Switzerland. 06.06.2004

B. C. G. 瑞士纳沙泰尔州福落里 Fleurier, Neuchatel, Switzerland. 12.06.2006

F. 坷路龙胆（*Gentiana clusii*）的盗蜜者；瑞士伯尔尼州莎斯哈尔山 Nectary robbery on flowers of Gentiana clusii , Chasseral, Bern, Switzerland. 06.06.2004

D
E
F
G
H

迪纳里德龙胆

Gentiana dinarica

基生叶阔椭圆形；萼齿长约萼筒长度的一半，狭披针形，基部收缩；花冠筒倒锥形，长 4~6 cm，喉部几无斑点，花冠裂片渐尖。

仅分布于巴尔干半岛的迪纳里德山区及意大利中北部。喜钙植物，生长在钙质岩山坡草地与林缘。海拔 1000~2000 m。花果期 6~9 月。

Rosette leaves broadly elliptic; calyx teeth about half as long as the tube, narrowly lanceolate, narrowed at the base; corolla obconical, 4-6 cm long, without green spots in the throat, lobes acuminate.

In Dinarides in Balkan and central to northern Italy only. Calcicole, in alpine pastures. Altitude: 1000-2000 m. Flowering & fruiting: June-September.

意大利科莫格锐尼亚山 Mt. Grigna, Como, Italy. 03.07.1996

A

B

西部龙胆

Gentiana occidentalis

基生叶椭圆形；萼齿长超过萼筒长度的一半，卵形到披针形，基部收缩；花冠筒倒锥形，长 4~5 cm，喉部几无绿色斑点，花冠裂片急尖到渐尖。

野生种仅在法国和西班牙的比利牛斯山西部和 Cordillera Cantábrica 山区。欧洲植物园常栽培。喜钙植物，生长在钙质岩山坡。海拔 800~1500 m。花果期 6~9 月。

Rosette leaves elliptic; calyx teeth more than half as long as the tube, ovate to lanceolate, narrowed at the base; corolla obconical, 4-5 cm long, nearly no green spots in the throat, lobes acute to acuminate.

In western Pyrenees and Cordillera Cantábrica only. Calcicole, in alpine pastures. Altitude: 800-1500 m. Flowering & fruiting: June-September.

A. 瑞士沃州托马斯山地植物园
La Thomasia, Jardin Alpin de Pont-de-Nant, Vaud, Switzerland. 25.06.2005

B. 瑞士沃州托马斯山地植物园：始建于 1891 年，绝对是世外桃源般的存在，深藏在阿尔卑斯山深谷里的瑰宝，一小半人工，一大半天然，完美诠释了瑞士的低调奢华。
Hidden in a deep valley in the Alps, the La Thomasia, Jardin Alpin de Pont-de-Nant, initiated in 1891, major half natural and minor half artificial, is absolutely a paradise for plants and visitors. Vaud, Switzerland. 25.06.2005.

秦艽 *Gentiana macrophylla*

中国甘肃省宕昌县 Tanchang, Gansu, China. 05.09.2023

秦艽组
sect. *Cruciata*

多年生草本；根和根状茎发达，稍肉质，须根扭结成发辫状结构；茎基部被残存的纤维状叶鞘包裹；基生叶形成发达的莲座叶丛；花顶生、轮生和轮状簇生成轮状聚伞花序；种子无翅，具细网纹。本组约 22 种，分布于亚欧大陆温带地区。

Perennial, roots and rhizome stout, slightly fleshy, small fibrous roots contiguous and twisted into a plait structure; caudex covered with conspicuous fibrous remains of leaf bases; basal leaves crowed to rosette; verticillaster, flowers terminal, axillary whorls or clusters; seed without wings, seedcoat minutely reticulate. About 22 species, widely spread in the temperate areas of the Eurasian continent.

粗茎秦艽

Gentiana crassicaulis

多年生草本，茎斜升，高 20~40 cm；基生莲座叶大型，叶片卵状椭圆形到狭椭圆形，5~7 条粗壮弧形脉；茎生叶 3~5 对，愈向顶端明显愈大；花无柄，常密集簇生于顶端；花萼筒向一侧开裂成佛焰苞状，1~5 裂片，齿状；花冠瓮状，喉部开口处稍收缩，浅黄绿色到蓝紫色，4~5 浅裂，裂片卵状三角形，先端急尖。

分布于中国西藏自治区、四川、云南等省区。生长在山坡草地、牧场和林缘。海拔 2500~4000 m。花果期 6~10 月。

Perennial, stem ascending, 20-40 cm high; basal rosette leaves large, blade ovate-elliptic to narrowly elliptic, with 5-7 strong longitudinal veins; cauline leaves 3-5 pairs, becoming distinctly larger toward apex; flowers sessile, crowded to terminal clusters; calyx tube split down to one side, spathe-like, calyx lobes 1-5, dentiform; corolla tube urceolate, slightly narrowed at the throat opening; pale greenish-yellow to light blue purple; 4-5 lobes to upper part, lobes ovate-triangular, apex acute.

In Xizang, Sichuan, and Yunnan provinces of China. In hill slopes, grassland, pasture and forest margin. Altitude 2500-4000 m. Flowering & fruiting: June-October.

A-B. 中国云南省香格里拉大雪山
Grand Snow Mountain, Yunnan, China. 04.08.2002

C. 中国四川省稻城县亚丁村
Yading, Daocheng, Sichuan, China. 09.07.2014

D-E. 中国云南省香格里拉大雪山
Grand Snow Mountain, Yunnan, China. 05.08.2002

F. 中国四川省稻城县亚丁村
Yading, Daocheng, Sichuan, China. 09.07.2014

G. 中国四川省九龙县
Jiulong, Sichuan, China. 14.09.2023

H. 中国云南省香格里拉市
Xianggelila, Yunnan, China. 18.09.2023

D
E
H
G

十字秦艽(欧洲秦艽)

(原亚种)

Gentiana cruciata subsp. *cruciata*

多年生草本，茎数枚，从莲座叶丛生出，斜升，高 10~40 cm，发紫；叶片条状卵状；茎生叶多对，近等大；花无柄，数枚轮生叶腋或顶生，花 4 基数，偶尔 5 数；花萼裂片微小齿状，比萼筒短；花冠瓮状到管状，长 20~25 mm，外面紫色，内面淡蓝色，4 裂，卵状长圆形，先端圆钝。

分散分布于欧洲东部、中部和南部及高加索和西亚部分地区。喜钙，生长在较潮湿的草地、牧场和林缘。海拔 800~2500 m。花果期 7~9 月。

Perennial, stems several from basal rosette, ascending, 10-40 cm, purplish; leaves blade oblong-ovate to narrowly elliptic; cauline leaves numerous pairs similar in size; flowers sessile, several verticillate in axillary and terminal clusters, 4-merous, occasionally 5-merous; calyx teeth triangular, much short than tube; corolla tube urceolate to tubular, purple outside light blue inside, 4 lobes ovate-oblong, apex obtuse.

Scattered in east, central and south European to Caucasus and west Asian countries. Calcicole, in wet grassland, pasture and forest margin. Altitude: 800-2500 m. Flowering & fruiting: July-September.

A

B

A-B. 保加利亚伯姬利卡 Beglika, Bulgaria. 22.07.1993

C-D. 瑞士沃州马切如山口 Col du Marchairuz, Vaud, Switzerland. 27.07.2023

C

D

福考秦艽（亚种）

Gentiana cruciata subsp. *phlogifolia*

该亚种与十字秦艽原亚种植物外部形态接近或重叠，唯花萼裂片更细长，与萼筒近等长，线形或线状披针形，可以区分。

野生种仅见于喀尔巴阡山东部及南部的罗马尼亚、捷克和斯洛伐克等地。喜钙，生长在较潮湿的草地、牧场和林缘。海拔1200~2200 m。花果期 7~9 月。欧洲各植物园多有栽培。

This subspecies highly resembles or overlap with the subspecies cruciata in morphology; major differences are the size and shape of the calyx teeth: the calyx teeth of this subspecies linear to linear-lanceolate, about as long as the tube.

Wildly in east and south Carpathians, in Romania, Czech and Slovakia. Calcicole, in damp grassland, pasture and forest margin. Altitude: 1200-2200 m. Flowering & fruiting: July-September. Often cultivated in botanical gardens in Europe.

A-B. 瑞士瓦莱州阿尔卑斯之花高山植物园

Flore-Alpe Botanical Garden，Champex-Lac, Valais, Switzerland. 12.07.2003.

B

达乌里秦艽

Gentiana dahurica

多年生草本，茎斜升，高 10~25 cm，发紫；莲座叶叶片条状披针形；茎生叶 3~5 对，顶部者更小；花有柄，数枚轮生叶腋或顶生，花 5 基数；花萼筒不裂或偶尔一侧开裂，萼齿不等大，线形；花冠漏斗状，蓝紫色，喉部有斑点，花冠裂片卵形或卵状披针形，先端圆钝或急尖，褶三角形到卵形。

分散分布于中亚及东北亚，包括中国北部、蒙古及邻近的俄罗斯。生长在向阳的山坡草地、牧场和林缘。海拔 800~3500 m。花果期 7~10 月。

Perennial, stems ascending, 10-25 cm, purplish; rosette leaves linear-lanceolate; cauline leaves 3-5 pairs becoming small toward apex; flowers pedicellate, several verticillate in axillary and terminal clusters, 5-merous; calyx tube entire or split on one side, calyx teeth unequal, linear; corolla funnelform, with spots in throat, blue-purple, lobes ovate or ovate-lanceolate, apex obtuse or acute, plicae triangular to ovate.
Scattered in central and northeast Asia, including northern China, Mongolia, and adjacent Russia. In sunny grassy slopes, pasture and forest margin. Altitude 800-3500 m. Flowering & fruiting: July-October.

A-D. 中国山西五台山
Wutaishan, Shanxi, China. 20.09.2004
E. 中国青海青海湖
Qinghai Lake, Qinghai, China. 02.08.2006
F. 中国甘肃省榆中县
Yuzhong, Gansu, China. 25.07.2011

D
E

斜升秦艽

Gentiana decumbens

多年生草本，茎斜升，高 20~50 cm，发紫；莲座叶叶片狭披针形，先端渐尖；茎生叶 2~3 对，顶部者渐小；花无柄或有短柄，数枚轮生上部叶腋或顶生；花萼筒一侧开裂至基部，萼齿不等大，钻形；花冠钟形，蓝紫色，花冠裂片卵形或卵状圆形，先端圆钝，褶短小，截形。

分散分布于东欧、中亚及东北亚，包括中国西北地区。生长在较潮湿的山坡草地和林缘。海拔 1500~2500 m。花果期 7~9 月。

Perennial, stems ascending, 20-50 cm, purplish; rosette leaves narrowly lanceolate, acuminate; cauline leaves 2-3 pairs becoming small toward apex; flowers sessile or with short pedicels, several verticillate in upper axillary and terminal clusters; calyx tube split to the base on one side, calyx teeth unequal, subulate; corolla campanulate, blue-purple, lobes ovate or orbicular, apex obtuse or rounded, plicae small, truncate.

Scattered in eastern Europe, central and northern Asia, including northwestern China. In damp grassy slopes and forest margin. Altitude: 1500-2500 m. Flowering & fruiting: July-September.

A–D. 中国新疆维吾尔自治区喀纳斯
Hanasi, Xinjiang, China. 08.08.1995

D

川西秦艽

Gentiana dendrologi

多年生草本；根肉质；茎斜升，高 20~40 cm；莲座叶叶片线状披针形，先端渐尖，叶面光滑；茎生叶 3~5 对，向上逐渐变小，最上部叶无柄，包裹花簇；花无柄，密集成头状簇生顶端，偶尔有少数花簇生在顶端以下的叶腋；花萼筒倒锥形，向一侧开裂，萼齿细小，线形；花冠管状钟状，绿白色到黄白色，花冠裂片卵形，先端圆钝，褶偏斜，截形，边缘齿状。

特产中国四川西北部。生长在高山草甸与灌丛间，海拔 3500~4500 m。花果期 7~9 月。

Perennial, root fleshy; stems ascending, 20-40 cm; rosette leaves linear-lanceolate, acuminate, glabrous; stem leaves 3-5 pairs becoming smaller toward apex; flowers sessile, crowded to a terminal cluster, occasionally subapical axillary; calyx tube obconic, split to one side, calyx teeth filiform; corolla tubular-campanulate, greenish white to yellowish white, lobes ovate, apex obtuse, plicae oblique, truncate, margin dentiform.

Endemic to northwestern Sichuan province of China. Growing in alpine meadows and bushes. Altitude: 3500-4500 m. Flowering & fruiting: July-September.

A-E. 中国四川省康定市

Kangding, Sichuan, China. 13.09.2023

C

D

E

何氏秦艽

Gentiana hoae

多年生草本；茎平铺或斜升，高 5~20 cm，发紫；莲座叶叶片线状狭披针形，先端渐尖；茎生叶 2~4 对，近等大；花有短柄，单生茎顶；花萼筒倒锥形，不开裂，萼齿近等大，狭椭圆形到匙形，基部稍变狭；花冠管状漏斗形，淡蓝色或蓝灰色，花冠裂片卵形，先端圆钝，褶偏斜三角形，具细齿。

中国青藏高原东部特产，见于西藏和青海等省区。生长在高山草甸。海拔 3500~4200 m。花果期 8~9 月。

Perennial, stems prostrate or ascending, 5-20 cm, purplish; rosette leaves linear to narrowly lanceolate, acuminate; cauline leaves 2-4 pairs subequal; flowers with short pedicels, solitary and terminal; calyx tube entire, obconic, calyx teeth nearly equal, narrowly elliptic; corolla tubular-funnelform, pale blue to grey-blue, lobes ovate, apex obtuse, plicae oblique triangular, denticulate.

Endemic to eastern Qinghai-Xizang Plateau of China, in Xizang and Qinghai provinces. Growing in alpine meadows. Altitude: 3500-4200 m. Flowering & fruiting: August-September.

A

B

A–D. F. 中国西藏自治区江达县
Jiangda, Xizang, China. 20.08.2006

E. G. 中国西藏自治区昌都市 摄影 /Photo：付鹏程
Changdu, Xizang, China. 16.08.2017

C
D
E
G
F

何氏秦艽（*Gentiana hoae*）的种名是为了纪念何廷农教授（1938—2011）。何教授是著名的龙胆科专家，毕生从事龙胆科植物的分类学研究，发现和描述了大量新种和一些新属，发表了许多论文及《中国植物志》中文版和英文版、*A worldwide monograph of Gentiana*、*A worldwide monograph od Swertia and its allies* 等多部著作，为龙胆科植物的研究做出了杰出贡献。她主导的 "中国龙胆科植物的研究" 荣获 2004 年中国国家自然科学二等奖。刘尚武教授（1934—2021）是何教授的丈夫，也是她最重要的研究伙伴，二人一起发表了大量著作，堪称楷模。

The species name *Gentiana hoae* was created after Ting-Ning Ho (1938-2011) to honor Prof. Ho for her great contribution on Gentianaceae. Prof. Ho is a well recognized expert on Gentianaceae family internationally. She spend nearly her whole life studying gentians, and discovered and described many new species and some new genera for this family. She published many papers and wrote some important works such as *the Flora of China*, *volumes on Gentianaceae* (both Chinese and English editions respectively), *A worldwide monograph of Gentiana*, and *A worldwide monograph od Swertia and its allies*, etc. In recognizing those great contributions on understanding the biodiversity of the family Gentianaceae, she worn a National Natural Science Award of China in 2004. Prof. Shang-Wu Liu, a well known Chinese botanist, was husband of Prof. Ho, and also the most important collaborator and coauthor of many publications.

A. 何廷农教授在野外工作 Prof. Ho working in the field

B. 刘尚武教授（左一）、何廷农教授（左二）与他们的研究组成员在野外工作（感谢刘源先生供图）Prof. Shang-Wu Liu (1st from left) and Prof. Ho (2nd from left) and their research team working in the field (photos courtesy from Yuan Liu)

中亚秦艽

Gentiana kaufmanniana

多年生草本；茎斜升，高 10~30 cm，发紫；莲座叶叶片狭披针形，先端渐尖；茎生叶 2~3 对，顶部者渐小；花有短柄，数枚生上部叶腋或顶生；花萼筒不开裂，萼齿不等大，线形；花冠钟状漏斗形，蓝色到紫色，花冠裂片卵形或卵状圆形，先端圆钝，褶偏斜，耳状。

分散分布于中亚，包括中国西北。生长在山坡草地。海拔 1800~3000 m。花果期 7~9 月。

Perennial, stems ascending, 10-30 cm, purplish; rosette leaves narrowly lanceolate, acuminate; cauline leaves 2-3 pairs becoming small toward apex; flowers with short pedicels, several in upper axillary and terminal clusters; calyx tube entire, calyx teeth unequal, linear; corolla campanulate-funnel form, blue to purple, lobes ovate or orbicular, apex obtuse, plicae oblique, auriculate.

Scattered in central Asia, including Afghanistan, Pakistan, Tajikistan, Kyrgyzstan, Kazakhstan, Russia (Altai) and northwestern China. In damp grassy slopes and forest margin. Altitude: 1800-3000 m. Flowering & fruiting: July-September.

A-C. 中国新疆维吾尔自治区和静县巴音布鲁克
Bayinbuluke, Xinjiang, China. 09.09.2004

全萼秦艽

Gentiana lhassica

多年生草本; 茎斜升，高 5~20 cm，发紫; 莲座叶叶片线状狭披针形，先端渐尖; 茎生叶 2~3 对，近等大; 花有短柄，单生茎顶，极个别 2~3 朵簇生; 花萼筒倒锥形，不开裂，萼齿近等大，狭椭圆形，基部稍变狭; 花冠管状漏斗形，蓝色或蓝紫色，花冠裂片卵状心形，先端圆钝，褶偏斜三角形，具细齿。

中国青藏高原特产，见于西藏、四川和青海等省区。生长在高山草甸。海拔 3500~4500 m。花果期 8~9 月。

Perennial, stems ascending, 5-20 cm, purplish; rosette leaves linear to narrowly lanceolate, acuminate; cauline leaves 2-3 pairs subequal; flowers with short pedicels, solitary and terminal, very occasionally 2-3 in terminal clusters; calyx tube entire, obconic, calyx teeth nearly equal, narrowly elliptic; corolla tubular-funnelform, blue to dark purple, lobes ovate orbicular, apex obtuse, plicae oblique triangular, denticulate.

Endemic to Qinghai-Xizang Plateau of China, in Xizang, Sichuan and Qinghai provinces. Growing in alpine meadows. Altitude: 3500-4500 m. Flowering & fruiting: August-September.

A-E. 中国西藏自治区隆子县

Longzi, Xizang, China. 13.08.2006

秦艽

Gentiana macrophylla

多年生草本；茎斜升，高 5~20 cm，发紫；莲座叶叶片线状狭披针形，先端渐尖；茎生叶2~3对，近等大；花有短柄，单生茎顶，极个别 2~3 朵簇生；花萼筒倒锥形，不开裂，萼齿近等大，狭椭圆形，基部稍变狭；花冠管状漏斗形，蓝色或蓝紫色，花冠裂片卵状心形，先端圆钝，褶偏斜三角形，具细齿。

中国青藏高原特产，见于西藏、四川和青海等省区。生长在高山草甸。海拔 3500~4500 m。花果期 8~9 月。

Perennial, root fleshy; stems ascending, 30-60 cm; rosette leaves ovate-lanceolate, acuminate, glabrous; cauline leaves 3-5 pairs becoming smaller toward apex; flowers sessile, crowded to a terminal cluster and few flowers in axillary clusters; calyx tube obconic, split to one side, calyx teeth equal, dentiform; corolla urceolate, blue, lobes ovate, apex obtuse, plicae triangular, margin entire.

In central and northern Asia, including northern China. Cultivated in Europe. Growing in slopes, grassland and forest margin. Altitude: 500-2500 m. Flowering & fruiting: July-October.

A

B

A–C. 中国甘肃省宕昌县
Tanchang, Gansu, China. 08.08.2012

D. 瑞士日内瓦植物园栽培
Cultivated in Geneva Botanical Garden, Switzerland. 30.07.2023

C

D

黄管秦艽

Gentiana officinalis

多年生草本；根肉质；茎斜升，高15~40 cm；莲座叶叶片卵状披针形，先端渐尖，叶面光滑；茎生叶3~5对，向上逐渐变小；花几无柄，密集成头状簇生顶端，时常有少数花簇轮生在顶端以下1~2节的叶腋；花萼筒倒锥形，向一侧开裂，萼齿细小，近等大，线形；花冠管状，白色到黄白色，花冠裂片卵形，先端圆钝，褶钻形，边缘整齐。

中国特产，见于青海、甘肃和四川。生长在向阳山坡、草甸与农田边缘。海拔2000~3500 m。花果期7~9月。

Perennial, root fleshy; stems ascending, 15-40 cm; rosette leaves ovate-lanceolate, acuminate, glabrous; cauline leaves 3-5 pairs becoming smaller toward apex; flowers sessile, crowded to a terminal cluster and few flowers in axillary clusters; calyx tube obconic, split to one side, calyx teeth equal, linear; corolla tubular, white to yellowish white, lobes ovate, apex obtuse, plicae subulate, margin entire.

Endemic to China, in Qinghai, Sichuan and Gansu provinces. Growing in sunny slopes, grassland and marginal agriculture fields. Altitude: 2000-3500 m. Flowering & fruiting: July-September.

A

B

A-B. 中国甘肃省榆中县 Yuzhong, Gansu, China. 25.07.2011

C. 中国青海省河南蒙古族自治县 Henan, Qinghai, China. 09.09.2023

D. 中国甘肃省卓尼县 Zhuoni, Gansu, China. 08.09.2023

粗壮秦艽

Gentiana robusta

多年生草本；根肉质；茎斜升，高 20~40 cm；莲座叶叶片椭圆状披针形；茎生叶 3~5 对，向上逐渐变小，最上部叶无柄，包裹花簇；花无柄，密集成头状簇生顶端，偶尔有少数花轮生在顶端以下的叶腋；花萼筒倒锥形，向一侧开裂，萼齿细小，不等大，线形；花冠钟状，绿白色到黄白色，花冠裂片卵状三角形，先端圆钝，褶偏斜，截形，边缘齿状。

喜马拉雅特产，见于尼泊尔、锡金和中国西藏。生长在向阳山坡、草甸与农田边缘。海拔 3500~4500 m。花果期 7~10 月。

Perennial, root fleshy; stems ascending, 20-40 cm; rosette leaves elliptic-lanceolate; stem leaves 3-5 pairs becoming smaller toward apex; flowers sessile, crowded to a terminal cluster, and sometimes few flowers in axillary clusters below apex; calyx tube obconic, split to one side, calyx teeth unequal, filiform; corolla campanulate, greenish white to yellowish white, lobes ovate-triangular, apex obtuse, plicae oblique, truncate, margin dentiform.

Limited to the Himalayas, in Nepal, Sikkim, and Xizang of China. Growing in sunny slopes, grassland and margins of farming land. Altitude: 3500-4500 m. Flowering & fruiting: July-October.

A–E. 中国西藏自治区浪卡子县
Langkazi, Xizang, China. 09.08.2006

F. 青稞田边的粗壮秦艽（Gentiana robusta）和迎亲仪仗队；中国西藏自治区浪卡子县
Gentiana robusta and the bride procession beside the barley field; Langkazi, Xizang, China. 09.08.2006

D

E

F

管花秦艽

Gentiana siphonantha

多年生草本；根肉质；茎纤细，直立，高 10~30 cm；莲座叶叶片线形到线状披针形；茎生叶 2~3 对，向上逐渐变小；花几无柄，密集成头状簇生顶端，时有少数花簇轮生在顶端以下的叶腋；花萼筒倒锥形，萼齿较小，近等大，钻状；花冠管状钟形，深蓝紫色，花冠裂片长圆形，先端圆钝，褶三角形，偏斜，边缘整齐或 2 裂。

中国西部特产，见于青海、甘肃和四川。生长在山坡草地与林缘。海拔 1500~3500 m。花果期 7~9 月。

Perennial, root fleshy; stems slender, erect, 10-30 cm; rosette leaves linear to linear-lanceolate; cauline leaves 2-3 pairs becoming smaller toward apex; flowers sessile, crowded to a terminal cluster and few flowers in lower axillary clusters; calyx tube obconic, calyx teeth equal, subulate; corolla tubular-campanulate, dark blue-purple, lobes oblong, apex obtuse, plicae triangular, oblique, margin entire or 2-cleft.

Endemic to China, in Qinghai, Gansu and Sichuan. Growing in slopes, grassland and forest margin. Altitude: 1500-3500 m. Flowering & fruiting: July- September.

A

B

A-D. 中国青海省门源回族自治县

Menyuan, Qinghai, China. 19.08.2015

C

D

麻花艽

Gentiana straminea

多年生草本；根肉质；茎斜升，高10~30 cm；莲座叶叶片卵状披针形到阔披针形；茎生叶3~5对，明显较小；花具柄，数枚顶生和腋生成较松散的聚伞花序；花萼筒向一侧开裂到接近基部，萼齿细小，不等大，钻形；花冠漏斗状，绿白色到黄白色，有绿色斑点，花冠裂片卵状三角形，先端圆钝，褶偏斜，三角形，边缘整齐。

中国西部特产，见于西藏、四川、青海、甘肃等地。生长在向阳干燥山坡、草甸与农田边缘。海拔2500~4500 m。花果期7~10月。

Perennial, root fleshy; stems ascending, 10-30 cm; rosette leaves ovate-lanceolate to broad lanceolate; stem leaves 3-5 pairs, much smaller; flowers pedicellate, several terminal and axillary as sparse cyme; calyx tube obconic, split on one side to base, calyx teeth unequal, subulate; corolla broadly funnelform, greenish white to yellowish white, with green spots, lobes ovate-triangular, apex obtuse, plicae oblique, triangular, margin entire.

Endemic to western China, in Xizang, Sichuan, Qinghai and Gansu. Growing in sunny and dry slopes, meadows and margins of farming land. Altitude: 2500-4500 m. Flowering & fruiting: July-October.

A

B

C

A-B. 中国青海省日月山
Riyueshan Mt., Qinghai, China. 02.08.2006
C. 中国青海省青海湖
Qinghai Lake, Qinghai, China. 02.08.2006
D. F. 中国四川省甘孜县 Ganzi, Sichuan, China. 21.08.2006
E. 中国青海省格尔木市沱沱河
Tuotuohe, Golmud, Qinghai, China. 05.08.2006
G. 中国甘肃省卓尼县 Zhuoni, Gansu, China. 08.09.2023

D

G

沐浴着阳光也沐浴着佛光的麻花艽，中国四川省阿坝藏族羌族自治州各莫寺

Gentiana straminea under the sunshine and the shine of Buddha;

Gemo Monastery, Aba, Sichuan, China. 11.09.2023

天山秦艽

Gentiana tianschanica

多年生草本，茎斜升，高 10~25 cm，发紫；莲座叶叶片条状椭圆形到椭圆状披针形；茎生叶 3~5 对线形，向上变小；花几无柄，数枚轮生叶腋或簇生顶端；花萼筒不裂或一侧稍开裂，萼齿不等大，线状钻形；花冠漏斗状，淡蓝色，花冠裂片卵形或卵状椭圆形，先端急尖，褶三角形，2 裂。

分散分布于中亚、克什米尔及喜马拉雅西部地区，包括中国西北。生长在向阳的山坡草地和牧场。海拔 1500~3500 m。花果期 7~9 月。

Perennial, stems ascending, 10-25 cm, purplish; rosette leaves linear-elliptic to elliptic-lanceolate; cauline leaves 3-5 pairs becoming small toward apex; flowers nearly sessile, several verticillate in axillary and terminal clusters; calyx tube entire or slightly split on one side, calyx teeth unequal, linear-subulate; corolla funnelform, light blue, lobes ovate or elliptic, apex acute, plicae triangular, 2-clift. Scattered in central Asia, Kashmir and western Himalayan areas including northwestern China. In sunny grassy slopes and pastures. Altitude: 1500-3500 m. Flowering & fruiting: July-September.

A

B

A–B. 中国新疆维吾尔自治区昭苏县
Zhaosu, Xinjiang, China. 05.09.2004

C–E. 中国新疆维吾尔自治区和静县巴音布鲁克
Bayinbuluke, Hejing, Xinjiang, China. 09.09.2004

中国新疆维吾尔自治区和静县巴音布鲁克 Bayinbuluke, Hejing, Xinjiang, China. 09.09.2004

西藏秦艽

Gentiana tibetica

多年生草本，茎粗壮，斜升，高 40~60 cm；基生莲座叶大型，叶片卵状椭圆形到狭椭圆形；茎生叶 3~5 对，愈向顶端明显愈大；花无柄，常密集簇生于顶端；花萼筒向一侧开裂成佛焰苞状，1~5 裂片，齿状；花冠阔管状到钟状，内面浅黄绿色到黄白色，外面淡紫褐色条纹，裂片三角形，偏斜，褶三角形。

喜马拉雅特产，见于尼泊尔、锡金、不丹及中国西藏自治区。生长在山坡草地、牧场和林缘。海拔 2500~4000 m。花果期 6~8 月。

Perennial, stem, stout, ascending, 40-60 cm; basal rosette leaves large, blade ovate-elliptic to narrowly elliptic; cauline leaves 3-5 pairs, becoming distinctly larger toward apex; flowers sessile, crowded to terminal clusters; calyx tube split down to one side, calyx lobes dentiform; corolla tube broad tubular to campanulate, inside pale greenish-yellow to yellowish white, outside with purple-brown stripes; lobes triangular, plicae oblique, triangular.

Endemic to the Himalayas, in Nepal, Sikkim, Bhutan, and Xizang of China. In hill slopes, grassland, pasture and forest margin. Altitude: 2500-4000 m. Flowering & fruiting: June-October.

A–D. 中国西藏自治区亚东县
Yadong, Xizang, China. 10.08.2006

E. 中国西藏自治区错那市
Cuona, Xizang, China. 14.08.2006

本种形态与粗茎秦艽（*Gentiana crassicaulis*）接近，但二者花的大小和颜色截然不同。This species highly resemble *Gentiana crassicaulis*, but the sizes and colors of flowers are distinct.

长梗秦艽

Gentiana waltonii

多年生草本；茎斜升到直立，高 10~30 cm，发紫；莲座叶叶片狭披针形，先端渐尖；茎生叶 2~3 对，显著更小；花有柄，数枚生上部叶腋或顶生；花萼筒紫色，一侧开裂至基部，萼齿近等大，钻形；花冠漏斗形，蓝紫色，花冠裂片卵形或卵状圆形，先端圆钝，褶偏斜，三角形。

中国西藏特产，见于西藏南部和东南部。生长在石质山坡。海拔 3000~4500 m。花果期 8~10 月。

Perennial, stems ascending to erect, 10-30 cm, purplish; rosette leaves narrowly lanceolate, acuminate; cauline leaves 2-3 pairs distinctly smaller; flowers with pedicels, several in upper axillary and terminal clusters; calyx tube purple open to base on one side, spathe-like, calyx teeth subequal, subulate; corolla funnelform, blue-purple, lobes ovate or orbicular, apex obtuse, plicae oblique, triangular.

Endemic to southern and southeastern Xizang of China. In stony and rocky slopes. Altitude: 3000-4500 m. Flowering & fruiting: August-October.

A–E. 中国西藏自治区曲水县
Qushui, Xizang, China. 20.09.2003

C

D

中国西藏自治区浪卡子县羊卓雍措湖边的石质山坡上生长着美丽的长梗秦艽（*Gentiana waltonii*）。

The Yangzhuoyongcuo Lake in Langkazi, Xizang, China, on the rocky slopes beside the lake grows the beautiful *Gentiana waltonii*. 20.09.2003.

新疆秦艽

Gentiana walujewii

多年生草本；根肉质；茎斜升，高20~40 cm；莲座叶叶片卵状椭圆形，先端渐尖，叶面光滑；茎生叶2~3对，向上逐渐变小；花无柄，密集成头状簇生顶端，偶尔有少数花簇轮生在顶端以下叶腋；花萼筒倒锥形，不开裂，萼齿较小，近等大，匙状；花冠管状漏斗形，外面淡蓝色，内面黄白色，花冠裂片卵形，先端圆钝，褶披针形，边缘整齐或2裂。

仅分布于哈萨克斯坦和中新疆国北部。生长在山坡草地。海拔2000~2500 m。花果期：7~9月。

Perennial, root fleshy; stems ascending, 20-40 cm; rosette leaves elliptic, acuminate, glabrous; cauline leaves 2-3 pairs becoming smaller toward apex; flowers sessile, crowded to a terminal cluster and few flowers in axillary clusters; calyx tube obconic, not split, calyx teeth equal, spathulate; corolla tubular funnelform, outside light blue, inside yellowish white, lobes ovate, apex obtuse, plicae lanceolate, margin entire or 2 clefts.
In Kazakhstan and northern Xinjiang of China. Growing in grassy slopes. Altitude: 2000-2500 m. Flowering & fruiting: July-September.

A

B

A-C. 中国新疆维吾尔自治区和静县巩乃斯
Gongnaisi, Xinjiang, China. 07.09.2004

C

泽库秦艽

Gentiana zekuensis

多年生草本；根肉质；茎纤细，斜升，高 10~30 cm；莲座叶叶片线形到线状披针形；茎生叶 2~3 对，明显更小；花几无柄，密集成头状簇生顶端，时有少数花簇轮生在顶端以下的叶腋；花萼筒倒锥形，一侧开裂到基部，萼齿细小，钻状；花冠筒管状，淡蓝紫色，喉部有蓝色斑点，花冠裂片卵圆形，先端圆钝，褶三角形，偏斜，边缘整齐。

特产中国青海和四川。生长在高山草甸和山坡。海拔 3000~3500 m。花果期 7~9 月。

Perennial, root fleshy; stems slender, ascending, 10-30 cm; rosette leaves linear to linear-lanceolate; cauline leaves 2-3 pairs distinctly smaller; flowers sessile, crowded to a terminal cluster and few flowers in lower axillary clusters; calyx tube obconic, split to base on one side, calyx teeth subulate; corolla tubular, light blue-purple, with dark blue spot in throat, lobes ovate, apex obtuse, plicae triangular, oblique, margin entire.

Endemic to Qinghai and Sichuan of China. In alpine slopes and meadows. Altitude: 3000-3500 m. Flowering & fruiting: July- September.

A

B

C

A–C. 中国四川省理塘县海子山
Haizishan, Litang, Sichuan, China. 08.08.2002

D. F. 中国青海省泽库县 Zeku, Qinghai, China. 09.09.2023

E. 中国四川省理塘县 Litang, Sichuan, China. 08.08.2002

D

F

E

高山龙胆 *Gentiana algida*
中国吉林长白山 Changbaishan Mt., Jilin, China.19.08.2000

高山龙胆组

sect. *Frigida*

多年生草本；根和根状茎稍肉质；茎合轴分支，有 1 至数枚莲座叶丛和 1 至数枚花枝，莲座叶丛直立，花枝基部具残余叶基部形成的鞘；花单生或簇生顶端；种皮具海绵状网纹。本组约 18 种，主要分布于中国西南部，个别种扩散到亚欧大陆及北美洲高纬度地区。

Perennial, roots and rhizome stout, slightly fleshy; stems branching sympodially, 1 to few basal rosettes and 1 to few flowering stems, rosettes erect, and bases of flowering stems sheathed with conspicuous fibrous remains of leaf bases; flowers solitary or clustered at terminal; seedcoat with spongy-like membranes. About 18 species highly concentrated in southwestern China, with few species spread to high latitude areas in Eurasia and North America.

高山龙胆

Gentiana algida

多年生草本，直立，高 5~15 cm；莲座叶丛 1~3，直立，叶片边缘向内卷，线状披针形；茎生叶 2~3 对，明显较小；花 1~3 枚顶生；花萼筒不裂或偶尔一侧浅裂，萼齿不等大，狭长圆形；花冠钟状到管状漏斗状，黄白色到黄绿色，外面有绿紫色条纹，花冠裂片三角形，先端圆钝，有绿色斑点，褶偏斜，截形。

分布于亚洲中部、北部到东北部及北美高山和高纬度地区。生长在高山冻原、高山草甸和高山石质山坡。海拔 2000~5000 m。花果期 6~9 月。

Perennial, stems erect, 5-15 cm; rosette 1-3, erect, basal leaves involute, linear-lanceolate; stem leaves 2-3 pairs, distinctly smaller; flowers 1-3, in terminal clusters; calyx tube entire or slightly split on one side, calyx teeth unequal, narrowly oblong; corolla campanulate to tubular-funnelform, yellowish white to greenish yellow, with purple stripes outside, lobes triangular, apex obtuse, with green spots, plicae oblique, truncate.

Scattered in central, northern to northeast Asia, and northern America. In alpine tundra, alpine meadows and alpine stony slopes. Altitude: 2000-5000 m. Flowering & fruiting: June-September.

B

A. 中国青海省拉脊山
Laji Mt., Qinghai, China. 02.09.1992

B. E. 中国四川省德格县雀儿山
Queershan, Dege, Sichuan, China. 21.08.2006

C. 美国犹他州尤因塔山
Uinta Mts, Utah, USA. 09.09.2019. Photo: Adrien Favre

D. 中国青海省唐古拉山口
Tanggula Pass, Qinghai, China. 05.08.2006

F. 中国新疆维吾尔自治区喀纳斯
Hanasi, Xinjiang, China. 09.08.1995

C

D

E

F

高山龙胆 *Gentiana algida* 与多刺绿绒蒿 *Meconopsis horridula*
中国青海省唐古拉山口 Tanggula Pass, Qinghai, China. 05.08.2006

太白龙胆

Gentiana apiata

多年生草本，直立，高 10~20 cm；莲座叶丛 2~4，直立，叶片条状披针形；茎生叶 2~3 对，明显较小；花数枚密集顶生；花萼筒不裂，萼齿近等大，狭长圆形，开展或反曲；花冠管状到漏斗状，黄色到黄绿色，外面有绿色条纹和斑点，花冠裂片卵形，先端圆钝，有绿色斑点，褶偏斜，截形。

中国陕西太白山特产。生长在高山草甸、林线边缘和山坡矮灌丛中，海拔 3000~3500 m。花果期 7~8 月。

Perennial, stems erect, 10-20 cm; rosette 2-4, erect, linear-lanceolate; stem leaves 2-3 pairs, distinctly smaller; flowers several in terminal clusters; calyx tube entire, calyx teeth subequal, narrowly oblong, spreading to reflexed; corolla tubular to funnelform, yellowish to greenish yellow, with purple stripes and spots outside, lobes ovate, apex obtuse, with green spots, plicae oblique, truncate.

Endemic to Taibaishan Mountain of Shaanxi province of China. In alpine meadows, margin of timberline and alpine bushes. Altitude: 3000-3500 m. Flowering & fruiting: July -August.

A

B

A-B. 中国陕西省太白山
Taibaishan Mt., Shaanxi, China. 15.08.1990

C-D. 中国陕西省太白山
Taibaishan Mt., Shaanxi, China. 02.09.2023

C

D

阿墩子龙胆

Gentiana atuntsiensis

多年生草本，直立，高 10~50 cm；莲座叶丛 5~10，直立，叶片长椭圆形到卵状披针形；茎生叶 3~5 对，明显较小；花多数，散生在茎上部，顶部较密集；花萼筒管状，不裂，檐部有膜，萼齿大小不等，钻形到披针形，开展或反曲；花冠管状到漏斗状，亮蓝色到深蓝色，花冠裂片卵形，先端圆钝，褶很小，偏斜，三角形到截形。

中国西南部云南和四川特产。生长在高山草甸、林缘和山坡灌丛中。海拔 2500~4000 m。花果期 7~10 月。

Perennial, stems erect, 10-50 cm; rosette 5-10, erect, oblong-elliptic to ovate-lanceolate; stem leaves 3-5 pairs, distinctly smaller; flowers many, sparsely clustered in upper part of stems, more flowers in terminal clusters; calyx tube entire, with membrane along limb, calyx teeth unequal, subulate to lanceolate, spreading to reflexed; corolla tubular to funnelform, light to dark blue, lobes ovate, apex obtuse, plicae tiny, oblique, triangular to truncate.

Endemic to Yunnan and Sichuan provinces in southwestern China. In alpine meadows, bushes and forest margins. Altitude: 2500-4000 m. Flowering & fruiting: July - October.

A

B

A-E. 中国云南省德钦县白马雪山

Baimaxueshan Mt., Deqin, Yunnan, China. 20.08.2002

德钦县旧称阿墩子，盛产龙胆，也盛产美景。在这里梅里雪山的倩影远在天边，近在眼前。

Deqin, formerly named Adunzi or Atuntsi, is a paradise for gentians, and for sight-seeing as well. The snow-covered Meili mountains are far in the horizon, but one feels they could be touched with ones fingers. 12.10.2003.

直萼龙胆

Gentiana erectosepala

多年生草本，直立，高 10~20 cm；莲座叶丛 1~3，叶片长圆状匙形；茎生叶 2~3 对，较基生叶稍小；花 3~5 枚簇生顶端；花萼筒不裂，萼齿不等大，狭条形到匙形，有紫色斑点；花冠倒锥形，白色到黄白色，外面有紫色斑点和条纹，花冠裂片三角形，先端圆钝，有绿色斑点，褶偏斜，截形，边缘齿状。

特产于中国西藏东南部。生长在高山草甸和灌丛中。海拔 3500~4500 m。花果期 7~9 月。

Perennial, stems erect, 10-20 cm; rosette 1-3, erect, basal leaves oblong- spathulate; stem leaves 2-3 pairs, slightly smaller; flowers 3-5, in terminal clusters; calyx tube entire, calyx teeth unequal, narrowly oblong to spathulate; corolla obconic, white to greenish yellow, with purple spots and stripes outside, lobes triangular, apex obtuse, with green spots, plicae oblique, truncate, margin dentiform.

Endemic to southeastern Xizang of China. In alpine meadows and bushes. Altitude: 3500-4500 m. Flowering & fruiting: July-September.

A–C. 中国西藏自治区错那市
Cuona, Xizang, China. 14.08.2006

D. 中国西藏自治区林芝市色季拉山
Sejilashan Mt. Linzi, Xizang, China. 17.08.2006

C

D

冷龙胆

Gentiana frigida

多年生草本，直立，高 5~10 cm；莲座叶丛 1~3，直立，叶片边缘向内卷，线状披针形；茎生叶 2~3 对，上部叶鞘状抱茎；花 1~3 枚顶生；花萼筒不裂，萼齿近等大，狭长圆形到匙形；花冠管状钟形，黄白色到黄绿色，外面有绿紫色条纹，花冠裂片三角形，先端圆钝，有绿色斑点，褶偏斜，三角形，边缘齿状。

仅见于阿尔卑斯山东部及喀尔巴阡山。嫌钙植物，生长在高山石质山坡。海拔 1800~2500 m。花果期 7~9 月。

Perennial, stems erect, 5-10 cm; rosette 1-3, erect, basal leaves involute, linear-lanceolate; Stem leaves 2-3 pairs, upper ones sheathing at base; flowers 1-3 in terminal clusters; calyx tube entire, calyx teeth subequal, narrowly oblong to spathulate; corolla tubular-campanulate, yellowish white to greenish yellow, with purple stripes and spots outside, lobes triangular, apex obtuse, with green spots, plicae oblique, triangular, margin denticulate.

In eastern Alps and the Carpathians. Calcifuge, in alpine stony slopes. Altitude: 1800-2500 m. Flowering & fruiting: July-September.

A

B

A–C. 保加利亚里拉山 Rila Mountains, Bulgaria. 18.07.1993

福丽奇龙胆

Gentiana froelichii

多年生草本，直立，高 5~15 cm；莲座叶丛 1~5，直立，叶片线状披针形；茎生叶 1~3 对；花 1~3 枚簇生顶端；花萼筒不裂，萼齿近等大，狭条形到线形，外卷或卷曲；花冠管状钟形到管状，纯净亮蓝色，花冠裂片卵形，先端急尖，有绿色斑点，褶偏斜，三角形，边缘近整齐。

特产于阿尔卑斯山东南部的奥地利、斯洛文尼亚和意大利交界地区。生长在高山石质山坡。海拔 1500~2000 m。花果期 7~9 月。该种染色体数目 2n=42，与本组其他种（2n=24）不同，DNA 分子证据也显示该种不应归属于高山龙胆组，而与龙胆组近缘。

Perennial, stems erect, 5-15 cm; rosette 1-5, erect, basal leaves linear- lanceolate; stem leaves 1-3 pairs; flowers 1-3, in terminal clusters; calyx tube entire, calyx teeth subequal, narrowly oblong to linear, reflexed or twisted; corolla tubular-campanulate to tubular, clear blue, lobes ovate, apex acute, with green spots, plicae oblique, triangular, margin nearly entire. Endemic to southeastern Alps in Austria, Slovenia and Italy. In alpine stony slopes. Altitude: 1500-2000 m. Flowering & fruiting: July-September.

A-D. 意大利波得龙省切莫莱斯

Cimolais, Friuli Venezia Giulia, Pordenone, Italy.

07.07.1996

C

D

E. 意大利波得龙省切莫莱斯地区的景观，这里有覆瓦叶龙胆（*Gentiana terglouensis*）和福丽奇龙胆（*Gentiana froelichii*）。

Landscape in the area of Cimolais, Friuli Venezia Giulia, Pordenone, Italy, where two unique gentians grow, *Gentiana terglouensis* and *Gentiana froelichii*. 07.07.1996.

蓝绿龙胆

Gentiana glauca

多年生草本，具纤细根状茎，茎直立，高 4~15 cm，光滑；莲座叶叶片椭圆形到卵形；茎生叶 2~4 对，卵形；花数枚簇生枝顶；花萼筒倒锥形，萼齿钻形；花冠常闭合，管状，蓝绿色，花冠裂片三角状卵形，先端圆钝，褶微小。

特产于东北亚和北美西北部高纬度高山地区；多生长在沼泽化草甸。海拔 5~2300 m。花果期 6~9 月。

Perennial herb with slender rhizome; stems erect, 4-15 cm, glabrous; rosette leaves elliptic to ovate; stem leaves 2-4 pairs, ovate; flowers several clustered at terminal; calyx tube obconic, calyx lobes subulate; corollas often closed, tubular, greenish-blue, lobes triangular-ovate, plicae tiny.

Endemic to mountains in high latitude regions in NE Asia and NW America; In moist bogs. Altitude 5-2300 m. Flowering & fruiting: June-September.

A–B. 美国阿拉斯加 摄影 /Photo: Erica Simonson
Flower Mountain, Alaska, USA. 23.08.2021

斑点龙胆

Gentiana handeliana

多年生草本，直立，高 10~15 cm，紫红色，具乳突；莲座叶丛 2~5，叶片狭椭圆到倒披针形；茎生叶 2~3 对，较小；花多数，密集簇生顶端和上部叶腋；花无梗，花萼筒漏斗形，不开裂或一侧开裂外面具乳突，萼齿大小不等，钻形到披针形，开展到反曲；花冠管状到漏斗状，淡黄色，外面具蓝紫色斑点和细条纹，花冠裂片卵形，先端圆钝，褶很小，偏斜，三角形到截形，边缘齿状。

特产中国西藏东南部和云南西北部。生长在高山草甸。海拔 3500~4500 m。花果期 7~10 月。

Perennial, stems erect, 10-15 cm, reddish purple, papillate; rosette 2-5, narrowly elliptic to oblanceolate; stem leaves 2-3 pairs, smaller; flowers many, densely clustered at terminal or upper axillary; flowers sessile; calyx tube funnelform, entire or split to one side, papillate outside, calyx teeth unequal, subulate to lanceolate, spreading to reflexed; corolla tubular to funnelform, pale yellow, lobes ovate, apex obtuse, plicae tiny, oblique, triangular to truncate, margin denticulate.

Endemic to southeastern Xizang and northwestern Yunnan of China. In alpine meadows. Altitude: 3500-4500 m. Flowering & fruiting: July-October.

A

B

A–B. 中国西藏自治区察隅县 摄影 /Photo：付鹏程
Chayu, Xizang, China. 29.08.2018

巴塘龙胆

Gentiana susanneae

多年生草本，茎直立，光滑，高 30~60 cm；莲座叶丛 1~2，叶片长圆状椭圆形到披针形；茎生叶 1~4 对，较基生叶小；花 3~5 枚簇生顶端，个别花腋生；花无柄，花萼筒倒锥形，不裂，萼齿不等大，直立，长圆状三角形；花冠管状钟形，白色到黄白色，外面有紫色条纹，花冠裂片三角形，内面有绿色斑点，褶偏斜，截形，边缘啮齿状。

仅见于中国四川省西部。生长在高山灌丛中。海拔 4400~4600 m。花果期 8~9 月。

Perennial, stems erect, glabrous, 30-60 cm; rosette 1-2, basal leaves oblong-elliptic to lanceolate; stem leaves 1-4 pairs, slightly smaller; flowers 3-5, in terminal clusters, few flowers axillary; flowers sessile, calyx tube obconic, entire, calyx teeth unequal, erect, oblong-triangular; corolla tubular-campanulate, white to greenish yellow, with purple stripes outside, lobes triangular, with green spots inside, plicae oblique, truncate, margin erose. Known only from western Sichuan of China. In alpine bushes. Altitude: 4400~4600m. Flowering & fruiting: August-September.

A–B. 中国四川省巴塘县 摄影 /Photo：付鹏程
Batang, Sichuan, China. 21.08.2016

小齿龙胆

Gentiana microdonta

多年生草本，直立，高 20~40 cm；莲座叶丛 1~5，叶片长圆状披针形；茎生叶 3~7 对，较小；花多数，排列成较为紧密狭细的圆锥花序；花萼筒倒锥形，一侧开裂，萼齿大小不等，钻形到披针形，开展到反曲；花冠管状到漏斗状，深蓝色，花冠裂片卵形，先端圆钝，褶很小，偏斜，三角形到截形，边缘齿状。

中国云南西北部特产。生长在高山林缘和山坡灌丛中。海拔 2500~3500 m。花果期 7~10 月。

Perennial, stems erect, 20-40 cm; rosette 1-5, erect, oblong-lanceolate; stem leaves 3-7 pairs, smaller; flowers many, densely clustered in upper part of stems forming a narrow panicle; calyx tube obconic, split to one side, calyx teeth unequal, subulate to lanceolate, spreading to reflexed; corolla tubular to funnelform, dark blue, lobes ovate, apex obtuse, plicae tiny, oblique, triangular to truncate, margin denticulate.

Endemic to northwestern Yunnan province of China. In alpine bushes and forest margins. Altitude: 2500-3500 m. Flowering & fruiting: July-October.

A

B

A–B. 中国云南省大理市 Dali, Yunnan, China. 02.08.2011

C–D. 中国云南省大理市 Dali, Yunnan, China. 23.09.2023

C

D

云雾龙胆

Gentiana nubigena

多年生草本，直立，高 10~20 cm；莲座叶丛 1~5，直立，叶片狭披针形到匙形；茎生叶 1~3 对，较基生叶稍小；花 2~5 枚簇生顶端；花萼筒不裂，萼齿不等大，狭条形到匙形，有紫色斑点；花冠管状钟形，淡蓝色到深蓝色，外面有深紫色条纹，花冠裂片三角形，先端圆钝，褶偏斜，截形，边缘齿状。

分布于尼泊尔、印度、锡金、不丹及西藏、青海、甘肃、四川等省。生长在高山草甸、沼泽草甸和灌丛中。海拔 3000~4500 m。花果期 7~9 月。

Perennial, stems erect, 10-20 cm; rosette 1-5, erect, basal leaves narrowly lanceolate to spathulate; stem leaves 1-3 pairs, slightly smaller; flowers 2-5, in terminal clusters; calyx tube entire, calyx teeth unequal, narrowly oblong to spathulate; corolla tubular-campanulate, light to dark blue, with purple stripes outside, lobes triangular, apex obtuse, with green spots, plicae oblique, truncate, margin denticulate.

Distributed in Nepal, India, Sikkim, Bhutan and Xizang, Sichuan, Qinghai and Gansu provinces of China. In bog meadows, alpine meadows and bushes. Altitude: 3000-4500 m. Flowering & fruiting: July-September.

A

B

A. B. E. 中国西藏自治区聂拉木县 Nielamu, Xizang, China. 16.09.1992

C-D. 中国甘肃省玛曲县 Maqu, Gansu, China. 10.08.1992

C
D
E

岷县龙胆

Gentiana purdomii

多年生草本，直立，高 10~30 cm；莲座叶丛 2~5，直立，叶片狭披针形到条形；茎生叶 2~5 对，较基生叶稍小；花 2~5 枚簇生顶端和顶端下部 1~2 节；花萼筒不裂，萼齿近等大，狭条形到匙形，有绿色斑点；花冠管状钟形，淡黄色到黄白色，外面有深紫色条纹，花冠裂片三角形，先端圆钝，褶偏斜，截形，边缘齿状。

分布于中国西藏、青海、甘肃、四川等省。生长在高山草甸、沼泽草甸和灌丛中。海拔 3000~4500 m。花果期 7~9 月。

Perennial, stems erect, 10-30 cm; rosette 2-5, erect, basal leaves narrowly lanceolate to linear; stem leaves 2-5 pairs, slightly smaller; flowers 2-5, in terminal clusters and the 1-2 nodes below apex; calyx tube entire, calyx teeth subequal, narrowly oblong to spathulate with green spots; corolla tubular-campanulate, yellow to greenish white, with purple stripes outside, lobes triangular, apex obtuse, with green spots, plicae oblique, truncate, margin denticulate.

In Xizang, Sichuan, Qinghai and Gansu provinces of China. In bog meadows, alpine meadows and bushes. Altitude 3000-4500 m. Flowering & fruiting: July-September.

A. E. F. 中国青海省久治县 Jiuzhi, Qinghai, China. 11.09.2023

B. 中国四川省理塘县 Litang, Sichuan, China. 08.08.2002

C. 中国四川省德格县 Dege, Sichuan, China. 21.08.2006

D. 中国青海省久治县 Jiuzhi, Qinghai, China. 06.09.1994

C
D
E
F

多花龙胆

Gentiana striolata

多年生草本，直立，高 20~40 cm；莲座叶丛 3~8，直立，叶片狭披针形到匙形；茎生叶 2~5 对，较基生叶稍小；花近无柄，5~15 枚簇生顶端和茎上部各节；花萼筒不裂，萼齿细小，近等大，狭条形到匙形，有绿色斑点；花冠管状漏斗形，淡蓝色到深蓝色，外面有深蓝色条纹和斑点，花冠裂片三角形，先端圆钝，褶偏斜，截形，边缘整齐。

仅分布于中国四川和云南。生长在高山草甸、灌丛和林缘。海拔 3000~4500 m。花果期 7~9 月。

Perennial, stems erect, 20-40 cm; rosette 3-8, erect, basal leaves narrowly lanceolate to spathulate; stem leaves 2-5 pairs, slightly smaller; flowers sessile, 5-15 in terminal clusters and the upper part of stems; calyx tube entire, calyx teeth tiny, subequal, narrowly oblong to spathulate with green spots; corolla tubular-funnelform, light to dark blue, with dark blue stripes and spots outside, lobes triangular, apex obtuse, plicae oblique, truncate, margin entire.

In Sichuan and Yunnan provinces of China. In alpine meadows, bushes and forest margins. Altitude 3000-4500 m. Flowering & fruiting: July-September.

A. B. D. 中国四川省雅江县
Yajiang, Sichuan, China. 14.08.2002

C. 中国云南省德钦县
Deqin, Yunnan, China. 14.08.2002

E. 中国四川省康定市
Kangding, Sichuan, China. 13.09.2023

C

D

E

三岐龙胆

Gentiana trichotoma

多年生草本，直立，高 15~30 cm；莲座叶丛 3~5，直立，叶片狭披针形到匙形；茎生叶 2~3 对，较基生叶稍小；花 3~5 枚簇生顶端；花萼筒不裂，萼齿较大而显著，不等大，伸展，狭条形到匙形，外面深紫色；花冠管状漏斗形，淡蓝色到深蓝色，外面有深蓝色条纹和斑点，花冠裂片三角形，先端圆钝，褶偏斜，截形，边缘整齐。

仅分布于中国青海、甘肃和四川。生长在高山草甸、灌丛和林缘。海拔 3000~4500 m。花果期 7~9 月。

Perennial, stems erect, 15-30 cm; rosette 3-5, erect, basal leaves narrowly lanceolate to spathulate; stem leaves 2-3 pairs, slightly smaller; flowers 3-5 in terminal clusters; calyx tube entire, calyx teeth conspicuous, unequal, spreading, narrowly oblong to spathulate, dark purple outside; corolla tubular-funnelform, light to dark blue, with dark blue stripes and spots outside, lobes triangular, apex obtuse, plicae oblique, truncate, margin entire.

In Qinghai, Gansu and Sichuan provinces of China. In alpine meadows, bushes and forest margins. Altitude: 3000-4500 m. Flowering & fruiting: July-September.

A-D. 中国四川省康定市
Kangding, Sichuan, China. 12.08.2004

D

川西龙胆

Gentiana wilsonii

多年生草本，直立，高 10~20 cm；莲座叶丛 3~8，直立，叶片狭披针形到狭椭圆形或匙形；茎生叶 1~2 对，较基生叶稍小；花 1~3 枚簇生顶端；花萼筒几不裂，萼齿不等大，伸展或直立，狭条形到匙形；花冠管状漏斗形，淡蓝色到深蓝色，外面有深蓝色条纹和斑点，花冠裂片三角形，先端圆钝，褶偏斜，截形，边缘整齐。

仅分布于中国四川和云南。生长在高山草甸、灌丛和林缘。海拔 3000~4500 m。花果期 7~9 月。

Perennial, stems erect, 10-20 cm; rosette 3-8, erect, basal leaves narrowly lanceolate to spathulate; stem leaves 1-2 pairs, slightly smaller; flowers 1-3 in terminal clusters; calyx tube entire, calyx teeth conspicuous, unequal, narrowly oblong to spathulate; corolla tubular-funnelform, light to dark blue, with dark blue stripes and spots outside, lobes triangular, apex obtuse, plicae oblique, truncate, margin entire. In Sichuan and Yunnan provinces of China. In alpine meadows, bushes and forest margins. Altitude 3000-4500 m. Flowering & fruiting: July-September.

A

B

A–C. 中国四川省康定市雅拉山
Yala Mt., Kangding, Sichuan, China. 03.07.2021

D. 中国四川省康定市雅拉山山口景观
Yala Pass, Kangding, Sichuan, China. 03.07.2021

C

D

黄龙胆 *Gentiana lutea*

瑞士纳沙泰尔拉渡恩 La Tourne, Neuchatel，Switzerland. 15.06.2018

龙胆组

sect. *Gentiana*

多年生高大草本；根和根状茎粗壮、肉质；叶片硕大，长可达 30 cm，形成莲座；轮状聚伞花序，花顶生、轮生和轮状簇生；花冠的褶减退，很小或无。本组约 7 种（含种下类群），分布主要限于欧洲。

Gigantic perennial, roots and rhizome stout, fleshy; leaves large, up to 30 cm, crowed to rosette; verticillaster, flowers terminal, axillary whorls or clusters; corolla plicae reduced or absent. About 7 species (including intraspecific taxa), almost in Europe only.

博色龙胆（维拉亚种）

Gentiana burseri ssp. *villarsii*

多年生草本，高 20~60 cm；基生莲座叶大型，叶片披针形到阔卵形，5~7 条粗壮弧形脉；茎上部叶无柄；花无柄，数量较少，顶生或簇生叶腋；花萼筒一侧开裂到，萼裂片 2~5 枚，大小不等；花冠钟状，黄色，有紫黑色斑点，5~7 浅裂，裂片近圆形，先端圆钝。

博色龙胆常被划分为两个地理亚种，原亚种 ssp. *burseri* 仅分布于比利牛斯山，维拉亚种 ssp. *villarsii* 仅分布于阿尔卑斯山西南部。生长在较潮湿的山坡、草地、牧场、林缘和林窗。海拔 1000~3000 m。花果期 6~9 月。

Perennial, stout, 20-60 cm high; basal rosette leaves large, blade lanceolate to broad ovate, with 5-7 strong longitudinal veins; upper cauline leaves sessile; flowers sessile, in terminal and axillary clusters; calyx tube split down to one side, calyx lobes 2-5, unequal; corolla campanulate, yellow, with purple-black spots; corolla divided to 5-7 lobes to upper part, lobes obtuse.

Gentiana burseri is often divided into 2 subspecies, ssp. *burseri* endemic to the Pyrenees, and ssp. *villarsii* endemic to southwestern Alps, in Europe. In damp hill slopes, grassland, pasture, and forest margin. Altitude: 1000-3000 m. Flowering & fruiting: June - September.

A–C. 法国阿尔卑斯山 摄影 /Photo: Adrien Favre
Col du Lautaret, Alps, France. 16.07.2005

黄龙胆

Gentiana lutea

高大多年生草本，有粗大的肉质根；高50~120 cm；叶大型，有发达的莲座叶，叶片披针形到椭圆形或阔卵形，表面有粉霜，5~7条粗壮弧形脉；花具短柄，多数，常密集顶生或腋生；花冠亮黄色，几乎全裂，5~9裂片，开展，条状披针形。

广泛分布于欧洲中部到南部的山区。生长在钙质岩山坡、草地、牧场、林缘和林窗。海拔500~2000 m。花果期6~8月。根富含黄酮类物质，用它制作多种酒精类饮品，历史悠久。

本种是龙胆属的命名模式，但花和植物体的形态与本属其他种相差甚远。

Gigantic perennial, 50-120 cm high; basal leaves large, well developed to rosette, blades lanceolate to elliptic or broad ovate, glaucous, with 5-7 strong longitudinal veins; flowers pedicellate, crowded in terminal and axillary cyme; corolla bright yellow, divided into 5-9 lobes nearly to the base, lobes spreading, oblong-lanceolate.

Widely spread in mountains in central to southern Europe. Calcicole, in hill slopes, grassland, pasture, forest margin and forest clearings. Altitude: 500-2000 m. Flowering & fruiting: June-August.

A

B

A-B. 瑞士纳沙泰尔州萨瓦聂
Savagnier, Neuchatel, Switzerland. 22.06.2003

C. E. 法国中央高原奥维尼火山
Volcans d'Auvergne, Massif Central, France. 01.08.2005

D. 瑞士沃州然燃
La Barillette, Gingins, Vaud, Switzerland. 27.07.2023

F. 黄龙胆 *Gentiana lutea* 的传粉者
瑞士纳沙泰尔州萨瓦聂
Savagnier, Neuchatel, Switzerland. 22.06.2003

C
D
E
F

G–H. 抽薹前苗期的黄龙胆（*Gentiana lutea*）形态与单子叶植物黑药花科白藜芦（*Veratrum album*）非常相似，而且在欧洲的草场上它们时常混生在一起，容易混淆。黄龙胆的叶片光滑，被白粉，而白藜芦的叶粗糙无粉可以区分。瑞士纳沙泰尔州萨瓦聂。

Young plants of Yellow Gentian are easily confused with young plants of *Veratrum album* (Melanthiaceae). Leaves of Yellow Gentian are glabrous and glaucous, whereas the leaves of the latter are scabrous. Savagnier, Neuchatel, Switzerland. 22.06.2003.

I. 瑞士沃州马切如山口 Col du Marchairuz, Vaud, Switzerland. 27.07.2023

马切如山口是瑞士沃州境内汝拉山的一个重要通道，海拔 1350 m 左右。远处的雪山是位于意大利、法国和瑞士三国交界处的欧洲最高峰勃朗峰，海拔 4807 m。

J. 法国中央高原奥维尼火山 Volcans d'Auvergne, Massif Central, France. 01.08.2005

匈牙利龙胆

Gentiana pannonica

多年生草本，高 20~60 cm；基生叶大型，有莲座叶，叶片长圆形到椭圆形或阔卵形，5~7 条粗壮弧形脉；茎上部叶较小无柄；花常密集顶生或簇生叶腋；花萼裂片反折；花冠钟状，紫红色，有紫黑色斑点，5~7 浅裂，裂片倒卵形，先端圆钝。

分布于欧洲中部到东部的山区，包括瑞士、德国、奥地利、捷克和斯洛伐克等。生长在山坡、草地、牧场、林缘和林窗。海拔 800~2000 m。花果期 6~9 月。

Perennial, 20-60 cm high; basal leaves large, crowded as rosette, blade oblong to elliptic or broad ovate, with 5-7 strong longitudinal veins; upper cauline leaves smaller and sessile; flowers crowded terminal and axillary clusters; calyx lobes recurved; corolla campanulate, reddish purple, with purple-black spots; corolla divided to 5-7 lobes to upper part, lobes obovate, apex obtuse.

In mountains from central to eastern Europe, including Switzerland, Germany, Austria, Czech, and Slovakia. In hill slopes, grassland, pasture, and forest margin. Altitude: 800-2000 m. Flowering & fruiting: June-September.

A–D. 德国巴伐利亚州贝切噶登
Berchtesgaden, Bavaria, Germany. 19.08.2005

C

D

欧洲斑点龙胆

Gentiana punctata

多年生草本，粗壮，高 20~60 cm；基生莲座叶大型，叶片长披针到阔卵形，5~7 条粗壮弧形脉；茎上部叶无柄；花无柄，常密集顶生或簇生叶腋；花萼筒不裂到基部，萼裂片 5~8 枚，直立；花冠钟状，浅黄绿色，有紫黑色斑点，5~8 浅裂，裂片卵形，先端圆钝。

广泛分布于欧洲中部到东南部的山区。生长在较潮湿的山坡、草地、牧场、林缘和林窗。海拔 1000~3000 m。花果期 6~9 月。

Perennial, stout, 20-60 cm high; basal rosette leaves large, blade long lanceolate to broad ovate, with 5-7 strong longitudinal veins; upper cauline leaves sessile; flowers sessile, crowded terminal and axillary clusters; calyx tube not split down to one side, calyx lobes 5-8, erect; corolla campanulate, pale greenish-yellow, with purple-black spots; corolla divided to 5-8 lobes to upper part, lobes obtuse.

Widely spread in mountains from central to southeastern Europe. In damp hill slopes, grassland, pasture, and forest margin. Altitude: 1000-3000 m. Flowering & fruiting: June-September.

A

B

A–D. 瑞士瓦莱州奥拜瓦尔德 Oberwald, Valais, Switzerland. 16.07.2005

E. 保加利亚里拉山 Rila Mountains, Bulgaria. 18.07.1993

C
D
E

紫龙胆

Gentiana purpurea

多年生草本，直立，高 20~60 cm；基生莲座叶大型，具柄，叶片披针到阔卵形，5~7 条粗壮弧形脉；茎上部叶无柄；花无柄，数量较少，顶生或簇生叶腋；花萼筒一侧开裂至基部，裂片 2~5 枚，不等大；花冠筒状，暗红紫色，有少数紫黑色斑点，5~7 裂至中部，裂片近圆形，先端圆钝。

分布于欧洲中部山区，包括挪威南部、奥地利，德国、瑞士、法国、意大利等地。生长在山坡草地、牧场和林缘。海拔 1000~2500 m。花果期 6~9 月。

Perennial, erect, 20-60 cm high; basal rosette leaves large, blade lanceolate to broad ovate, with 5-7 strong longitudinal veins; upper cauline leaves sessile; flowers sessile, in small terminal and axillary clusters; calyx tube split on one side to the base, lobes 2-5, unequal in size; corolla cylindrical, dark purple, with few purple-black spots; corolla divided to 5-7 lobes to middle part, lobes obtuse.

In the mountains of central Europe, including southern Norway, Austria, Germany, Switzerland, France and Italy. In grassland, pasture, and forest margin. Altitude: 1000-2500 m. Flowering & fruiting: June-September.

A

B

A–C. 意大利莎奴泽高山植物园 Chanouzia Alpine Botanical Garden, Italy. 09.08.2005

D. 法国夏慕尼 Chamonix, France. 08.08.2005

E. 法国夏慕尼冰海冰川 The "Mer de Glace" Glacier, Chamonix, France. 08.08.2005

F. 瑞士瓦莱 摄影 /Photo：Adrien Favre Valais, Switzerland. 05.08.2020

C
D
E
F

一片紫龙胆 *Gentiana purpurea* 花海，法国夏慕尼附近；摄影 /Photo：Regula Naef
A flower carpet of *Gentiana purpurea*, nearby Chamonix, France. 08.2000

黄紫龙胆（杂交种）

Gentiana × *hybrida*

(*Gentiana lutea* × *purpurea*)

据文献报道，龙胆组种间天然杂种比较常见，不同种间均可自然或人工杂交。其中黄龙胆与紫龙胆的天然杂种很早就被人们注意到。

植物体的形态接近黄龙胆，但花色介于黄龙胆与紫龙胆之间，花冠外面橘红色，内面则黄里透红。

野生个体见于法国。生长在山坡草地。海拔 1000~1500 m。花果期 6~9 月。

According to research reports, natural interspecific hybrids among most species of the sect. *Gentiana* commonly occur. Most species of the section can be hybridized naturally or artificially. Among the hybrids, *Gentiana* × *hybrida* was named and suggested being the hybrid between *Gentiana lutea* and *Gentiana purpurea*.

Plant morphology highly resemble that of *Gentiana lutea*. Flower color is obviously between the two parents: reddish purple outside of corolla, while reddish yellow inside.

Wildly in France. In grassland and pasture. Altitude: 1000-1500 m. Flowering & fruiting: June-September.

A

B

A–B. 黄紫龙胆 *Gentiana* × *hybrida*
法国中央高原奥维尼火山
Volcans d'Auvergne, Massif Central, France. 01.08.2005

C. 黄紫龙胆与亲本黄龙胆在野外混生
法国中央高原奥维尼火山
Gentiana lutea × *purpurea* and *Gentiana lutea* grow together in nature; Volcans d'Auvergne, Massif Central, France. 01.08.2005

D. 黄紫龙胆 *Gentiana lutea* × *purpurea*
意大利莎奴泽高山植物园
Chanouzia Alpine Botanical Garden, Italy. 09.08.2005

C

D

位于勃朗峰南侧小圣贝尔纳山口意大利与法国交界处的莎奴泽高山植物园。它历史悠久，地理位置特殊，小巧玲珑，景色优美，高山植物收藏丰富，栽培了多种龙胆，尤其是紫龙胆及其他与黄龙胆的杂交种，令人印象深刻。

To the southern side of the Mont Blanc, at the Pass of Petit-Saint-Bernard and the border between France ad Italy, located a tiny botanical garden, i.e. the Chanouzia Alpine Botanical Garden. It is small but very unique, with long history, special location, beautiful scenery, and rich collection of alpine plants, including many gentians, among them *Gentiana purpurea* and its hybrid with *Gentiana lutea* are extraordinary. 09.08.2005.

滇西龙胆 *Gentiana georgei*
中国甘肃省卓尼县 Zhuoni, Gansu, China. 08.09.2023；摄影 /Photo：马惠

匍茎组

sect. *Isomeria*

多年生草本，合轴分枝，有发达的匍匐茎；叶片常具软骨质边缘；花茎单一，从匍匐茎上生出；花无梗，单花顶生或少数几朵花簇生茎顶。本组约 10 种，分布限于喜马拉雅山地区。

Perennial, stem branching sympodially, stoloniferous; leaves with whitish cartilaginous margins; flowering stems arising from slender stolon; flowers sessile, solitary or few in a terminal cluster. About 10 species, limited to the Himalayan areas only.

硕花龙胆

Gentiana amplicrater

多年生草本，根状茎短缩，直立，茎直立，多单生，高 7~15 cm；叶基生，密集覆瓦状排列，卵形到卵状椭圆形，边缘软骨质；花 1~4 朵，顶生，包围于上部叶丛中；花萼筒钟状，花萼裂片长圆椭圆卵，大小不等，具短小尖头，边缘软骨质；花冠筒钟状，蓝色到淡紫色，花冠裂片卵圆形，褶卵圆形，整齐。

分布于喜马拉雅山地区，包括锡金、尼泊尔和中国西藏南部。生长在高山沼泽化草甸。海拔 4000~4800 m。花果期 8~10 月。

Perennial, rhizome short, dense to erect caudex, Stem simple, erect, 7-15 cm; leaves mostly basal, densely imbricate, margins cartilaginous; flowers 1-4, surrounded by upper leaves; calyx tube campanulate, calyx lobes ovate-elliptic, unequal, mucronate, margins cartilaginous; corolla tube campanulate, blue to pale purple, corolla lobes round ovate, plicae rounded, entire.

In the Himalayan regions including Sikkim, Nepal, and southern Xizang of China. In alpine swamp meadows. Altitude: 4000-4800 m. Flowering & fruiting: August-October.

A

B

C

A-C. 中国西藏自治区扎囊县 摄影 /Photo：付鹏程
Zhanang, Xizang, China. 19.09.2020

平龙胆

Gentiana depressa

匍匐茎发达，多分枝，连接多个营养莲座叶丛及花茎；营养枝莲座叶卵形，草质，交互对生；花单生叶丛中，花枝上的叶更为宽大；花萼筒钟状；花冠筒阔钟状，淡蓝色，蓝白相间。

分布于喜马拉雅山地区，包括印度、尼泊尔、锡金、不丹合中国西藏南部。生长在山地石质山坡、草地和林缘。海拔3000~4500 m。花果期 8~10 月。

Stolon well developed, much branched, connecting numerous vegetative rosettes and flowering stems; leaves of vegetative rosette ovate, herbaceous, decussate, leaves on flowering stems broader; flowers solitary, terminal; calyx tube campanulate; corolla tube broad campanulate, light blue with white stripes.

In the Himalayan regions including India, Nepal, Sikkim, Bhutan, and southern Xizang of China. In stony alpine slopes, meadows and forest margin. Altitude: 3000-4500 m. Flowering & fruiting: August-October.

A–C. 中国西藏自治区聂拉木县樟木镇
Zhangmu, Nielamu, Xizang, China. 18.09.1992

扇叶龙胆

Gentiana emodii

匍匐茎发达，多分枝，连接多个营养莲座叶丛及个别花茎；营养枝莲座叶扇形，交互对生，折叠套合，边缘软骨质，有细睫毛；花单生叶丛中；花萼筒钟状，花萼裂片扇形；花冠筒钟状，黄白色到淡蓝色，有紫色斑纹。

分布于喜马拉雅山地区，包括锡金、不丹和中国西藏南部。生长在高山石质山坡和高山冻原。海拔 4000~5000 m。花果期 6~10 月。

Stolon well developed, much branched, connecting numerous vegetative rosettes and few flowering stems; leaves of vegetative rosette flabelliform, decussate, overlapping, margins cartilaginous and ciliolate; flowers solitary, terminal; calyx tube campanulate, calyx lobes flabelliform; corolla tube campanulate, light blue to greenish white with purple speckles and stripes.

In the Himalayan regions including Sikkim, Bhutan, and southern Xizang of China. In stony alpine slopes and alpine tundra. Altitude: 4000-5000 m. Flowering & fruiting: June-October.

A-B. 中国西藏自治区定日县
Dingri, Xizang, China. 15.09.1992

B

滇西龙胆（滇藏型）

Gentiana georgei(Yunnan-Xizang Type)

多年生草本，有较粗壮且短小的主根，连接营养莲座叶丛及 1~3 个花茎；莲座叶披针形，密集交互对生，折叠套合，边缘软骨质；花枝斜升，花单生枝顶；花萼筒钟状，花萼裂片匙形；花冠筒钟状，红紫色，喉部白色，有紫色条纹。

特产于中国西南部云南、四川和西藏等省区交界的横断山南部地区。生长在高山草甸及石质山坡。海拔 3000~4500 m。花果期 8~10 月。

Perennial, with stout and short taproot, connecting vegetative rosette and 1-3 flowering stems; leaves of vegetative rosette lanceolate, decussate, densely overlapping, margins cartilaginous; flowering stems ascending, flowers solitary, terminal; calyx tube campanulate, calyx lobes spathulate; corolla tube campanulate, ruddy purple, throat white with purple stripes.

Endemic to southern Hengduan Mountains connecting Xizang, Sichuan and Yunnan provinces of China. In alpine meadows and stony slopes. Altitude: 3000-4500 m. Flowering & fruiting: August-October.

A

B

A–C. 中国西藏自治区察隅县
Chayu, Xizang, China. 08.10.2003

D–E. 中国云南省德钦县白马雪山
Baimaxueshan Mt., Deqin, Yunnan, China. 29.09.1992

C

D

E

中国西藏自治区波密县米堆冰川

Midui Glacier, Bomi, Xizang, China. 08.10.2003

滇西龙胆（甘青型）

Gentiana georgei (Gansu-Qinghai Type)

本类型的植物形态与“滇藏型”非常相似，但花冠筒狭钟状到漏斗状，花冠裂片蓝紫色，褶红紫色，喉部颜色与裂片和褶的颜色一致（非白色），分布区独立，且与邻近分布的白花类型大花龙胆（*Gentiana szechenyii*）的花色、花型和分布区显著不同，很可能代表了一个独立类型，有必要深入研究界定。

特产于中国青海、甘肃和四川交界的横断山北部地区，生长在高山草甸。海拔 3000~4000 m。花果期 8~10 月。

Highly resemble the “Yunnan-Xizang Type” in gross morphology, however, minor but stable differences include narrower corolla tube (narrowly campanulate to funnelform), corolla color patterns (corolla lobes blue-purple, plicae ruddy purple, throat not white), and disjunct distribution. It is also different from *Gentiana szechenyii* which occurs in the same regions but with white flowers. Thus, this type may represent a distinct species and deserve deeper studies.

Endemic to northern Hengduan Mountains connecting Qinghai, Gansu and Sichuan provinces of China. In alpine meadows. Altitude: 3000-4000 m. Flowering & fruiting: August-October.

A–C. 中国青海省河南蒙古族自治县
Henan, Qinghai, China. 09.09.2023

D. 滇西龙胆 *Gentiana georgei* + 蓝玉簪龙胆 *Gentiana veitchiorum* + 黄管秦艽 *Gentiana officinalis*
中国青海省河南蒙古族自治县
Henan, Qinghai, China. 10.09.2023

E. 滇西龙胆 *Gentiana georgei* + 线叶龙胆 *Gentiana lawrencei* var. *farreri*
中国青海省河南蒙古族自治县
Henan, Qinghai, China. 10.09.2023

D

E

撕裂边龙胆

Gentiana lacerulata

匍匐茎发达，多分枝，连接多个营养莲座叶丛及数枚花茎；营养枝莲座叶扇形，草质波浪状，交互对生，较疏松折叠套合，边缘软骨质，撕裂状；花单生叶丛中；花萼筒钟状，花萼裂片扇形，边缘撕裂状；花冠筒钟状，蓝紫色。

分布于喜马拉雅山地区，包括印度、锡金、不丹和中国西藏南部。生长在高山草甸及石质山坡。海拔 3500~4500 m。花果期 9~10 月。

Stolon well developed, much branched, connecting numerous vegetative rosettes and several flowering stems; leaves of vegetative rosette flabelliform, herbaceous, undulate, decussate, sparsely overlapping, margins cartilaginous and lacerate; flowers solitary, terminal; calyx tube campanulate, calyx lobes flabelliform, margins lacerate; corolla tube campanulate, blue-purple.

In the Himalayan regions including India, Sikkim, Bhutan, and southern Xizang of China. In alpine meadows and alpine slopes. Altitude: 3500-4500 m. Flowering & fruiting: September-October.

A-C. 中国西藏自治区错那市
Cuona, Xizang, China. 14.08.2006

c

短柄龙胆

Gentiana stipitata

多年生草本，有较粗壮且短的主根，连接多个密集的营养莲座叶丛和花茎；莲座叶短小椭圆形；花枝斜升，茎生叶密集在顶在花周围，花单生枝顶，无柄；花萼筒钟状，花萼裂片匙形；花冠筒管状，白色到蓝灰色，喉部有紫色条纹，花冠裂片先端有小尖头。

分布于喜马拉雅周边，包括印度、尼泊尔及中国云南、四川和西藏等省区。生长在高山草甸及干山坡。海拔 2800~4500 m。花果期 6~10 月。

Perennial, with stout and short taproot, connecting numerous densely arranged vegetative rosette and flowering stems; leaves of vegetative rosette narrowly elliptic; flowering stems ascending, stem leaves crowded at terminal surrounding flowers; flowers solitary, sessile; calyx tube campanulate, calyx lobes spathulate; corolla tube tubular, white pale to blue-grey, throat with purple stripes.

In the Himalayas, India, Nepal and the Yunnan, Xizang and Qinghai provinces of China. In alpine meadows and stony slopes. Altitude: 2800-4500 m. Flowering & fruiting: June-October.

A

B

C

A. 中国四川省石渠县 Shiqu, Sichuan, China. 25.09.1993

B. 中国青海省久治县 Jiuzhi, Qinghai, China. 08.09.1994

C. 中国西藏自治区丁青县 Dingqing Xizang, China. 25.09.1992

D. 中国四川省甘孜县 Ganzi, Sichuan, China. 29.09.1993

E. 短柄龙胆 *Gentiana stipitata* + 蓝玉簪龙胆 *Gentiana veitchiorum* 中国四川省石渠县 Shiqu, Sichuan, China. 25.09.1993

D

E

大花龙胆

Gentiana szechenyii

本种与滇西龙胆（*Gentiana georgei*）形态相似，主要区别在于花色和花冠裂片形状和大小不同: 本种花冠白色，外面有紫色条纹，花冠裂片较短，长约 5 mm，而滇西龙胆花冠紫红色，花冠裂片较长，可达 10 mm。

特产于中国西南部云南、四川、西藏和甘肃等省区交界的横断山周边地区。生长在高山草甸及石质山坡。海拔 3000~4500 m。花果期 6~10 月。

Gentiana szechenyii highly resemble Gentiana georgei, major differences include flower color and size of corolla lobes: flowers of *Gentiana szechenyii* white, corolla lobes triangular ~5 mm long, whereas flowers of Gentiana georgei ruddy purple, lobes reaches 10 mm.

Endemic to the regions around Hengduan Mountains connecting Xizang, Sichuan, Yunnan and Gansu provinces of China. In alpine meadows and stony slopes. Altitude: 3000-4500 m. Flowering & fruiting: June-October.

A

B

C

A–B. 中国青海省共和县
Gonghe, Qinghai, China. 22.09.1993

C–D. 中国青海省久治县 Jiuzhi, Qinghai, China. 11.09.2023

E. 中国青海省河南蒙古族自治县
Henan, Qinghai, China. 09.09.2023

D

E

提宗龙胆

Gentiana tizuensis

本种与短柄龙胆（*Gentiana stipitata*）相似，常被合并或作为种下单元（亚种或变种）处理。我们在野外注意到本种即提宗龙胆（*Gentiana tizuensis*）的花时常有四基数出现，花冠紫色到粉白色，包围花的茎生叶较宽，而短柄龙胆（*Gentiana stipitata*）的个体为纯五基数花，花色为白色到浅蓝灰色，花下的茎生叶较窄，可以区别。或许这两个种间存在杂交或种质渐渗。

仅见于中国四川省西部和西藏自治区东部。生长在较为潮湿的山坡。海拔2800~3500 m。

Current species *Gentiana tizuensis* highly resembles *Gentiana stipitata*, and were often merged together or treated as intraspecific taxon as subspecies or variety. We noticed in our field observations that *Gentiana tizuensis* frequently have 4-merous flowers with corolla ruddy purple to purple-white, the leaves surrounding flowers wider, whereas *Gentiana stipitata* got only white to light blue-grey flowers 5-merous only, the leaves surrounding flowers narrower. In western Sichuan and eastern Xizang of China only. In moist alpine slopes. Altitude: 2800-3500 m. Flowering & fruiting: July-October.

A. 中国西藏自治区左贡县
Zuogong, Xizang, China. 27.09.1992

B-C. 中国四川省道孚县 Daofu, Sichuan, China. 30.09.1993

D. 中国四川省炉霍县 Luhuo, Sichuan, China. 12.09.2023

乌奴龙胆

Gentiana urnula

匍匐茎发达，多分枝，连接多个营养莲座叶丛及个别花茎；营养枝莲座叶扇形，交互对生，折叠套合，边缘软骨质，光滑无睫毛；花单生或 2~5 枚簇生叶丛中；花萼筒钟状，花萼裂片扇形；花冠筒阔钟状，黄白色到红紫色，有紫色条纹。

分布于喜马拉雅山地区，包括尼泊尔、锡金、不丹和中国西藏南部及青海西南部。生长在高山流石滩和高山陡坡。海拔 3100~6150 m。花果期 8~10 月。

Stolon well developed, much branched, connecting numerous vegetative rosettes and few flowering stems; leaves of vegetative rosette flabelliform, decussate, overlapping, margins cartilaginous but not ciliolate; flowers solitary or 2-5 in terminal clusters; calyx tube campanulate, calyx lobes flabelliform; corolla tube broad campanulate, greenish white to ruddy purple with dark purple stripes.

In the Himalayan regions including Nepal, Sikkim, Bhutan, and southern Xizang and southwestern Qinghai of China. In alpine screes and steep stony slopes. Altitude: 3100-6150 m. Flowering & fruiting: August-October.

A

B

A–B. 中国西藏自治区羊八井镇
Yangbajing, Xizang, China. 11.09.1992

C–D. 中国西藏自治区浪卡子县
Langkazi, Xizang, China. 12.09.1992

C

D

线叶龙胆 *Gentiana lawrencei* var. *farreri*

中国青海省河南蒙古族自治县 Henan, Qinghai, China. 09.09.2023

多枝组

sect. *Kudoa*

多年生草本，营养枝和根状茎不明显；营养叶密集成莲座叶丛与数枚花枝从根头上生出；花通常单生花枝顶端，花枝上的茎生叶发达。本组约 25 种，分布于喜马拉雅山地区及相邻的中国西部和西南部。

Perennial, vegetative stem and rhizome inconspicuous; vegetative leaves crowded to rosette which arising from a caudex together with multiple flowering stems; flowers solitary and terminal. About 25 species, limited to the Himalayan and adjacent western and southwestern provinces of China.

椭圆叶龙胆

Gentiana altigena

多年生草本；茎斜升或平铺，表面具乳突；莲座叶丛缺；茎生叶椭圆形到匙形，边缘具细睫毛；花单生枝顶，花冠檐部以上深蓝色，基部黄白色，花冠筒管状，较小，长不及 3 cm。

分布于印度阿萨姆及中国西藏和云南。生长于较潮湿的高山草甸和石岩边。海拔 3000~4000 m。花果期 8~10 月。

Perennials, stems ascending or decumbent, papillate; basal rosette absent; stem leaves elliptic to spathulate, margin ciliolate; flowers solitary at terminal; corolla upper part blue, lower part yellowish white, corolla tube narrowly tubular, less than 3 cm long.

In Assam of India and Xizang and Yunnan of China. In damp alpine meadows and rock sides. Altitude: 3000-4000 m. Flowering & fruiting: August-October.

A. 中国云南省贡山独龙族怒族自治县
Gongshan, Yunnan, China. 04.08.2000

B. 中国西藏自治区墨脱县
Motuo, Xizang, China. 24.09.2003

C-E. 中国云南省贡山独龙族怒族自治县
摄影 /Photo：付鹏程
Gongshan, Yunnan, China. 27.08.2018

C

D

E

道孚龙胆

Gentiana altorum

多年生草本，茎斜升；莲座叶丛不发达；茎生叶对生，基部叶片卵形，中部叶片椭圆形，上部叶片渐狭；花单生枝顶，5 基数，花萼裂片狭椭圆形；花冠亮蓝色到深蓝色，外面具蓝紫色与黄绿色相间的条纹，花冠筒阔管状到漏斗状，长 4~6 cm，直径 2~3 cm。

分布于中国青藏高原东部的西藏、青海和四川交界地区。生长于较潮湿向阳的高山冻原、高山草甸和灌丛边。海拔 4000~4500 m。花果期 8~10 月。

Perennials, stems ascending; basal rosette inconspicuous; stem leaves opposite, basal leaf blades ovate, middle leaf blades elliptic, upper leaf blade becoming narrower; flowers solitary at terminal, 5-merous; calyx lobes narrowly elliptic; corolla bright blue to dark blue, with dark blue and yellowish green stripes, corolla tube broad tubular to funnelform, 4~6 cm long and 2~3 cm in diameter.

In the eastern Qinghai-Xizang Plateau delta connecting Xizang, Qinghai and Sichuan provinces of China. In damp and sunny alpine tundra, alpine meadows and bushes. Altitude: 4000-4500 m. Flowering and fruiting: August-October.

B

C

A–D. 中国四川省康定市 Kangding, Sichuan, China.13.09.2023

D

川东龙胆

Gentiana arethusae

多年生草本，茎斜升或平铺；莲座叶丛不发达；茎生叶 6 枚轮生，叶片线形，先端渐尖；花单生枝顶，花 6~8 基数，花冠亮蓝色，外面具深蓝紫色条纹，花冠筒管状到管状钟形，长 5~7 cm，直径约 2 cm。

特产中国陕西和甘肃南部、青海东南部及四川北部山区。生长于较潮湿的高山草甸和灌丛。海拔 3000~4000 m。花果期 8~10 月。本种常与六叶龙胆混淆，其实本种花冠筒更细，管状，茎生叶片也更窄细，而与其不同。

Perennials, stems ascending or decumbent; basal rosette inconspicuous; stem leaves verticillate by 6, blades linear, apex acuminate; flowers solitary at terminal, 6-8-merous; corolla bright blue, outside with dark blue stripes, corolla tube tubular, 5~7 cm long and ~2 cm in diameter.

Endemic to the mountains in southern Shaanxi, southern Gansu, southeastern Qinghai and northern Sichuan provinces of China. In damp alpine meadows and bushes. Altitude 3000-4000 m. Flowering & fruiting: August - October.

A-C. 中国陕西省太白山
Taibaishan Mt., Shaanxi, China.10.08.1990

D. 中国青海省久治县 Jiuzhi, Qinghai, China. 08.09.1994

E-F. 中国陕西省太白山
Taibaishan Mt., Shaanxi, China. 02.09.2023

G. 中国四川省若尔盖县
Ruoergai, Sichuan, China. 07.08.1992

D
E
F
G

天蓝龙胆

Gentiana caelestis

多年生草本，茎斜升；有较明显的莲座叶丛；茎生叶对生，叶片狭披针形，先端渐尖；花单生枝顶，5 基数，花冠亮蓝色到深蓝色，外面具蓝紫色与黄绿色相间的条纹，花冠筒阔管状钟形到管状漏斗形，显著宽大，长 5~7 cm，直径 4~5 cm。

见于中国西藏、云南和四川相邻接的高山地区。生长于较潮湿向阳的高山草甸和灌丛。海拔 3000~4000 m。花果期 8~10 月。本种常与蓝玉簪龙胆等混淆，唯本种花冠更宽阔硕大而与其不同。

Perennials, stems ascending; basal rosette conspicuous; stem leaves opposite, blades narrowly lanceolate, apex acuminate; flowers solitary at terminal, 5-merous; corolla bright blue to dark blue, outside with dark blue and yellowish green stripes, corolla tube broad tubular-campanulate, 5~7 cm long and 4~5 cm in diameter.

In the mountains connecting Xizang, Yunnan and Sichuan provinces of China. In damp and sunny alpine meadows and bushes. Altitude 3000-4000 m. Flowering & fruiting: August-October.

A. 中国云南省丽江市 摄影 / Photo: Adrien Favre
Lijiang, Yunnan, China.18.10.2004

B. F. 中国西藏自治区曲水县
Qushui, Xizang, China. 20.09.2003

C. 中国西藏自治区林芝市
Linzhi, Xizang, China. 21.09.2003

D. 莲座叶丛明显；中国云南省香格里拉市
Xianggelila, Yunnan, China. 03.10.1999

E. 中国云南省香格里拉市
Xianggelila, Yunnan, China. 03.10.1999

C
D
E
F

长萼龙胆

Gentiana dolichocalyx

多年生草本，茎斜升，高 10~15 cm；花期莲座叶丛处于地下，不明显；茎生叶对生，叶片狭线形，先端尖；花单生枝顶，5 基数，花萼裂片与叶同形，长达花冠中部；花冠淡蓝色到蓝色，外面具蓝紫色与黄绿色相间的条纹，花冠筒狭倒锥状筒形，长有 5~7 cm，直径约 1.5 cm，花冠裂片圆钝。

特产于中国甘肃西南部、青海东南部和四川西北部。生长于向阳的高山草甸。海拔 3000~3800 m。花果期 8~9 月。

Perennials, stems ascending, 10-15 cm; basal rosette mostly underground and inconspicuous at anthesis; stem leaves opposite, blades narrowly linear, apex acuminate; flowers solitary at terminal, 5-merous; calyx lobes similar to leaves, as long as the half of corolla; corolla pale blue to blue, with dark blue and yellowish green stripes outside, corolla tube narrowly obconic-tubular, 5-7 cm long and -1.5 cm in diameter, corolla lobes obtuse.

Endemic to southwestern Gansu, southeastern Qinghai and northwestern Sichuan. In alpine meadows. Altitude: 3000-3800 m. Flowering & fruiting: August-September.

A

B

A–B. 中国四川省阿坝州 摄影 /Photo：付鹏程
Aba, Sichuan, China. 22.08.2017

C. 中国青海省久治 Jiuzhi, Qinghai, China. 11.09.2023

D–E. 中国甘肃省卓尼县 Zhuoni, Gansu, China. 08.09.2023

C

D

E

青藏龙胆

Gentiana futtereri

多年生草本，茎斜升；有较明显的莲座叶丛；茎生叶对生，叶片狭披针形，先端渐尖；花单生枝顶，5 基数，花冠亮蓝色到深蓝色，具蓝紫色与黄绿色相间的条纹和深紫色斑点，花冠筒管状漏斗形到倒锥形，显著宽大，长 5~7 cm，直径 4~5 cm。

见于中国西藏和青海的高原地区。生长于较潮湿向阳的高山草甸和灌丛边。海拔 3000~4000 m。花果期 8~10 月。

Perennials, stems ascending; basal rosette conspicuous; stem leaves opposite, blades narrowly lanceolate, apex acuminate; flowers solitary at terminal, 5-merous; corolla bright blue to dark blue, with dark blue and yellowish green stripes and dark blue spots, corolla tube tubular-funnelform to obconic, large, 5~7 cm long and 4~5 cm in diameter.

In Xizang and Qinghai provinces of China. In damp and sunny alpine meadows and bushes. Altitude: 3000-4000 m. Flowering & fruiting: August-October.

A–C. 中国青海省共和县 Gonghe, Qinghai, China. 22.09.1993

喜湿龙胆

Gentiana helophila

多年生草本，茎斜升；莲座叶丛不明显；茎生叶对生，叶片狭披针形，先端渐尖；花单生枝顶，5 基数，花冠亮蓝色，具蓝紫色与黄绿色相间的条纹，花冠筒狭倒锥形到漏斗状狭钟形，长可达 7 cm，直径约 2 cm，花冠裂片三角形。

仅见于中国云南西北部。生长于向阳的高山沼泽草甸。海拔 3000~3500 m。花果期 8~10 月。

Perennials, stems ascending; basal rosette inconspicuous; stem leaves opposite, blades narrowly lanceolate, apex acuminate; flowers solitary at terminal, 5-merous; corolla bright blue, with dark blue and yellowish green stripes, corolla tube narrowly obconic to narrowly funnelform-campanulate, large, up to -7 cm long and -2 cm in diameter, corolla lobes triangular. In northwestern Yunnan of China only. In alpine bog meadows and bushes. Altitude: 3000-3500 m. Flowering &fruiting: August-October.

A-C. 中国云南省丽江市 Lijiang, Yunnan, China. 13.10.2003

中国云南省丽江市老君山的喜湿龙胆群落
Population of *Gentiana helophila*, Laojunshan Mt., Lijiang, Yunnan, China

七叶龙胆

Gentiana heptaphylla

多年生草本，茎斜升；莲座叶丛不显著；茎生叶 6~7 枚轮生，叶片线形，先端渐尖；花单生枝顶，花 6~7 基数，花冠灰蓝色，具深蓝紫色条纹和斑点，花冠筒钟状，长 4~6 cm，直径 3~4 cm。

特产中国云南西北部及四川西北部山区。生长于高山草甸和灌丛中。海拔 3000~4000 m。花果期 8~10 月。

本种常常被归入川东龙胆，作为它的一个变种 *Gentiana arethusae* var. *delicatula*。本种花冠筒更阔，更短，钟状，与其不同。

Perennials, stems ascending; basal rosette inconspicuous; stem leaves verticillate by 6-7, blades linear, apex acuminate; flowers solitary at terminal, 6-7-merous; corolla grey-blue, with dark blue stripes and spots, corolla tube campanulate, 4-6 cm long and 3-4 cm in diameter.

Endemic to the mountains in northwestern Yunnan and northwestern Sichuan provinces of China. In alpine meadows and bushes. Altitude: 3000-4000 m. Flowering and fruiting: August-October.

A

B

C

A–C. 中国四川省小金县巴郎山
Balangshan Mt., Xiaojin, Sichuan, China. 23.09.1994

D–E. 花基数在 5~7 间变化
中国四川省小金县巴郎山
Corolla 5-7-merous; Balangshan Mt., Xiaojin, Sichuan, China. 23.09.1994

F–G. 中国云南省德钦县白马雪山
Baimangxueshan Mt., Deqin, Yunnan, China. 25.09.1999

H–I. 中国云南省德钦县 Deqin, Yunnan, China. 19.09.2023

D
E
F
G
H
I

六叶龙胆

Gentiana hexaphylla

多年生草本，茎斜升；莲座叶丛不显著；茎生叶 6~7 枚轮生，叶片线状匙形，先端圆钝；花单生枝顶，花 6~7 基数，花冠灰蓝色，具深蓝紫色条纹，花冠裂片三角形，具尾，有斑点，花冠筒钟状，长 4~5 cm，直径 2~4 cm。

特产中国云南西北部及四川西北部山区。生长于高山草甸和灌丛中。海拔 3000~4000 m。花果期 8~10 月。本种变异较大，与七叶龙胆易混淆。本种叶片更宽大，匙形，先端圆钝，花冠筒更长，钟状，裂片三角形。

Perennials, stems ascending; basal rosette inconspicuous; stem leaves verticillate by 6-7, blades linear-spathulate, apex obtuse; flowers solitary at terminal, 6-7-merous; corolla grey-blue, with dark blue stripes, corolla lobes triangular, caudate, corolla tube campanulate, 4~6 cm long and 3~4 cm in diameter.

Endemic to the mountains in northwestern Yunnan and northwestern Sichuan provinces of China. In alpine meadows and bushes. Altitude: 3000-4000 m. Flowering & fruiting: August-October. This species is highly variable. It can be distinguished from G. heptaphylla by leaf shapes.

A

B

C

A-C. 中国云南省香格里拉市 Xianggelila, Yunnan, China. 21.09.1999

D-E. 中国云南省德钦县 Deqin, Yunnan, China. 21.08.2002

D

E

线叶龙胆

Gentiana lawrencei var. *farreri*

多年生草本，茎斜升；莲座叶丛不发达；茎生叶对生，叶片细长，细线形，先端渐尖；花单生枝顶，5 基数，花萼裂片细长，长超过萼筒 1.5 倍，花冠亮蓝色到深蓝色，外面具蓝紫色与黄绿色相间的条纹，花冠筒阔管状到漏斗状，长 5~6 cm，直径 2~3 cm。

分布于中国青藏高原的西藏、青海、甘肃和四川。生长于较潮湿向阳的高山草甸和灌丛边。海拔 3000~4500 m。花果期 8~10 月。

Perennials, stems ascending; basal rosette inconspicuous; stem leaves opposite, narrow and long, blades narrowly linear, apex acuminate; flowers solitary at terminal, 5-merous; calyx lobes fine and long, 1.5 times as long calyx tube, corolla bright blue to dark blue, with dark blue and yellowish green stripes, corolla tube broad tubular to funnelform, 5-6 cm long and 2-3 cm in diameter.

In the Qinghai-Xizang Plateau, including Xizang, Qinghai, Gansu and Sichuan provinces of China. In damp and sunny alpine meadows and bushes. Altitude: 3000-4500 m. Flowering & fruiting: August-October.

A

B

A. 中国四川省甘孜县 Ganzi, Sichuan, China. 29.09.1993

B. 中国甘肃省合作市 Hezuo, Gansu, China. 19.10.2018

C–F. 中国青海省河南蒙古族自治县 Henan, Qinghai, China. 09.09.2023

D
E
F

中国青海省河南蒙古族自治县
Henan, Qinghai, China. 09.09.2023

聂拉木龙胆

Gentiana nyalamensis

多年生草本，茎斜升；莲座叶丛明显，叶片线状披针形；茎生叶多对，狭卵形，先端渐尖；花单生枝顶，5 基数，花冠亮蓝色，外面具蓝紫色与黄绿色相间的条纹，花冠筒钟形，长 3~4 cm，直径 2~3 cm。

仅见于中国西藏。生长于较潮湿向阳的高山草甸和灌丛边。海拔 3500~4500 m。花果期 8~10 月。

Perennials, stems ascending; basal rosette conspicuous, blades linear lanceolate; stem leaves several pairs opposite, narrowly ovate, apex acuminate; flowers solitary at terminal, 5-merous; corolla bright blue, with dark blue and yellowish green stripes, corolla tube campanulate, 3-4 cm long and 2-3 cm in diameter.

In Xizang of China only. In damp and sunny alpine meadows and bushes. Altitude: 3500-4500 m. Flowering and fruiting: August-October.

A-B. 中国西藏自治区聂拉木县
Nielamu, Xizang, China.16.09.1992

山景龙胆

Gentiana oreodoxa

多年生草本，较矮小，茎斜升，高 3~5 cm，有不育枝；莲座叶丛不显著；茎生叶数对生，叶片细线形到钻形，先端渐尖，枝顶部叶较长；花单生枝顶，5 基数，花冠淡蓝色，具深紫色斑点，狭倒锥形，长 3~4 cm，直径 1~2 cm。

分布于喜马拉雅东部地区的不丹、缅甸及中国西南部（西藏东南部、云南西北部和四川西南部）。生长于向阳的高山草甸。海拔 3000~5000 m。花果期 8~10 月。

Perennials, plants short and small; stems ascending, 3-5 cm, often with vegetative shoots; basal rosette inconspicuous; stem leaves opposite, blades narrowly linear to subulate, apex acuminate; flowers solitary at terminal, 5-merous; corolla pale blue with park purple spots, corolla tube narrowly obconic, 3-4 cm long and 1-2 cm in diameter.

In the eastern Himalayan areas including Bhutan, Myanmar and southwestern China (southeastern Xizang, northwestern Yunnan and southwestern Sichuan). In sunny alpine meadows. Altitude: 3000-5000 m. Flowering & fruiting: August-October.

A–B. 中国西藏自治区浪卡子县 摄影 /Photo：付鹏程
Langkazi, Xizang, China. 19.09.2020

倒锥花龙胆

Gentiana obconica

多年生草本，有时具匍匐茎，茎斜升，有不育营养枝；莲座叶丛不显著；茎生叶数对生，叶片卵形到钻形，先端渐尖；花单生枝顶，5 基数，花冠纯净的深蓝色，具蓝紫色条纹，花冠筒阔钟状倒锥形，长 3~4 cm，直径 2~3 cm。

分布于喜马拉雅地区的尼泊尔、不丹及中国西藏南部和东南部。生长于向阳的高山草甸和高山冻原。海拔 4000~5000 m。花果期 8~9 月。

Perennials, sometimes with stolon; stems ascending, often with vegetative shoots; basal rosette inconspicuous; stem leaves opposite, blades ovate to subulate, apex acuminate; flowers solitary at terminal, 5-merous; corolla clear dark blue, with dark blue stripes, corolla tube broad campanulate-obconic, 3~4 cm long and 2~3 cm in diameter.

In the Himalayan areas including Nepal, Bhutan, and south and southeastern Xizang of China. In sunny alpine meadows and alpine tundra. Altitude: 4000-5000 m. Flowering & fruiting: August-September.

A–G. 中国西藏自治区隆子县

Longzi, Xizang, China. 15.08.2006

D
E
F
G

观赏龙胆

Gentiana prolata

多年生草本，茎斜升或平铺，表面具乳突；莲座叶丛不显著；茎生叶圆形、椭圆形到匙形，向上增大；花单生枝顶，花冠檐部以上亮蓝色，基部黄绿色，花冠筒狭管状到漏斗状，长 3~5 cm，直径不及 1 cm。

分布于喜马拉雅地区的印度阿萨姆、尼泊尔、锡金、不丹及中国的西藏南部和东南部。生长于较潮湿的高山草甸和山坡。海拔 3500~4500 m。花果期 8~10 月。

Perennials, stems ascending or decumbent, papillate; basal rosette inconspicuous; stem leaves orbicular or elliptic to spathulate, becoming larger toward apex; flowers solitary at terminal; corolla upper part bright blue, lower part greenish yellow, corolla tube narrowly tubular to funnelform, 3-5 cm long, diameter less than 1 cm.

In the Himalayan areas including Assam of India, Nepal, Sikkim, Bhutan and southern and southeastern Xizang of China. In damp alpine meadows and slopes. Altitude: 3500-4500 m. Flowering and fruiting: August-October.

A-E. 中国西藏自治区错那市
Cuona, Xizang, China. 14.08.2006

D

E

类华丽龙胆

Gentiana sinoornata

多年生草本，茎斜升；莲座叶丛不发达；茎生叶对生，外弯曲，叶片狭披针形，先端渐尖，向上增大；花单生枝顶，无柄，花萼裂片细长，与萼筒近等长，花冠淡蓝色，基部黄白色，具蓝紫色与黄绿色相间的条纹，花冠筒阔管状到漏斗状，长 5~6 cm，直径 2~3 cm。

分布于中国青藏高原的西藏、青海、云南和四川。生长于较潮湿向阳的高山草甸和灌丛边。海拔 3000~4500 m。花果期 7~10 月。

Perennials, stems ascending; basal rosette inconspicuous; stem leaves opposite, narrow and incurved, blades narrowly lanceolate, apex acuminate; flowers solitary at terminal, sessile; calyx lobes fine, about as long as the tube; corolla pale blue, base yellowish white, with dark blue and yellowish green stripes, corolla tube broad tubular to funnelform, 5-6 cm long and 2-3 cm in diameter.

In the Qinghai-Xizang Plateau, including Xizang, Qinghai, Yunnan and Sichuan provinces of China. In damp and sunny alpine meadows and bushes. Altitude: 3000-4500 m. Flowering & fruiting: July-October.

A

B

C

A. 中国青海省久治县 Jiuzhi, Qinghai, China. 07.09.1994

B. E. 中国西藏自治区林芝市 Linzi, Xizang, China. 21.09.2003

C. 中国四川省阿坝州 Aba, Sichuan, China. 09.09.1994

D. 中国西藏自治区米拉山 Milashan Mt., Xizang, China. 20.09.2003

F-G. 中国西藏自治区左贡县 Zuogong, Xizang, China. 10.10.2003

D
E
F
G

三叶龙胆

Gentiana ternifolia

多年生草本，茎斜升；莲座叶丛不发达；茎生叶 3~4 枚轮生，叶片线形，先端渐尖；花单生枝顶，花 5~6 基数，花冠蓝色，外面具深蓝紫色条纹，花冠筒漏斗形到管状钟形，长 5~6 cm，直径 1~2 cm，花冠裂片先端无尾状尖头。

仅见于中国四川北部，文献报道云南也有。生长于较潮湿的高山草甸和灌丛。海拔 3000~4000 m。花果期 9~10 月。本种可能是川东龙胆与蓝玉簪龙胆的杂交后代，花的形态介于二者之间，兼具二者特征，分布区也相互重叠。

Perennials, stems ascending; basal rosette inconspicuous; stem leaves verticillate by 3-4, blades linear, apex acuminate; flowers solitary at terminal, 5-6-merous; corolla blue, outside with dark blue stripes, corolla tube funnelform to tubularcampanulate, 5-6 cm long and 1-2 cm in diameter, lobes not caudate.

In northern Sichuan and reported from Yunnan of China. In damp alpine meadows and bushes. Altitude: 3000-4000 m. Flowering and fruiting: September - October. This species is probably a hybrid between *Gentiana arethusae* and *Gentiana veitchiorum*, as it is morphologically in between of them.

A–C. 中国四川省松潘县
Songpan, Sichuan, China. 12.09.1994

D. 川东龙胆 *Gentiana arethusae*，可能是三叶龙胆的一个亲本；中国四川省松潘县
A putative parent of *Gentiana ternifolia*; Songpan, Sichuan, China. 12.09.1994

E. 蓝玉簪龙胆 *Gentiana veitchiorum*，可能是三叶龙胆的另一个亲本；中国四川省松潘县
Another putative parent of *Gentiana ternifolia*; Songpan, Sichuan, China. 12.09.1994

D

E

蓝玉簪龙胆

Gentiana veitchiorum

多年生草本，茎斜升，高 5~8 cm；基生莲座叶发达；茎生叶较大，多集中于顶端；花单生枝顶，花冠深蓝色，基部具浅黄白色和深色条纹，有时具斑点，花冠筒狭漏斗形，4~6 cm 长。

分布于青藏高原南部和东部及其邻近地区常见，包括尼泊尔、不丹以及中国大陆的西藏、四川、云南、青海、甘肃等地。生长于高山河滩、草甸、灌丛、山坡草地及林缘。海拔 2500~4500 m。花果期 7~10 月。

Perennials, stems ascending 5-8 cm tall; basal rosette leaves well developed; stem leaves larger and more crowded toward apex; flowers solitary at terminal; corolla intense blue, with pale yellow-white base and black streaks and spots, narrowly funnelform, 4-6 cm long.
Commonly seen in the southern and eastern Qinghai-Xizang Plateau and adjacent areas, including Nepal and the Xizang, Sichuan, Yunnan, Qinghai, and Gansu provinces of China. In alpine river banks, meadows, bushes and grassland and forest margins. Altitude: 2500-4500 m. Flowering & fruiting: July-October.

A. 中国西藏自治区左贡县
Zuogong, Xizang, China. 19.08.2006

B. 中国西藏自治区波密县
Lulang, Bomi, Xizang, China. 04.10.2003

C. 中国四川省阿坝州 Aba, Sichuan, China. 09.09.1994

D. 中国四川省石渠县 Shiqu, Sichuan, China. 25.09.1993

E. 蓝玉簪龙胆 *Gentiana veitchiorum* 的白花变异个体
中国青海省泽库县 A mutant with white flowers; Zeku, Qinghai, China. 31.08.1994

F. 中国四川省阿坝州 Aba, Sichuan, China. 11.09.2023

G-H. 中国青海省河南蒙古族自治县
Henan, Qinghai, China. 09.09.2023

D

E

F

G

H

I. 蓝玉簪龙胆 *Gentiana veitchiorum* + 短柄龙胆 *Gentiana stipitata*（白色花）编织的鲜花地毯　A carpet made with fresh flowers of *Gentiana veitchiorum* + *Gentiana stipitata* (white)；中国四川省石渠县 Shiqu, Sichuan, China. 25.09.1993

J. 蓝玉簪龙胆 *Gentiana veitchiorum* + 短柄龙胆 *Gentiana stipitata*；中国四川省石渠县 Shiqu, Sichuan, China. 25.09.1993

K. 并列生长的蓝玉簪龙胆 *Gentiana veitchiorum* 和大花龙胆 *Gentiana szechenyii*　*Gentiana veitchiorum* and *Gentiana szechenyii* growing side by side；中国青海省河南蒙古族自治县至泽库县间　between Henan and Zeku, Qinghai, China. 31.08.1994

L. 蓝玉簪龙胆 *Gentiana veitchiorum* 和它的传粉者　*Gentiana veitchiorum* and its pollinator；中国青海省达日县 Dari, Qinghai, China. 06.09.1994

M. 蓝玉簪龙胆 *Gentiana veitchiorum* 和它的邻居旱獭　*Gentiana veitchiorum* and its neighbor marmot；中国青海省玛沁县 Maqin, Qinghai, China. 05.09.1994

N. 蓝玉簪龙胆 *Gentiana veitchiorum* 和它的邻居猫头鹰　*Gentiana veitchiorum* and its neighbor owl；中国青海省达日县 Dari, Qinghai, China. 06.09.1994

O. 牧场上的蓝玉簪龙胆 *Gentiana veitchiorum*　*Gentiana veitchiorum* on a yak pasture；中国青海省达日县 Dari, Qinghai, China. 06.09.1994

P. 中国四川省石渠县虾扎镇附近牧场　Yak farm nearby Xiaza, Shiqu, Sichuan, China. 25.09.1993

Q. 在由蓝玉簪龙胆 *Gentiana veitchiorum* 和大花龙胆 *Gentiana szechenyii* 鲜花编织的地毯上休闲的藏族家庭
中国青海省泽库县
Ahappy family resting on the carpet made with fresh flowers of *Gentiana veitchiorum* & *Gentiana szechenyii*
Zeku, Qinghai, China. 31.08.1994

R. 在由蓝玉簪龙胆 *Gentiana veitchiorum* 和大花龙胆 *Gentiana szechenyii* 鲜花编织的地毯上拍摄的瑞士植物学家 Philippe Kuepfer 教授
中国青海省泽库县
Swiss botanist Prof. Philippe Kuepfer exploring the carpet made with fresh flowers of *Gentiana veitchiorum* & *Gentiana szechenyii*
Zeku, Qinghai, China. 31.08.1994

蓝玉簪龙胆 *Gentiana veitchiorum* + 大花龙胆 *Gentiana szechenyii*
中国青海省泽库县 Zeku, Qinghai, China. 09.09.2023

川藏公路北线与南线的分路口：中国西藏自治区左贡县邦达镇及附近的业拉山景观

The crossroad of the southern and northern Chuang-Zang highway: Bangda and the landscape of Mt. Yelashan nearby the village Bangda, Zuogong, Xizang, China. 19.08.2006

中国四川省塔公草原远眺雅拉雪山

Yala Snow Mountain seen from Tagong, Sichuan, China. 13.09.2023

微籽组

sect. *Microsperma*

一年生草本；茎多分枝，3 到多花；花萼裂片增大，叶状；种皮具有蜂窝状纹饰构造。本组约 8 种，分布于喜马拉雅山到相邻的中国西南部。

Annual; stem much branched with 3 to many flowers; calyx lobes enlarged, leaf-like; seedcoat alveolate with a honeycomb-like sculpture. About 8 species, limited to the Himalayan and adjacent southwestern provinces of China.

微籽龙胆

Gentiana delavayi

一年生草本，茎直立，高 5~15 cm，多分枝，全株密被紫色乳突；茎生叶绿色或紫色，叶片椭圆形到披针形；花无梗，密集生枝端；花冠淡蓝色到蓝紫色或紫红，花冠筒阔管状钟形到管状漏斗形。

仅见于中国云南西北部和四川西南部山区。生长于较干燥向阳的高山裸露山坡。海拔 2000~3500 m。花果期 9~11 月。

Annual, stems erect, 5-15 cm, much branched, densely purple papillate; stem leaves green or purple, blades elliptic to lanceolate; flowers sessile, crowded terminal; corolla light blue to dark blue or ruddy purple, corolla tube tubular-campanulate to tubular-funnelform.

In the mountains of northwestern Yunnan and southwestern Sichuan provinces of China. In dry and sunny open slopes. Altitude 2000-3500 m. Flowering & fruiting: September-November.

A–C. 中国云南省香格里拉市
Xianggelila, Yunnan, China. 21.09.1999

D. 中国云南省丽江市 Lijiang, Yunnan, China. 04.10.1992

E. 中国云南省大理白族自治州
Dali, Yunnan, China. 04.10.1999

F–H. 中国云南省香格里拉市
Xianggelila, Yunnan, China. 21.09.1999

G. 中国云南省香格里拉市
Xianggelila, Yunnan, China. 12.10.2003

D
E
F
G
H

着色龙胆

Gentiana picta

一年生草本，茎直立，高 5~15 cm，紫色；基生叶在花期枯萎，茎生叶叶片椭圆形到匙形；花 1~3 朵簇生枝端；花萼筒倒锥形，花萼裂片增大，叶状，匙形；花冠淡蓝色，外面具紫色条纹，花冠筒阔管状钟形到管状漏斗形。

仅见于中国云南西北部和四川西南部山区。生长于较潮湿向阳的裸露山坡。海拔 2000~3500 m。花果期 8~11 月。

Annual, stems erect, 5-15 cm, tinged purple; basal leaves withered at anthesis; stem leaves elliptic to spathulate; flowers 1-3 clustered terminal; calyx tube obconic, calyx lobes leaf-like, spathulate; corolla light blue with purple stripes outside, corolla tube tubular-campanulate to tubular-funnelform.

In the mountains of northwestern Yunnan and southwestern Sichuan provinces of China. In damp and sunny open slopes. Altitude: 2000-3500 m. Flowering & fruiting: August-November.

A–B. 中国四川省雅江县 Yajiang, Sichuan, China. 13.08.2003

圆萼龙胆

Gentiana suborbisepala

一年生草本，高 5~15 cm，茎平卧或斜升，多分枝，紫色；茎生叶叶片圆形、倒卵形到心形；花单生分枝顶端；花萼筒管状，紫红色，花萼裂片增大，叶状，心形到圆形；花冠黄白色，外面具紫色条纹。

仅见于中国云南西北部和四川西南部山区。生长于较干燥向阳的山坡草甸。海拔 2000~3500 m。花果期 8~11 月。

Annual, 5-15 cm, stems decumbent to ascending, much branched, purple; stem leaves orbicular, obovate to cordate; flowers solitary and terminal on each branch apex; calyx tube tubular, reddish purple, calyx lobes enlarged, leaf-like, cordate to orbicular; corolla yellowish white.

In the mountains of northwestern Yunnan and southwestern Sichuan provinces of China. In dry and sunny slopes meadows. Altitude: 2000-3500 m. Flowering & fruiting: August-November.

A–B. 中国云南省香格里拉市
Xianggelila, Yunnan, China. 05.08.2002

匙萼龙胆

Gentiana spathulisepala

一年生草本，高 5~15 cm，茎平卧或斜升，多分枝，紫色，密被紫色乳突；茎生叶叶片圆形、倒卵形到心形；花单生分枝顶端；花萼筒管状，紫红色，花萼裂片增大，叶状，匙状、心形到圆形；花冠紫红色，外面具紫色条纹，花冠筒管状。

仅见于中国云南西北部。生长于较干燥向阳的裸露山坡。海拔 2000~3500 m。花果期 8~11 月。

该种是何廷农和刘尚武二位先生于 2010 年发表的新种，与圆萼龙胆非常相似，但花色和花冠裂片形状不同。

Annual, 5-15 cm, stems decumbent to ascending, much branched, tinged purple, papillate; stem leaves orbicular, obovate to cordate; flowers solitary and terminal on each branch apex; calyx tube tubular, reddish purple, calyx lobes enlarged, leaf-like, cordate to orbicular; corolla ruddy blue, with purple stripes outside, corolla tube tubular.

In the mountains of northwestern Yunnan of China. In dry and sunny open slopes. Altitude: 2000-3500 m. Flowering & fruiting: August-November.

A-E. 中国云南省香格里拉大雪山
Grand Snow Mountain, Yunnan, China. 05.08.2002
F. 果期的匙萼龙胆 *Gentiana spathulisepala*
中国云南省香格里拉
Xianggelila, Yunnan, China. 28.09.1999

D

E

F

东俄洛龙胆

Gentiana tongolensis

一年生草本，高 5~10 cm，茎平卧或斜升，多分枝，紫色，密被紫色乳突；茎生叶叶片卵形到心形；花单生分枝顶端；花萼筒管状，紫红色，花萼裂片增大，叶状，心形；花冠亮黄色，花冠筒高脚杯状，花药亮黄色。

仅见于中国西藏东部、云南西北部和四川西南部山区。生长于较干燥向阳的裸露山坡。海拔 2000~3500 m。花果期 8~10 月。

Annual, 5-15 cm, stems decumbent to ascending, much branched, tinged purple, densely papillate; stem leaves ovate to cordate; flowers solitary and terminal on each branch terminal; calyx tube tubular, reddish purple, calyx lobes enlarged, leaf-like, cordate; corolla bright yellow, corolla tube salverform; anthers bright yellow.

In the mountains of eastern Xizang, northwestern Yunnan and southwestern Sichuan provinces of China. In dry and sunny open slopes. Altitude: 2000-3500 m. Flowering & fruiting: August-October.

A. B. F. 中国四川省雅江县
Yajiang, Sichuan, China. 13.08.2002

C. 中国四川省道孚县
Daofu, Sichuan, China. 30.09.1993

D. E. G. 中国四川省康定市
Kangding, Sichuan, China. 13.09.2023

D
E
F
G

从折多山口眺望东俄洛方向：中国四川省康定市

View of Tongole area at Zheduoshan Pass, Kangding, Sichuan, China. 03.07.2021

露蕊龙胆

Gentiana vernayi

一年生草本，高 3~10 cm，茎平卧或斜升，多分枝，紫色，被紫色乳突或光滑；茎生叶叶片卵形到心形；花单生分枝顶端；花萼筒管状，具紫红色条纹，花萼裂片增大，叶状，卵形到心形；花冠亮蓝色，喉部白色，花冠筒高脚杯状，花药紫红色。

见于喜马拉雅山地区的印度、尼泊尔、不丹和中国西藏南东部及云南西北部。生长于较干燥向阳的草甸边缘裸露山坡。海拔 3500~4500 m。花果期 7~9 月。

Annual, 3-10 cm, stems decumbent to ascending, much branched, tinged purple, papillate or glabrous; stem leaves ovate to cordate; flowers solitary and terminal on each branch terminal; calyx tube tubular, with reddish purple stripes, calyx lobes enlarged, leaf-like, ovate to cordate; corolla bright blue, corolla tube salverform; anthers ruddy red.
In the Himalayan mountains of India, Nepal, Bhutan, and southeastern Xizang and northwestern Yunnan of China. In dry and sunny open slopes alone alpine meadows. Altitude: 3500-4500 m. Flowering & fruiting: July-September.

A-H. 中国西藏自治区隆子县 Longzi, Xizang, China. 13.08.2006

D
E
F
G
H

云南龙胆（白冠绿萼型）

Gentiana yunnanensis
(white corolla green calyx type)

一年生草本，高 5~30 cm，直立，紫色，密被紫色乳突；茎生叶叶片倒卵形到匙形；花 1 到数枚，簇生；花萼筒管状，绿色，基部具淡紫色条纹，花萼裂片增大，叶状，大小不等，通常 3 大 2 小，心形到匙形；花冠白色，外面具紫色斑点，花冠筒管状，花冠裂片卵形，褶不整齐，啮齿状。

仅见于中国西藏东南部、云南西北部和四川西南部山区。生长于向阳的山坡、草甸和灌丛边。海拔 2500~4000 m。花果期 8~11 月。

Annual, 5-30 cm, stems erect, tinged purple, papillate; stem leaves obovate to spathulate; flowers 1 to a few, clustered; calyx tube tubular, green with pale purple stripes, calyx lobes enlarged, leaf-like, unequal in size, often 3 larger and 2 smaller, cordate to spathulate; corolla white, with purple spots outside, corolla tube tubular; plicae irregular, erose.
In the mountains of southeastern Xizang, northwestern Yunnan and southwestern Sichuan provinces of China. In alpine slopes; meadows, and bushes. Altitude: 2500-4000 m. Flowering & fruiting: August-November.

A

B

A–B. 中国云南省香格里拉市
Xianggelila, Yunnan, China. 18.09.2023
C. 中国四川省盐源县
Yanyuan, Sichuan, China. 16.09.2023

云南龙胆（蓝冠紫萼型）

Gentiana yunnanensis
(blue corolla purple calyx type)

本类型与白冠绿萼型的生境相似，分布区基本重叠，外形也较相似，但植株体型纤细高挑，花数量多，排列成密集的穗状，花萼筒外面深紫色，花冠淡蓝色，外面紫色斑点更密集，而明显可以区别，也未见有中间类型。二者是否属于同种有待深入研究。

This type highly resembles the “white corolla green calyx type”, with similar gross morphology, similar habitat and overlapping distribution. However, it can be easily distinguished due to its slender and higher size, much more flowers densely arranged in a compact spike, darker purple calyx tube dark purple, and corolla pale blue with more purple spots on corolla tube. No intermediate type between these two types was found in the nature. It is necessary to be further studied if these two types are conspecific.

A–D. 中国云南省德钦县 Deqin, Yunnan, China. 19.09.2023

C

D

中国西藏自治区八宿县业拉山 Mt. Yela, Basu, Xizang, China. 19.08.2006

单轴组

sect. *Monopodiae*

多年生草本；莲座叶丛发达，主轴单轴分枝；1 到多花簇生花枝顶端；花萼裂片不等大，花冠的褶偏斜不对称。本组约 15 种，分布于喜马拉雅山到相邻的东南亚和中国西南部。

Perennial; vegetative leaves forming a well developed rosette; core axes branching monopodially; 1 to many flowers clustered at apex terminal; calyx lobes unequal in size; corolla plicae asymmetrical. About 15 species, in the Himalayan and adjacent southeastern Asian countries and southwestern provinces of China.

头花龙胆

Gentiana cephalantha

多年生草本，营养主茎短，生成莲座叶丛，从其基部生出斜升花枝；莲座叶狭椭圆形到椭圆状披针型；花枝基部木质；花枝上的茎生叶向上愈大，在枝顶端围托着花簇；花无梗，3~10 花密集簇生枝端成头状；花冠淡蓝色到蓝紫色，花冠筒管状钟形，花冠裂片卵形到三角形，先端具绿色斑点。

分布于缅甸、泰国、越南和中国西南各省（云南、广西、贵州、四川）。生长在山坡草地、灌丛和林缘。海拔 1500~3500 m。花果期 8~11 月。

Perennial, vegetative caudex short, forming well developed rosette, from the base arising the flowering stems; blades of rosette leaves narrowly elliptic to elliptic-lanceolate; flowering stem woody at base, stem leaves becoming larger toward apex, apically crowded and subtending terminal flowers; flowers sessile, 3-10 flowers densely clustered and capitate; corolla light blue to dark blue, corolla tube tubular-campanulate, corolla lobes ovate to triangular, with green spots. In Myanmar, Thailand, Vietnam and southwestern provinces of China (Yunnan, Sichuan, Guizhou and Guangxi). In slopes grassland, bushes, and forest margins. Altitude: 1500-3500 m. Flowering & fruiting: August-November.

A. 中国云南省大理白族自治州苍山
Cangshan Mt., Dali, Yunnan, China. 06.10.1999

B. 中国云南省昆明市
Kunming, Yunnan, China. 29.11.1997

C. G. 中国云南省丽江市
Lijiang, Yunnan, China. 13.10.2003

D. 中国云南省宾川县
Binchuan, Yunnan, China. 29.11.2007

E-F. 中国云南省大理白族自治州苍山
摄影 /Photo: Adrien Favre
Cangshan Mt., Dali, Yunnan, China. 06.10.1999

D

中国龙胆

Gentiana chinensis

多年生草本，匍匐茎发达，多分枝，无明显莲座叶丛，花枝斜升；茎生叶卵圆形，在花基部密集簇生；花无梗，1~3 花簇生枝端；花冠淡蓝色，花冠筒管状，花冠裂片卵形到三角形。

特产中国云南和四川。生长在山坡林间和灌丛岩石上等向阳处。海拔 2000~3000 m。花果期 7~10 月。

Perennial, stolon well developed, much branching; rosette inconspicuous, flowering stems ascending; stem leaves ovate to ovate-orbicular, clustered at apex surrounding bases of flower clusters; flowers sessile, 1-3 in clusters; corolla bright blue, corolla tube tubular, corolla lobes ovate to triangular.

Endemic to Sichuan and Yunnan provinces of China. On rocky and sunny places in forest and bush margins and cleared places. Altitude: 2000-3000 m. Flowering & fruiting: July-October.

A. C. 中国四川省峨眉山
Emeishan Mt., Sichuan, China. 08.10.1993
B. 中国四川省峨眉山
Emeishan Mt., Sichuan, China. 24.08.2006
D. 中国四川省峨眉山
Emeishan Mt., Sichuan, China. 07.08.2004

五岭龙胆

Gentiana davidii

多年生草本，花枝斜升；莲座叶丛发达，叶卵线状披针形；花无梗，1~3 花簇生枝端；花萼筒倒锥形，紫色，花萼裂片大小不等，花冠淡蓝色，花冠筒管状漏斗形到管状钟形，花冠裂片卵形到三角形。

特产中国中部和西南部山区。生长在山坡林间和灌丛间向阳处空地和草丛。海拔 300~2000 m。花果期 8~11 月。

Perennial; rosette well developed, flowering stems arising from base of rosette, ascending; leaves linear lanceolate; flowers sessile, 1-3 clustered terminal; calyx tube obconic, purple, calyx teeth unequal in size; corolla light blue, corolla tube tubular, corolla lobes ovate to triangular.

Endemic to the mountains in central to southwestern China. On open and sunny places in grassland, forest margins and bush margins. Altitude: 300-2000 m. Flowering & fruiting: August-November.

B

A. B. E. 中国浙江省龙泉市黄茅尖
Huangmaojian Mt., Zhejiang, China. 17.10.2022

C. 蕾期的五岭龙胆 *Gentiana davidii*
中国福建省黄冈山
Huanggangshan Mt., Fujian, China. 21.07.2020

D. 果期的五岭龙胆 *Gentiana davidii*
中国浙江省龙泉市黄茅尖
Huangmaojian Mt., Zhejiang, China. 17.10.2022

D

E

昆明龙胆

Gentiana duclouxii

多年生草本，花枝斜升；莲座叶丛发达，叶片匙形；花无梗，1~3 花簇生花枝顶端；花萼筒倒锥形，紫色，花萼裂片大小不等，花冠玫瑰红色，有时淡蓝灰色，花冠较大，直径可达 4 cm，花冠筒管状漏斗形到管状钟形，花冠裂片卵形到三角形。

特产中国云南中部和西北部低山区。生长在山坡林间和灌丛间向阳处空地和岩石上。海拔 1500~2000 m。花果期 3~5 月。

Perennial; rosette well developed, flowering stems arising from base of rosette, ascending; leaves spathulate; flowers sessile, 1-3 clustered terminal; calyx tube obconic, purple, calyx teeth unequal in size; corolla pink or rose-colored, corolla large, up to ~4 cm diameter, corolla tube funnelform, corolla lobes ovate to triangular.

Endemic to the low mountains in central to northwest Yunnan of China. On open and sunny places of forest margins and bushes. Altitude: 1500-2000 m. Flowering & fruiting: March-May.

A

B

C

A-B. 中国云南省武定县 Wuding, Yunnan, China. 11.04.2020

C-E. 中国云南省昆明市 Kunming, Yunnan, China. 18.04.1999

D

E

壶冠龙胆

Gentiana elwesii

多年生草本，高 10~15 cm，茎单一，纤细；叶片长圆形；花无梗，簇生花枝顶端；花萼筒倒锥形，紫色；花冠坛状，多闭合，蓝紫色。分布于喜马拉雅山地区。生长高山草甸及灌丛边缘。海拔 3500~4500 m。花果期 7~10 月。

Perennial; 10-15 cm high, stems simple, slender; leaves oblong; flowers sessile, clustered terminal; calyx tube obconic, purple; corolla urceolate, often closed. In the Himalayas. Alpine meadows and bush margins. Altitude: 3500-4500 m. Flowering & fruiting: July-October.

中国西藏亚东 摄影/Photo：曾佑派

Yadong, Xizang, China. 11.08.2019

露萼龙胆

Gentiana emergens

多年生草本，高 3~5 cm，垫状植物；匍匐茎发达，不育营养枝叶密集；叶片圆形到倒卵形；花无梗，单生低矮的花枝顶端；花萼筒钟状，花冠淡蓝到深蓝色，花冠筒钟状，多闭合；花谢后花梗基本不伸长或只略微伸长。

特产中国西南部（西藏、云南、四川）。生长高山草甸及冰缘山坡。海拔 3500~4500 m。花果期 8~10 月。

本种有时被作为变种并入矮龙胆，区别是本种叶片圆形或倒卵形，而矮龙胆的叶为狭条形；本种也易与叶萼龙胆混淆，区别是本种花萼外露出叶片，花梗在花谢后不伸长。

Perennial; 3-5 cm high, cushion plant; stolon and vegetative sterile branches well developed, crowded as cushion; leaves orbicular to obovate; flowers sessile, solitary terminal on short flowering stems; calyx tube campanulate; corolla pale blue to dark blue, corolla tube campanulate, often closed; pedicel not elongated or only slightly elongated after anthesis.

Endemic to southwestern China (Xizang, Yunnan and Sichuan). On alpine meadows and gravel slopes. Altitude: 3500-4500 m. Flowering & fruiting: August-October.

A-B. 中国西藏自治区米林多雄拉山
Duoxiongla Mt., Milin, Xizang, China. 24.09.2003

C. 花期的露萼龙胆 *Gentiana emergens*；中国西藏自治区米林市多雄拉山 Duoxiongla Mt., Milin, Xizang, China. 24.09.2003

D. 果期的露萼龙胆 *Gentiana emergens*；中国云南省德钦县白马雪山 Baimaxueshan Mt., Deqin, Yunnan, China. 26.09.1999

丝柱龙胆

Gentiana filistyla

多年生草本，匍匐茎较发达，连接营养枝与花枝；莲座叶丛发达，叶片卵形到长卵形，具尾尖，肉质，密集覆瓦状排列；花无梗，单生花枝顶端；花萼筒紫色，花萼裂片长披针型，花冠蓝紫色，花冠筒狭漏斗形，花冠裂片卵状三角形。

特产东喜马拉雅地区的缅甸和中国西藏和云南的高山区。生长高草甸、山坡和岩石上。海拔 3000~4500 m。花果期 7~9 月。

Perennial; stolon well developed connecting flowering branches and vegetative rosettes; rosette leaves ovate to oblong-ovate, fleshy, crowded and imbricate; flowers sessile, solitary and terminal; calyx tube purple, calyx teeth lanceolate; corolla blue-purple, corolla tube narrowly funnelform, corolla lobes ovate-triangular.

Endemic to the east Himalayan mountains including Myanmar and Xizang and Yunnan provinces of China. On alpine meadows, slopes and rocky places. Altitude: 3000-4500 m. Flowering & fruiting: July-September.

A-B. 中国西藏自治区错那市
Cuona, Xizang, China. 14.08.2006

C–F. 中国西藏自治区错那市 Cuona, Xizang, China. 14.08.2006

E

F

黄条纹龙胆

Gentiana gilvostriata

多年生草本，高 5~8 cm，莲座叶丛和匍匐茎发达；花枝平铺或斜升；叶片倒卵状匙形；花无梗，单生花枝顶端；花萼筒钟状，黄绿色，花冠蓝紫色，下部具条纹和斑点，花冠筒管状，裂片卵形。

分布于喜马拉雅东部，包括中国西藏及云南。生长于高山草地及山坡林下。海拔 3000~4000 m。花果期 8~11 月。

Perennial; 5-8 cm high; rosette & stolon well developed; flowering stem decumbent to ascending; leaves obovate-spathulate; flowers sessile, solitary terminal; calyx tube campanulate, yellowish green; corolla blue purple, with stripes & spots, corolla tube tubular, lobes ovate.

In eastern Himalayan region, including Xizang and Yunnan of China. On alpine grassland and under forest. Altitude: 3000-4000 m. Flowering & fruiting: August-November.

A–B. 中国云南省德钦县 Deqin, Yunnan, China; 26.09.1999

扭果柄龙胆

Gentiana harrowiana

多年生草本，高 3~6 cm，匍匐茎发达，连接营养枝与花枝；莲座叶丛较发达，叶片卵圆形到椭圆形，先端圆钝；花无梗，1~3 朵簇生花枝顶端；花萼筒红紫色，花萼裂片三角形，花冠淡蓝紫色或黄白色，花冠筒狭管状。

特产东喜马拉雅地区的缅甸和中国西藏和云南高山区。生长高山草丛、林缘、山坡和岩石等向阳处。海拔 3000~4500 m。花果期 8~10 月。

本种在野外有花冠蓝色和白色两种类型，都较常见，混合生长，它们之间的关系尚不清楚，值得进一步观察研究。

Perennial; 3-6 cm high, stolon well developed connecting flowering branches and vegetative rosettes; rosette leaves round ovate to elliptic-ovate; flowers sessile, solitary and terminal; calyx tube ruddy purple, calyx teeth triangular; corolla pale blue or yellowish white, corolla tube narrowly tubular.

Endemic to the east Himalayan mountains including Myanmar and Xizang and Yunnan provinces of China. On sunny places in alpine grassland, forest margin, slopes and rocky places. Altitude: 3000-4500 m. Flowering & fruiting: August-October.

A–C. 中国云南省大理白族自治州 Dali, Yunnan, China. 07.10.1999

D

E

D. 蓝花型扭果柄龙胆 *Gentiana harrowiana*；中国云南省大理白族自治州 Blue flower type, Dali, Yunnan, China. 23.09.2023

E. 白花型扭果柄龙胆 *Gentiana harrowiana*；中国云南省大理白族自治州 White flower type, Dali, Yunnan, China. 23.09.2023

F. 两种花色的类型在一起生长；中国云南省大理白族自治州 Two types growing together; Dali, Yunnan, China. 23.09.2023

G. 果期的扭果柄龙胆 *Gentiana harrowiana*；中国云南省大理白族自治州 Dali, Yunnan, China. 07.10.1999

女娄菜叶龙胆

Gentiana melandriifolia

多年生草本，高 5~7 cm，花枝平卧或斜升；莲座叶丛发达，叶片倒卵形到匙形；花无梗，1~3 花簇生花枝顶端；花萼筒钟状，花冠亮蓝色到深蓝色，花冠较大，直径可达 3 cm，花冠筒漏斗状到钟状，花冠裂片卵形，具绿色斑点。

特产中国云南中部和西北部山区。生长在山坡林间和灌丛间向阳处空地和岩石上。海拔 2000~3000 m。花果期 7~10 月。

Perennial; 5-7 cm high; rosette well developed, flowering stems arising from base of rosette, decumbent or ascending; leaves obovate to spathulate; flowers sessile, 1-3 clustered terminal; calyx tube campanulate; corolla bright to dark blue, corolla large, up to ~3 cm diameter, corolla tube funnelform to campanulate, corolla lobes ovate, with green spots.

Endemic to the mountains in central to northwest Yunnan of China. On open and sunny places of forest margins, bushes and rocky places. Altitude: 2000-3000 m. Flowering & fruiting: July-October.

A

B

C

A–G. 中国云南省大理白族自治州
Dali, Yunnan, China. 06.10.1999

D
E
F
G

坚龙胆

Gentiana rigescens

多年生草本，高 10~50 cm，无莲座叶丛；花枝直立坚挺，基部木质；茎生叶椭圆形到椭圆状披针型；花无梗，5~10 朵花簇生花枝端和花枝上部；花冠粉红色到紫红色，花冠筒管状钟形，花冠裂片卵形到三角形，先端具绿色斑点。

分布于缅甸和中国西南各省（云南、贵州、四川）。生长在较干燥的山坡草地、灌丛和林缘。海拔 1200~3500 m。花果期 8~11 月。

Perennial, 10-50 cm high; rosette absent; flowering stems erect, rigid, woody at base, stem leaves elliptic to elliptic-lanceolate; flowers sessile, 5-10 flowers clustered terminal and axillary on upper part of flowering stems; corolla pink to dark purple-rec, corolla tube tubular-campanulate, corolla lobes ovate to triangular, with green spots.

In Myanmar and southwestern provinces of China (Yunnan, Sichuan, and Guizhou). On dry places in grasses, bushes, and forest margins. Altitude: 1200-3500 m. Flowering & fruiting: August-November.

A

B

A–D. 中国云南省昆明市
Kunming, Yunnan, China. 08.10.1992

E. 中国云南省大理白族自治州
Dali, Yunnan, China. 06.10.1999

C
D
E

锡金龙胆

Gentiana sikkimensis

多年生草本，高 5~20 cm，花枝直立或斜升；匍匐茎较发达，有不育营养枝，叶片卵圆形到匙形；花无梗，密集簇生花枝顶端及花枝上部叶腋；花萼筒钟状，花冠淡蓝色，有蓝绿色条纹，花冠筒管状到钟状，花冠裂片卵形。

分布于喜马拉雅到中国西南部西藏及云南。生长在山坡林缘和灌丛间处空地和岩石上。海拔 3000~4000 m。花果期 8~11 月。

Perennial; 5-20 cm high; rosette inconspicuous; stolon and vegetative sterile branches well developed; leaves ovate orbicular to elliptic; flowers sessile, crowded in clusters at terminal and in axillary of upper part of stems; calyx tube campanulate; corolla pale blue with green-blue stripes, corolla tube tubular to campanulate, corolla lobes ovate.

In Himalayan regions to the mountains in southwestern China (Xizang and Yunnan). On open and sunny places of forest margins, bushes and rocky places. Altitude: 3000-4000 m. Flowering & fruiting: August-November.

A

B

A-B. 中国西藏自治区米林市多雄拉山

Duoxiongla Mt., Milin, Xizang, China. 24.09.2003

维纳斯龙胆

Gentiana venusta

多年生草本，高 3~5 cm，匍匐茎发达，连接营养枝与花枝；花枝直立，叶聚生于不育营养枝顶端成莲座状，叶片卵圆形；花无梗，密集簇生花枝顶端；花萼筒倒锥形，紫红色，花萼裂片线形到狭披针型；花冠蓝紫色，有蓝绿色条纹，花冠筒管状，花冠裂片卵形。

分布于克什米尔到喜马拉雅地区，包括中国西藏南部。生长在石质山坡及灌丛边缘空地。海拔 3500~5000 m。花果期 8~10 月。本种与扭果柄龙胆接近，叶与萼片形状不同。

Perennial; 3-5 cm high; stolon and vegetative sterile branches well developed; leaf rosette on terminal of vegetative branches, leaf blade ovate orbicular; flowers sessile, crowded in clusters and terminal; calyx tube obconic, reddish purple, calyx lobes linear to narrowly lanceolate; corolla blue-purple, with green-blue stripes, corolla tube tubular, corolla lobes ovate.

In Kashmir to Himalayan regions, including southern Xizang of China. On stony slopes and bushes bush margins. Altitude: 3500-5000 m. Flowering & fruiting: August-October. This species is similar to *Gentiana harrowiana*, but shapes of leaves and calyx lobes different.

A–B. 中国西藏自治区米林市多雄拉山
Duoxiongla Mt., Milin, Xizang, China. 24.09.2003

C-D. 中国西藏自治区米林市多雄拉山 Duoxiongla Mt., Milin, Xizang, China. 24.09.2003

矮龙胆

Gentiana wardii

多年生草本，高 3~5 cm，垫状植物；匍匐茎发达，不育营养枝密集；叶片卵状条形；花无梗，单生低矮的花枝顶端；花萼筒钟状，花冠淡蓝到深蓝色，花冠筒钟状，多闭合。

分布于东喜马拉雅的缅甸及中国西南部（西藏、云南、四川）。生长高山草甸及冰缘山坡。海拔 3500~4500 m。花果期 8~10 月。

Perennial; 3-5 cm high, cushion plant; Stolon and vegetative sterile branches well developed, crowded as cushion; leaves linear-ovate to elliptic; flowers sessile, solitary terminal on short flowering stems; calyx tube campanulate; corolla pale blue to dark blue, corolla tube campanulate, often closed.

In the eastern Himalayan regions Myanmar and southwestern China (Xizang, Yunnan and Sichuan). On alpine meadows and gravel slopes. Altitude: 3500-4500 m. Flowering & fruiting: August-October.

A-B. 中国西藏自治区米林市多雄拉山
Duoxiongla Mt., Milin, Xizang, China. 24.09.2003

C–E. 中国西藏自治区米林市多雄拉山 Duoxiongla Mt., Milin, Xizang, China. 24.09.2003

中国西藏自治区米林市多雄拉山 Duoxiongla Mt., Milin, Xizang, China. 24.09.2003

叶萼组

sect. *Phyllocalyx*

多年生草本；主轴合轴分枝，有匍匐茎；花萼膜质半透明，完全包裹在最顶部的一对叶里。本组仅 1 种，分布于喜马拉雅山中部和东部到相邻的中国西南部地区。

Perennial; core axes branching sympodially, stoloniferous; calyx membranous, subtranslucent, entirely surrounded by upper pair of leaves. Only 1 species known so far, found in the high mountains from central and eastern Himalayan to adjacent southwestern provinces of China.

叶萼龙胆

Gentiana phyllocalyx

多年生草本，高 3~15 cm，匍匐茎较发达；不育营养枝的叶密集成莲座，叶片圆形到倒卵形；花枝的茎生叶 2~3 对；花无梗，单生或偶尔 2~3 朵簇生花枝顶端；花萼较小，膜质半透明，完全包藏于顶生的一对叶片中；花冠淡蓝到深蓝色，花冠筒钟状，开花或花谢后花梗多明显伸长。

分布于喜马拉雅到中国西南部（西藏、云南、四川）。生长在石质高山山坡。海拔 3500~4500 m。花果期 6~9 月。本种在花期易与露萼龙胆混淆，区别是本种花萼小，膜质，完全被最顶部的一对叶片包裹其中，而露萼龙胆的花萼较大，草质，至少部分外露在顶部叶片之外。

Perennial; 3-15 cm high with stolon; leaves of vegetative sterile branches crowded as rosettes, leaf blades orbicular to obovate; stem leaves 2-3 pairs; flowers sessile, solitary or occasionally 2-3 clustered and terminal; calyx small, membranous, semitranslucent, embodied inside the upper most pair of leaves; corolla pale to dark blue, corolla tube campanulate; pedicels elongated after anthesis.

In the regions from the Himalayas to southwestern China (Xizang, Yunnan and Sichuan). On alpine stony slopes. Altitude: 3500-4500 m. Flowering & fruiting: June-September.

A. 中国云南省贡山独龙族怒族自治县 摄影 /Photo：曾佑派
Gongshan, Yunnan, China. 27.07.2003

B. 中国西藏自治区多雄拉山
Duoxiongla Mt., Xizang, China. 24.09.2003

C. 花后期的叶萼龙胆 *Gentiana phyllocalyx*
中国云南省高黎贡山
Gaoligong Mt., Yunnan, China. 04.08.2000

D. 果期的叶萼龙胆 *Gentiana phyllocalyx*
中国西藏自治区多雄拉山
Duoxiongla Mt., Xizang, China. 24.09.2003

C

D

狭叶龙胆 *Gentiana manshurica*

中国安徽省明光市 Mingguang, Anhui, China. 10.10.2022

龙胆草组

sect. *Pneumonanthe*

多年生草本，根状茎多少肉质；茎合轴分枝，有 1 到数个直立的花枝，无营养莲座叶丛；花枝基部有鳞片状叶；种子常有翅，种皮具粗网纹。本组约 42 种，分布于北半球温带亚欧到北美的低山区，以北美东部和西部山地种类最为丰富。

Perennial, rhizome more or less fleshy; stems branching sympodially, with 1 to several erect flowering stems, vegetative rosette absent; seeds often winged; seed coat coarsely reticulate. About 42 species, widely spread in northern temperate areas from Asia, Europe to North America; the mountains in eastern and western North America harbor the most species.

草原褶龙胆

Prairie Pleated Gentian

Gentiana affinis

多年生草本，高 10~60 cm；茎数枝丛生，直立或斜升；叶无柄，卵形到长圆状披针形；花数枚，簇生顶端和茎上部叶腋；萼筒管状，萼齿钻形，大小不等；花冠漏斗状，蓝色，花冠裂片卵形，褶小，不规则分裂，齿状。

主要分布于北美西北部美国和加拿大太平洋沿岸山区。生长在潮湿草地、灌丛与牧场边缘。海拔 1000~2300 m。花果期 7~8 月。

Perennial, 10-60 cm high; stems several clumps, erect or decumbent; leaves sessile, blade ovate to oblong-lanceolate; flowers numerous, clustered at terminal and in axillaries of upper parts of stems; calyx tube tubular, teeth subulate; corolla tubular-funnelform, blue, corolla lobes ovate, plicae small, irregular clefted, denticulate.

In the mountains alone the Pacific coast of northwestern America in Canada and USA. In moist grassland, bushes and pastures. Altitude: 1000-2300 m. Flowering & fruiting: July-August.

A–C. 美国犹他 摄影 /Photo: Adrien Favre
Fishlake Utah, USA. 06.09.2019

闭花瓶龙胆

Closed Bottle Gentian

Gentiana andrewsii

多年生草本，高 30~60 cm；茎单一或基部分枝成丛生，直立或斜升；叶无柄，卵形到卵状披针形；花数枚，簇生顶端和顶部少数叶腋；萼筒管状，萼齿钻形，大小不等；花冠管状钟形，常闭合状，淡蓝色到蓝紫色，花冠裂片卵形，褶小，不规则分裂，齿状。

美国东北部大部分地区和相邻的加拿大东部地区较常见。生长在潮湿向阳的草地、林缘与牧场边缘。海拔 800~2000 m。花果期 8~10 月。

Perennial, 30-60 cm high; stems simple or clump branched at base, erect or decumbent; leaves sessile, blade ovate to ovate-lanceolate; flowers numerous, clustered at terminal and in few axils of top parts of stems; calyx tube small, tubular, teeth subulate; corolla campanulate, mostly closed, pale blue to blue-purple, corolla lobes ovate, plicae small, irregular clefted, denticulate. In the mountains of northeastern USA and adjacent eastern Canada. In moist and sunny grassland, forest margins and pastures. Altitude: 800-2000 m. Flowering & fruiting: August-October.

A

B

A–B. 美国密苏里 摄影 /Photo: Adrien Favre
St Louis Missouri, USA. 26.09.2019

马利筋龙胆

Gentiana asclepiadea

多年生草本，高 20~60 cm；茎单一，纤细，直立或外弯斜升；叶无柄，披针形到卵形，渐尖，叶脉明显；萼筒管状，萼齿钻形，极短；花冠钟状，长 3~5 cm，蓝色，内面有紫色斑点。

主要分布于欧洲中部到东南部的山区。生长在潮湿阴凉溪边，林缘与牧场边缘。海拔 300~2000 m。花果期 7~10 月。

Perennial, 20-60 cm high; stems simple, slender, erect or excurved ascending; leaves sessile, blade lanceolate to ovate, acuminate, veins strong; calyx tube obconic, teeth small, subulate; corolla campanulate, 3-5 cm, blue, with purple spots inside.

In the mountains from central to southeastern Europe. In damp and shady stream sides, forest margins, and pastures. Altitude: 300-2000 m. Flowering & fruiting: July-October.

A-B. 意大利波得龙省切莫莱斯

Cimolais, Friuli-Venezia Giulia, Pordenone, Italy. 07.07.1996

C. 瑞士瓦莱州埃莫森 摄影 /Photo: Adrien Favre
Emosson Valais, Switzerland. 23.08.2005

D. 德国法兰克福植物园
Frankfurt Botanical Garden, Germany. 23.08.2005

E. 瑞士瑞吉山 Rigi Mt., Switzerland. 20.09.2004

蓝岭龙胆

Blue Ridge Gentian

Gentiana austromontana

多年生草本，高 30~50 cm；茎单一或基部分枝成丛生，直立或斜升，密被短绒毛；叶无柄，长圆状披针形；花数枚，簇生顶端；萼筒管状，萼齿倒卵形，开展；花冠管状钟形，常闭合状，淡蓝色到紫红色，花冠裂片卵形，褶三角形，齿裂。

仅见于美国东南部蓝岭山脉的西弗吉尼亚、弗吉尼亚、北卡罗来纳和田纳西等州。生长在森林与草地的空地及边缘。海拔 600~2000 m。花果期 9~10 月。

Perennial, 30-50 cm high; stems simple or clump branched at base, erect or decumbent, densely puberulent; leaves sessile, blade oblong-lanceolate; flowers clustered at top of stems; calyx tube small, tubular, teeth obovate, extending; corolla campanulate, mostly closed, pale blue to reddish purple, corolla lobes ovate, plicae small, triangular, denticulate.

In the Blue Ridge Mountains in West Virginia, Virginia, North Carolina, and Tennessee of southeastern USA. In mountain forests and grassy windows and margins. Altitude: 600-2000 m. Flowering & fruiting: September-October.

A-B. 美国弗吉尼亚州大马士革 摄影 /Photo: Adrien Favre
DamascusVirginia, USA. 04.10.2019.

松林龙胆

Pine Barren Gentian

Gentiana autumnalis

多年生草本，高 20~35 cm；茎纤细单一，直立或斜升；叶无柄，线状倒披针形到线形；花单生枝顶；萼筒管状，萼齿细长，针状，开展；花冠阔漏斗形，淡蓝色到蓝紫色，喉部有绿色条纹，花冠裂片卵形，褶细小三角形，2 齿状裂。

仅见于美国东南部山区的特拉华、弗吉尼亚、北卡罗莱纳、南卡罗莱纳、佐治亚、马里兰和新泽西等州。生长在沿海平原及山地松林中。海拔 100~1000 m。花果期 9~12 月。

Perennial, 20-35 cm high; stems slender, simple, erect or decumbent, densely puberulent; leaves sessile, blade linear-oblanceolate to linear; flowers solitary at top of stems; calyx tube small, tubular, teeth linear, extending; corolla broad funnelform, pale blue to blue purple with greenish stripes in throat, corolla lobes ovate, plicae small, triangular, 2-cleft, denticulate.

In the mountains of southeastern USA, including the states of Delaware, Virginia, North Carolina, South Carolina, Georgia, Maryland and New Jersey. In longleaf pine forest forests. Altitude: 100-1000 m. Flowering & fruiting: September-December.

A-C. 美国新泽西 摄影 /Photo: Adrien Favre
Franklin Parker Preserve, New Jersey, USA. 08.10.2019

C

山沼龙胆

Mountain Bog Gentian

Gentiana calycosa

多年生草本，高 10~45 cm；茎数枚丛生，纤细单一，直立或斜升；叶无柄，卵状椭圆形，与节间近等长；花单生（偶尔 2~3）枝顶；萼筒倒锥形，萼齿钻形，大小不等；花冠阔漏斗形，淡蓝色到蓝紫色，有绿色斑点，花冠裂片卵形，褶细小三角形，2 裂，齿状。

主要分布于美国西北部与加拿大西部的内华达山脉及喀斯喀特山脉。生长在高山和亚高山潮湿及沼泽化草甸及溪流边。海拔 1500~2900 m。花果期 7~9 月。

Perennial, 10-45 cm high; stems several branched from caudex, slender, simple, erect or decumbent; leaves sessile, blade ovate-elliptic, as long as internode; flowers solitary (occasionally 2-3) at top of stems; calyx tube obconic, teeth subulate, unequal; corolla broad funnelform, pale blue to blue purple with greenish spots, corolla lobes ovate, plicae triangular, 2-cleft, denticulate.
Mainly in the Sierra Nevada and Cascade mountains of northwestern USA and western Canada. In moist and wet meadows and stream sides in alpine and subalpine zones. Altitude: 1500-2900 m. Flowering & fruiting: July-September.

A

B

A–C. 美国犹他尤盈塔山 摄影 /Photo: Adrien Favre
Uintah Mts, Utah, USA. 09.09.2019

瓶龙胆

Bottle Gentian

Gentiana clausa

多年生草本，高 30~60 cm；茎单一或基部分枝成丛生，直立或斜升；叶无柄，卵形到卵状披针形；花数枚，簇生顶端；萼筒管状，萼齿倒卵形到圆形，大小不等，开展；花冠管状钟形，常闭合状，淡蓝色到蓝紫色，花冠裂片卵形，褶小，不规则分裂，齿状。

分布于以阿帕拉契亚山区为主的美国东北部和相邻的加拿大东部地区。生长在潮湿的草地和林下。海拔 1000~3000 m。花果期 8~10 月。

Perennial, 30-60 cm high; stems simple or clump branched at base, erect or decumbent; Leaves sessile, blade ovate to ovate-lanceolate; flowers clustered at terminal; calyx tube obconic, calyx teeth obovate to orbicular, unequal, spreading; corolla campanulate, mostly closed, pale blue to blue-purple, corolla lobes minute ovate, plicae small, irregular clefted, denticulate.

Mostly in the Appalachians of northeastern USA and adjacent eastern Canada. In moist meadows and forest. Altitude: 1000-3000 m. Flowering & fruiting: August-October.

A–B. 美国弗吉尼亚杰斐逊国家森林
摄影 /Photo: Adrien Favre
Jefferson National Forest, Virginia, USA. 05.10.2019

美花龙胆

Showy Gentian

Gentiana decora

多年生草本，高 30~60 cm；茎单一或基部分枝成丛生，直立或斜升，密被细绒毛；叶无柄，卵状倒披针形；花数枚，簇生顶端；萼筒管状，密被细绒毛，萼齿线状钻形；花冠管状钟形，淡蓝色到蓝紫色，有深色条纹，花冠裂片卵形，褶小，2 裂。

特产于美国阿帕拉契亚山脉南部地区。生长在潮湿的草地和林下。海拔 800~2000 m。花果期 9~11 月。

Perennial, 30-60 cm high; stems simple or clump branched at base, erect or decumbent, densely puberulent; leaves sessile, blade ovate-oblanceolate; flowers clustered at terminal; calyx tube obconic, densely puberulent, calyx teeth linear-subulate; corolla campanulate, pale blue to blue-purple with darker stripes, corolla lobes minute ovate, plicae small, 2-cleft.

Mostly in the southern Appalachians of northeastern USA. In moist meadows and forest. Altitude: 800-2000 m. Flowering & fruiting: September-November.

A-B. 美国佐治亚州查茨沃斯 摄影 /Photo: Adrien Favre
Chatsworth, Georgia, USA. 30.09.2019

C. 美国佐治亚州 摄影 /Photo: Adrien Favre; Andrews, Georgia, USA. 01.10.2019

香脂山龙胆

Balsam Mountain Gentian

Gentiana latidens

与瓶龙胆（*Gentiana clausa*）非常接近，主要区别在于本种的花更大，闭合时更疏松，花萼裂片和花冠裂片更大、形态更多变。

特产于美国北卡罗莱纳州西部香脂山及邻近高山地区。生长在潮湿渗水开发向阳石质山坡。海拔 1300~2000 m。花果期 9~10 月。

Highly resemble the Bottle Gentian *Gentiana clausa*; major differences are larger flowers, corolla less tightly closed, larger corolla lobes, larger and variable calyx lobes.

Endemic to Balsam Mountains and nearby ranges in western North Carolina in USA. In moist, often seeping, more or less open sites on rocky slopes. Altitude: 1300-2000 m. Flowering & fruiting: September-October.

A–B. 美国北卡罗莱纳州蓝岭山观景公路
摄影 /Photo: Adrien Favre
Blue Ridge Parkway, North Carolina, USA. 03.10.2019

条叶龙胆

Gentiana manshurica

多年生草本，高 20~40 cm；茎单一，纤细，直立或斜升；叶无柄，叶片线形到线状披针形，长 3~8 cm，宽不及 1 cm；花无柄或有短柄，单生或偶尔 2 花生顶端及叶腋；萼筒管状，萼齿钻形；花冠管状钟形，长 4~5 cm，蓝色，内面长有绿色斑点。

主要分布于中国南部、东部、东北及朝鲜。生长在潮湿阴凉溪边，草地、林缘与牧场边缘，分布区严重片断化。海拔 100~1000 m。花果期 8~10 月。

Perennial, 20-40 cm high; stems simple, slender, erect or ascending; leaves sessile, blade linear to linear-lanceolate, 3-8 cm length, width less than 1cm; flowers sessile or with short pedicels, solitary or 2 in cluster, terminal and axillary; calyx tube obconic, teeth subulate; corolla tubular-campanulate, 4-5 cm, blue, with green spots inside.

In southern, eastern, and northeastern China, and Korea. In damp and shady stream sides, forest margins, and pastures, distribution fragmented severely. Altitude: 100-1000 m. Flowering & fruiting: July-October.

A

B

A-D. 中国安徽省明光市
Mingguang, Anhui, China. 10.10.2022

E. 狭叶龙胆（*Gentiana manshurica*）白花个体
中国安徽省明光市
White flowered, Mingguang, Anhui, China. 10.10.2022

牛百丽龙胆（原变种）

Newberry's Gentian

Gentiana newberryi var. *newberryi*

多年生小草本，高 10~15 cm；茎单生或少数丛生，纤细单一，斜升或铺地；叶无柄，披针形到匙形，弯曲；花单生或偶尔数枚簇生枝顶；萼筒倒锥形，萼齿匙形；花冠阔漏斗形，蓝色，有绿色斑点，花冠裂片卵形，褶细小，三角形，不规则分裂，齿状。

主要分布于美国西部加利福尼亚州和俄勒冈州。生长在潮湿高山草甸。海拔 1200~3500 m。花果期 7~9 月。

Perennial, 10-15 cm high; stems single or several branched from caudex, slender, simple, decumbent or prostrate; leaves sessile, blade lanceolate to spathulate, curved; flowers solitary or occasionally a few clustered at top of stems; calyx tube obconic, teeth spathulate; corolla broad funnelform, blue with greenish spots, corolla lobes ovate, plicae small, triangular, irregularly clefted, denticulate.

Mainly in California and Oregon states of USA. In wet meadows. Altitude: 1200-3500 m. Flowering & fruiting: July-September.

A

B

A–C. 美国加利福尼亚州 Gumboot 湖
摄影 /Photo: Adrien Favre
Gumboot Lake, California, USA. 15.08.2019

C

泰奥加龙胆（变种）

Tioga Gentian

Gentiana newberryi var. *tiogana*

本变种与原变种形态接近，区别在于原变种的花蓝色，而本变种的花为黄白色到很淡的蓝色，花冠裂片背面及其下部为紫褐色。

特产于美国西部加利福尼亚州和内华达州。生长在潮湿高山草甸。海拔最高 3000 m。花果期 7~9 月。

This variety is very similar to the original variety var. *newberryi*. Difference is the color of corolla: var. *newberryi* with blue corolla, whereas var. *tiogana* with yellowish white corolla which is abaxially dark brown-purple on, below lobes. Endemic to California and Nevada states of USA. In wet meadows. Altitude ~3000 m. Flowering & fruiting: July-September.

A–B. 美国加利福尼亚州 摄影 /Photo: Adrien Favre
Mineral to Chester, California, USA. 16.08.2019

帕丽山地龙胆

Parry's Mountain Gentian

Gentiana parryi

多年生草本，高 15~40 cm；茎数枚丛生，纤细单一，直立或斜升；叶无柄，卵形，排列较密集，较节间稍长；花簇生枝顶；萼筒倒锥形，萼齿钻形；花冠管状钟形，蓝紫色，有绿色斑点，花冠裂片卵形，褶三角形，2 裂。

本种与山沼龙胆相似，主要区别在于叶片和花萼裂片都明显更小。主要分布于美国落基山脉南部地区的怀俄明、科罗拉多、犹他和新墨西哥州。生长在潮湿高山草甸。海拔 1900~2500 m。花果期 7~9 月。

Perennial, 15-40 cm high; stems several branched from caudex, simple, erect or decumbent; leaves sessile, blade ovate, densely spaced, slightly longer than internode; flowers a few clustered at top of stems; calyx tube obconic, teeth subulate; corolla broad funnelform, blue purple with greenish spots, corolla lobes ovate, plicae triangular, 2-cleft.

Highly resemble Mountain Bog Gentian, leaves and calyx lobes smaller. Mainly in the states in the southern Rocky Mountains in USA. In moist alpine meadows. Altitude: 1900-2500 m. Flowering & fruiting: July-September.

A-B. 美国科罗拉多州阿拉帕霍国家森林公园
摄影 /Photo: Adrien Favre
Arapaho National Forest, Colorado, USA. 18.09.2019.

C–D. 美国科罗拉多州乌雷 摄影 /Photo: Adrien Favre; Ouray, Colorado, USA. 16.09.2019

E. 美国科罗拉多州梅萨 摄影 /Photo: Adrien Favre; Mesa, Colorado, USA. 13.09.2019

克拉玛斯龙胆

Klamath Gentian

Gentiana plurisetosa

多年生草本，高 10~40 cm；茎 2 到数枚丛生，直立或斜升；叶无柄，椭圆形到近圆形，先端圆钝；花数枚，簇生顶部叶腋；萼筒管状，萼齿披针状椭圆形，叶状；花冠管状，蓝色，花冠裂片卵形，褶三角形，不规则分裂，流苏状。

特产于美国西部加利福尼亚州和俄勒冈州山区。生长在潮湿高山草甸。海拔 1200~1900 m。花果期 7~9 月。

Perennial, 10-40 cm high; stems 2-several, erect or decumbent; leaves sessile, blade elliptic to suborbicular, apex obtuse; flowers several clustered at terminal and in axillaries of upper parts of stems; calyx tube tubular, teeth lanceolate-elliptic, leaf-like; corolla tubular, blue, corolla lobes ovate, plicae triangular, irregular divided, fimbriate.

Endemic to the mountains in California and Oregon states of USA. In wet mountain meadows. Altitude: 1200-1900 m. Flowering & fruiting: July-September.

A–C. 美国加利福尼亚州六河国家森林
摄影 /Photo: Adrien Favre
Six Rivers National Forest, California, USA. 10.08.2019

C

C

肺花龙胆

Gentiana pneumonanthe

多年生草本，高 10~50 cm；茎单一，纤细，直立或斜升；叶无柄，叶片线形到线状披针形，长 1~5 cm，宽 2~4 mm；花有短柄，单生或偶尔 2~3 花生顶端及上部叶腋；萼筒管状，萼齿线形；花冠漏斗状，长 4~5 cm，蓝色到蓝紫色，内面长有绿色斑点。

广泛分散分布于欧洲各地。生长在潮湿阴凉溪边，草地与牧场边缘，分布区严重片段化。海拔 100~2000 m。花果期 7~10 月。

Perennial, 10-50 cm high; stems simple, slender, erect or ascending; leaves sessile, blade linear to linear-lanceolate,1-5 cm length, width 2-4 mm; flowers with short pedicels, solitary or 2-3 in cluster, terminal and axillary on upper part of stems; calyx tube tubular, teeth linear; corolla funnelform, 4-5 cm, blue to blue-purple, with green spots inside.

Widely scattered in Europe. In damp and shady stream sides, grassland and pastures, distribution fragmented severely. Altitude: 100-2000 m. Flowering & fruiting: July-October.

A

B

A–B. 瑞士苏黎世植物园
Zurich Botanical Garden, Switzerland. 16.08.2005

C. 意大利波得龙省切莫莱斯 摄影 /Photo: Philippe Kuepfer Cimolais, Friuli Venezia Giulia, Pordenone, Italy. 07.07.1996

C

草原（柔毛）龙胆

Prairie (Downy) Gentian

Gentiana puberulenta

多年生草本，高 30~50 cm；茎数枚丛生，直立或斜升；叶无柄，长圆状披针形；花数枚，簇生顶端和茎上部叶腋；萼筒管状，萼齿钻形，开展；花冠管状，蓝色，花冠裂片卵形，褶小，不规则分裂，齿状。

特产于美国东南部印第安纳、肯塔基、马里兰、密西根、俄亥俄及田纳西等州的山区。生长在草地、灌丛与牧场边缘。海拔 1000~2300 m。花果期 8~11 月。

Perennial, 30-50 cm high; stems several clumps, erect or decumbent; leaves sessile, blade oblong-lanceolate; flowers numerous, clustered at terminal and in axillaries of upper parts of stems; calyx tube tubular, teeth subulate, spreading; corolla tubular, blue, corolla lobes ovate, plicae small, irregular clefted, denticulate.

Endemic to the mountains in the Indiana, Kentucky, Maryland, Michigan, Ohio, and Tennessee states of USA. In moist grassland, bushes and pastures. Altitude: 1000-2300 m. Flowering & fruiting: August-November.

A–B. 美国密苏里州埃尔多拉多 - 斯普林斯 摄影 /Photo: Adrien Favre
El Dorado Springs, Missouri, USA. 25.09.2019

C. 美国密苏里州 摄影 /Photo: Adrien Favre
Camdenton, Missouri, USA. 25.09.2019

C

肥皂草龙胆

Soapwort Gentian

Gentiana saponaria

多年生草本，高 30~60 cm；茎单一或基部分枝成丛生，直立或斜升；叶无柄，线状椭圆形；花数枚簇生顶端；萼筒管状，萼齿线状匙形；花冠管状钟形，淡蓝色到蓝紫色，有深色条纹，花冠裂片卵形，褶小，2 裂。

散布于北美东部的五大湖以南地区，从威斯康星到纽约州，向南到得克萨斯和佛罗里达。生长在沼泽等潮湿的草地和林下湿地。海拔 800~2000 m。花果期 8~11 月。

Perennial, 30-60 cm high; stems simple or clump branched at base, erect or decumbent; leaves sessile, blade linear-elliptic; flowers several clustered at terminal; calyx tube obconic, calyx teeth linear-spathulate; corolla campanulate, pale blue to blue-purple with darker stripes, corolla lobes minute ovate, plicae small, 2-cleft.

Widely scattered in eastern North America south of the Great Lakes, from Wisconsin to New York, and south to Texas and Florida. In bogs, marshes, wet hardwood forests, other moist to wet habitats. Altitude: 800-2000 m. Flowering & fruiting: August-November.

A–B. 美国田纳西 摄影 /Photo: Adrien Favre
Moore, Tennessee, USA. 29.09.2019

海洋之心龙胆

Gentiana scabra 'Blue Heart'

多年生草本，高 20~40 cm；茎从基部多分枝，纤细，直立或斜升；叶无柄，披针形；花单生枝顶，萼筒管状；花冠管状钟形，长 3~5 cm，蓝色，内面长有绿色斑点。

人工培育的较为耐热的园艺品种，生长旺盛，开花整齐，深受园艺爱好者喜爱，世界各地广泛栽培。近年来被引进中国，是中国新晋网红植物。花期 7~12 月。

Perennial, 20-40 cm high; stems branching vigorously from base, slender, erect or ascending; leaves sessile, blade lanceolate; flowers solitary terminal; calyx tube tubular, teeth small, subulate; corolla tubular- campanulate, 3-5 cm, blue, with greenish spots inside.

Artificially bred cultivar, tolerate to heat, vigorous, blooming uniformly, well welcome by growers and widely cultivated worldwide. It is becoming very popular in China in recent years. Flowering: July-December.

A–B. 中国上海 Shanghai, China. 24.10. 2022

岩间钻石龙胆 *Gentiana scabra* 'Rocky Diamond Blue Heart'
瑞士日内瓦超市花卉 Flower Shop, Geneva, Switzerland. 10.07.2023

权杖龙胆

King's Scepter Gentian

Gentiana sceptrum

多年生草本，高 10~80 cm；茎 1~ 数枚，单生，直立或斜升；叶无柄，椭圆形到近卵形，先端圆钝；花 1~ 数枚，单生或簇生顶部叶腋；萼筒管状，萼齿卵状椭圆形，叶状；花冠管状钟形，蓝色，外面有深色斑点，花冠裂片卵状长圆形，褶几近无。

特产于北美西海岸加利福尼亚至英属哥伦比亚间低山区。生长在沼泽和湿草甸。海拔 80~1200 m。花果期 7~9 月。

Perennial, 10-80 cm high; stems 1-several, simple, erect or decumbent; leaves sessile, blade elliptic to subovate, apex obtuse; flowers 1-several clustered at terminal and upper axillaries; calyx tube tubular, teeth ovate-elliptic, leaf-like; corolla tubular-campanulate, blue, corolla lobes ovate-oblong, plicae nearly not present.

Endemic to the mountains in California and Oregon states of USA. In wet mountain meadows. Altitude: 80-1200 m. Flowering & fruiting: July-September.

A–B. 美国加利福尼亚州曼彻斯特 摄影 /Photo: Adrien Favre
Manchester, California, USA. 06.08.2019

七瓣龙胆

Gentiana septemfida

多年生草本，高 10~30 cm；茎单一，直立或斜升；叶无柄，节间短，叶排列密集，卵形到卵状披针形；花密集簇生枝顶端及花枝上部；萼筒管状，萼齿钻形；花冠管状钟形，长 3~4 cm，蓝色，具蓝紫色条纹；褶流苏状。

分布于中亚及高加索一带，欧洲植物园常有栽培。生长在潮湿阴凉溪边，草地与牧场边缘。海拔 1000~3000 m。花果期 7~10 月。

Perennial, 10-30 cm high; stems simple, slender, erect or ascending; leaves sessile, internode short so leaves crowded, blade ovate to ovate-lanceolate; flowers crowded at terminal and upper part of stems; calyx tubular; corolla tubular-campanulate, 3-4 cm, blue, with blue-purple spots stripes.

In central Asia and Caucasus regions; often cultivated in botanical gardens in Europe. In damp and shady stream sides, grassland, and pastures. Altitude: 1000-3000 m. Flowering & fruiting: July-October.

A

B

A–C. 瑞士莱森植物园

Leysin Botanical Garden, Switzerland. 11.08.2005

门多西诺龙胆

Mendocino Gentian

Gentiana setigera

多年生草本，高 20~45 cm；茎单生或少数丛生，斜升；叶无柄，椭圆形到卵状椭圆形，先端圆钝；花单生或偶尔 2~3 簇生枝顶；萼筒倒锥形，萼齿卵状长圆形；花冠阔漏斗形，蓝色，有绿色斑点，花冠裂片椭圆状倒卵形，褶小，三角形，不规则 2~3 分裂。

特产于美国西部加利福尼亚州北部（门多西诺地区）和相邻的俄勒冈州西南部。生长在潮湿山地草甸。海拔 1100~1800 m。花果期 7~9 月。

Perennial, 20-45 cm high; stems single or several branched from caudex, decumbent; leaves sessile, blade elliptic to ovate-elliptic; flowers solitary or occasionally 2-3 clustered at top of stems; calyx tube obconic, teeth ovate-oblong; corolla broad funnelform, blue with greenish spots, corolla lobes elliptic-obovate, plicae small, triangular, irregularly 2-3 clefted.

Endemic to northern California (Mendocino county) and adjacent southwestern Oregon in USA. In wet meadows. Altitude: 1100-1800 m. Flowering & fruiting: July-September.

B

A-C. 美国加利福尼亚州门多西诺市
摄影 /Photo: Adrien Favre
Mendocino, California, USA. 08.08.2019

三花龙胆

Gentiana triflora

多年生草本，高 30~80 cm；茎单 1 至数枚，纤细，直立或斜升；叶无柄，叶片线形到线状披针形，长 2~10 cm，宽 2~8 mm；花无柄，单生或偶尔 2~3 花簇生顶端及上部叶腋；萼筒管状，萼齿钻形；花冠管状钟形，长 3~5 cm，蓝色到蓝紫色，偶有白色变异。

分散分布于东北亚的中国东北，俄罗斯远东，朝鲜半岛及日本。生长在潮湿阴凉溪边，草地与林缘和林间空地。海拔 600~1000 m。花果期 8~10 月。

Perennial, 30-80 cm high; stems simple to several, slender, erect or ascending; leaves sessile, blade linear to linear-lanceolate, 2-10 cm long, 2-8 mm wide; flowers sessile, solitary or 2-3 in cluster, terminal and axillary on upper part of stems; calyx tube tubular, teeth subulate; corolla tubular, 3-5 cm, blue to blue-purple, occasionally white.

Scattered in northeastern Asia, including northeastern China, Russia Far East, Korea Peninsula and Japan. In damp and shady stream sides, grassland, forest margins and cleared open places. Altitude: 600-1000 m. Flowering & fruiting: August-October.

A

B

A-C. 中国吉林省长白山
Changbaishan Mt., Jilin, China. 19.08.2000

D. 白花变异；中国吉林省长白山
White flowered mutant; Changbaishan Mt., Jilin, China. 19.08.2000

C

D

中国云南省白马雪山 Baimaxueshan Mt., Yunnan, China. 19.09.2023

四数组

sect. *Tetramerae*

一年生草本；花四基数；花萼裂片三角形，中脉呈龙骨状隆起并向萼筒延伸成翅。本组 2 种，特产自中国云南和四川。

Annual; flowers 4-merous; calyx lobes triangular, mid-veins prominent keeled and decurrent on the tube as wings. 2 species, endemic to Yunnan and Sichuan provinces of China.

四数龙胆

Gentiana lineolata

一年生草本，茎直立，高5~10 cm，多分枝，铺散斜升，全株密被紫色乳突；叶绿色或紫色，叶片卵状三角形到披针形；花四数，无梗，单生枝端；花冠蓝色到紫色，花冠筒高脚杯状，长2~3 cm，褶长约为花冠裂片的一半。

仅见于中国云南西北部和四川西南部山区。生长于较干燥向阳的高山裸露山坡。海拔2000~3000 m。花果期8~12月。

Annual, stems erect, 5-10 cm, much branched, branches lax, spreading, densely purple papillate; leaves green or purple, blades ovate-triangular to lanceolate; flowers 4-merous, sessile, solitary and terminal; corolla blue to dark purple, corolla salverform, 2-3 cm long, plicae about ½ as long as corolla lobes.

In the mountains of northwestern Yunnan and southwestern Sichuan provinces of China. In dry and sunny open slopes. Altitude: 2000-3000 m. Flowering and fruiting: August-December.

A

B

A. C. D. 中国云南省丽江市
Lijiang, Yunnan, China. 09.10.1999

B. 中国云南省丽江市 摄影 /Photo: Adrien Favre
Lijiang, Yunnan, China. 14.10.2004

C

D

E. 中国云南省大理白族自治州 Dali, Yunnan, China. 29.11.2007

肴萼龙胆

Gentiana praeclara

一年生草本，茎直立，高 5~10 cm，多分枝，铺散斜升，全株密被紫色乳突；叶绿色或紫色，叶片卵状三角形到披针形；花四数，无梗，单上生枝端；花冠蓝色到紫红色，花冠筒高脚杯状，长 4~5 cm，褶长约为花冠裂片的四分之一。

仅见于中国云南西北部和四川西南部山区。生长于较干燥向阳的高山裸露山坡。海拔 2000~3500 m。花果期 8~11 月。

Annual, stems erect, 5-10 cm, much branched, branches lax, spreading, densely purple papillate; leaves green or purple, blades ovate-triangular to lanceolate; flowers 4-merous, sessile, solitary and terminal; corolla blue to dark purple, corolla salverform, 4-5 cm long, plicae about ¼ as long as corolla lobes.

In the mountains of northwestern Yunnan and southwestern Sichuan provinces of China. In dry and sunny open slopes. Altitude: 2000-3500 m. Flowering and fruiting: August-November.

A. 中国云南泸沽湖 摄影 /Photo: Adrien Favre
Luguhu, Yunnan, China.

B-D. 中国云南省宁蒗彝族自治县
Ninglang, Yunnan, China. 17.09.2023

C

D

中国云南省松赞林寺 Songzanlin Monastory, Yunnan, China. 11.07.2014

4.1.3 耳褶属 *Kuepferia*

中国西藏自治区多雄拉山，耳褶属植物的典型生境

Typical habitat of *Kuepferia* plants, Duoxiongla Mountain, Xizang, China. 24.09.2003

多年生草本；根稍肉质；单轴分枝；花冠的褶减小退化成花冠裂片一侧的一个小齿或耳状结构，较花冠裂片明显小许多。

本属 12 种，分布于喜马拉雅山东部地区的尼泊尔、不丹、印度东部及中国西南部的西藏、云南和四川等省区。

属名 *Kuepferia* 是为了致敬瑞士植物学家 Philippe Kuepfer，以纪念他长期对龙胆科的研究。

Perennial; root slightly fleshy; stem branching monopodially. Plicae of corolla reduced to a small tooth or auricle on side of a lobe, much smaller than the lobe.

12 species, spread alone the eastern Himalayas in Nepal, Bhutan, northeastern India, and the Xizang, Yunnan, and Sichuan provinces of southwestern China.

The generic name *Kuepferia* was created in honor of the Swiss botanist Philippe Kuepfer, to acknowledge his continued researches on gentians.

瑞士植物学家 Philippe Kuefper 教授（左）和他的老朋友——意大利园艺学家 Egidio Anchisi 在中国青海省久治县考察龙胆科等高山植物

Swiss botanist Prof. Philippe Kuepfer (left) and his friend Italian horticulturist Egidio Anchisi exploring alpine plants in Jiuzhi, Qinghai, China. 07.09.1994.

深裂耳褶

Kuepferia damyonensis

多年生草本，高 5~10 cm；主根肉质，较粗壮；花枝丛生，斜升；莲座叶狭披针形到狭倒披针形；茎生叶卵状披针形；花无梗，多单生花枝顶端；花萼筒杯状，花萼裂片钻形，不等大；花冠筒浅钟形，较裂片短许多，黄绿色，具深绿色条纹和斑点，花冠裂片卵状椭圆形，先端圆钝；褶小，狭三角形到披针形。

特产于中国的西藏东南部和云南西北部及缅甸东北部。生长在高山草地与杜鹃林下。海拔 3700~5000 m。花果期 8~10 月。

Perennial; 5-10 cm high; taproot robust, fleshy; flowering stems multiple, ascending; rosette leaf blades narrowly lanceolate to narrowly oblanceolate, stem leaves ovate-lanceolate; flowers sessile, often solitary at terminal; calyx tube cupular, calyx lobes subulate, unequal; corolla tube shortly campanulate, much shorter than lobes, greenish yellow with dark green stripes and spots; corolla lobes ovate-elliptic, apex obtuse; plicae small, narrowly triangular to lanceolate.

Endemic to southwestern China in southeastern Xizang and northwestern Yunnan, and northeastern Myanmar. In alpine grassland and azalea scrub. Altitude: 3700-5000 m. Flowering & fruiting: August-October.

A-B. 中国西藏自治区察隅县 摄影 /Photo：付鹏程
Chayu, Xizang, China. 29.08.2018

美丽耳褶

Kuepferia decorata

多年生草本，高 4~6 cm；主根肉质，圆柱状；茎平卧或斜升，基部木质化；茎生叶较密集，稍肉质，卵形到匙形，先端圆钝；花单生花枝顶端，5 基数，偶尔 4 或 6 基数；花萼筒杯状，花萼裂片匙形；花冠筒杯状，淡蓝色到蓝紫色，有深色条纹，花冠裂片卵形；褶小，三角形。

分布于喜马拉雅山东部地区的中国西藏东南部和云南西北部及邻近的缅甸。生长在高山草甸与灌丛边缘。海拔 3500~4500 m。花果期 8~10 月。

Perennial; 4-6 cm high; taproot cylindric, fleshy; stems prostrate to ascending, woody at base; stem leaves crowded, slightly fleshy, blades ovate to spathulate, apex obtuse; flowers solitary and terminal, 5-merus, occasionally 4- or 6-merus; calyx tube cupular, calyx lobes spathulate; corolla tube cupular, pale blue to dark blue purple, with darker stripes; corolla lobes ovate; plicae small, triangular.

Scattered in the eastern Himalayas, in southeastern Xizang and northwestern Yunnan provinces of China adjacent Myanmar. On alpine meadows and bush margins. Altitude: 3500-4500 m. Flowering and fruiting: August-October.

A-B. 中国西藏自治区多雄拉山口

Duoxiongla Pass, Xizang, China. 24.09.2003

C

D

C–E. 中国西藏色季拉山 Sejilashan Mountain, Xizang, China. 17.08.2006
F–G. 中国西藏多雄拉山口 Duoxiongla Pass, Xizang, China. 24.09.2003

多雄拉耳褶

Kuepferia doxiongshangensis

多年生草本，高 5~10 cm；主根肉质，圆柱状；茎斜升；花期有明显莲座叶丛，莲座叶片倒披针形到狭椭圆形；茎生叶疏松排列，较节间短，叶片椭圆形，先端钝；花单生花枝顶端，无柄，生于最上部一对叶片中；花萼筒紫褐色，杯状，花萼裂片钻形；花冠筒钟状，蓝紫色，常闭合状，花冠裂片卵形，先端钝；褶小，耳状。

特产于喜马拉雅山东部地区的中国西藏南部和东南部、邻近的不丹及印度东北部高山地区。生长在高山草甸。海拔 3800~4200 m。花果期 8~10 月。

Perennial; 5-10 cm high; taproot cylindric, fleshy; stems ascending; rosette leaves present at anthesis, blade oblanceolate to narrowly elliptic; stem leaves sparsely spaced, shorter than internode, blades elliptic, apex obtuse; flowers solitary and terminal, surrounded by the top most pair of leaves; calyx tube cupular, dark purple, calyx lobes subulate; corolla tube campanulate, dark blue purple, often closed; corolla lobes ovate, apex obtuse; plicae small, auricular.
Endemic to the high mountains in the eastern Himalayas in south & southeastern Xizang adjacent Bhutan and northeastern India. On alpine meadows. Altitude: 3800-4200 m. Flowering and fruiting: August-October.

A

B

A–G. 中国西藏多雄拉山口
Duoxiongla Pass, Xizang, China. 24.09.2003

C

D

E

F

G

中国西藏自治区多雄拉山山口景观：曾经最美、最原始的徒步山路——派墨路，
2023 年通车以前是连接米林派区和墨脱的重要通道。

The famous hiking trial at the pass of Mt. Duoxiongla, Xizang, China. 24.09.2003.

耳褶

Kuepferia otophora

多年生草本，高 5~15 cm；主根肉质，圆柱状；营养枝的叶密集成发达的莲座叶丛，莲座叶片长圆状匙形或匙状倒卵形；茎生叶卵状椭圆形；花无梗，单生或数朵簇生花枝顶端或叶腋，成聚伞花序；花萼筒杯状，花萼裂片钻形、披针形或卵形，不等大；花冠筒管状钟形，黄色，具深绿色条纹和斑点，花冠裂片卵状椭圆形，先端钝；褶小，耳状。

特产于中国的西藏东南部和云南西北部。生长在高山草地与杜鹃林下。海拔 3500~4000 m。花果期 8~10 月。

Perennial; 5-15 cm high; taproot cylindric, fleshy; vegetative leaves crowded as well developed rosettes, leaf blades oblong-spathulate to spathulate-obovate, stem leaves ovate-elliptic; flowers sessile, several clustered at terminal and in axillary as cymes; calyx tube cupular, calyx lobes subulate, lanceolate or ovate, unequal; corolla tube tubular-campanulate, yellow with dark green stripes and spots; corolla lobes ovate-elliptic, apex obtuse; plicae small, auricular.
Endemic to southwestern China in southeastern Xizang and northwestern Yunnan. In alpine grassland and azalea scrub. Altitude: 3500-4000 m. Flowering and fruiting: August-October.

A

B

A-D. 中国云南省大理白族自治州
Dali, Yunnan, China. 02.08.2011

E. 中国云南省贡山独龙族怒族自治县
Gongshan, Yunnan, China. 03.08.2000

F. 中国云南省大理白族自治州
Dali, Yunnan, China. 23.09.2023

C
D
E
F

G. 中国云南省大理白族自治州 Dali, Yunnan, China. 23.09.2023

H. 乳黄杜鹃林下的耳褶 *Kuepferia otophora*；中国云南省大理白族自治州
Kuepferia otophora under forest of *Rhododendron lacteum*, Dali, Yunnan, China. 23.09.2023

拟耳褶

Kuepferia otophoroides

多年生草本，高 5~10 cm；主根肉质，圆柱状；营养枝的叶密集成莲座，叶片披针形到倒披针形；花枝平卧或斜升；花无梗，单生或偶尔 2 朵簇生花枝顶端；花萼筒杯状，花萼裂片披针形，多少平展；花冠筒管状，外面紫色，基部白色，花冠裂片椭圆形，白色到黄白色，有紫色斑点；褶小，三角形。

分布于中国西南部的西藏、云南和四川及邻近的缅甸。生长在高山草甸与灌丛边缘。海拔 3000~4000 m。花果期 8~10 月。

Perennial; 5-10 cm high; taproot cylindric, fleshy; vegetative leaves crowded as rosettes, leaf blades lanceolate to oblanceolate; flowering stems prostrate to ascending; flowers sessile, solitary or occasionally 2 clustered and terminal; calyx tube cupular, calyx lobes lanceolate, somewhat spreading; corolla tube tubular, outside purple, white at base ; corolla lobes elliptic, white to yellowish white, with purple spots; plicae small, triangular.

Scattered in the Xizang, Yunnan and Sichuan provinces in southwestern China and adjacent Myanmar. On alpine meadows and bush margins. Altitude: 3000-4000 m. Flowering and fruiting: August-October.

A

B

A-D. 中国云南省贡山独龙族怒族自治县
Gongshan, Yunnan, China. 03.08.2000

C

D

在中国云南省丽江市的山地与深谷，狭蕊属植物的种类最为丰富

High mountains and deep valleys in Lijiang, Yunnan, China, where highest diversity of *Metagentiana* occurs.18.09.2023

4.1.4
狭蕊属
Metagentiana

一年生或多年生草本；茎直立，基部以上多分枝；花单生分枝顶端；花萼筒具棱或翅；花冠的褶不对称，极度偏斜；雄蕊先端弯曲；花柱细长。

本属 12 种，主要集中在中国西南部的西藏、云南和四川等省区及相邻的缅甸、泰国和越南等国家。

Annual or perennial; stems erect, branching from base and above; flowers solitary on terminal of branches; calyx tubes angular or winged; Plicae of corolla asymmetrical, very oblique; stamens apically recurved; styles filiform.

12 species, highly concentrated in Xizang, Yunnan, and Sichuan provinces of southwestern China, and adjacent countries including Myanmar, Thailand and Vietnam.

高雅狭蕊

Metagentiana gentilis

一年生草本，高 10~30 cm，基部以上多分枝；茎直立或斜升，基部稍木质化，有棱角，沿棱角具刚毛；茎生叶稀疏，阔卵形到心形，沿叶脉密生刚毛；花单生花枝顶端；花萼筒倒锥形，花萼裂片钻形，中脉突起并下延成萼筒的翅，密生刚毛；花冠高脚杯状，亮蓝色到蓝紫色，有深色条纹，花冠裂片卵形；褶长圆形，边缘啮齿状。

特产中国云南省中部。生长在向阳干燥的山坡草地与灌丛边缘。海拔 2000~2500 m。花果期 10~11 月。

Annual; 10-30 cm high, much branched from base and above; stems erect to ascending, woody at base, angulate, hispid along the angles; stem leaves widely spaced, blades broad ovate to cordate, densely hispid along veins; flowers solitary and terminal; calyx tube obconic, calyx lobes subulate, midveins keeled and decurrent into wings of calyx tube, densely hispid; corolla tube salverform, bright blue to blue purple, with darker stripes; corolla lobes ovate; plicae oblong, margins erose.

Endemic to central Yunnan provinces of China. On sunny and dry grassy slopes and bush margins. Altitude: 2000-2500 m. Flowering and fruiting: October-November.

A

B

A–D. 中国云南省昆明市 摄影 /Photo: Adrien Favre
Kunming, Yunnan, China. 29.10.2004

C

D

蔓枝狭蕊

Metagentiana leptoclada

一年生草本，高 10~30 cm，基部以上多分枝；主茎直立或斜升，基部稍木质化，分枝平铺；茎生叶较密集，叶片阔卵形到心形，边缘锯齿状；花单生花枝顶端；花萼筒倒锥形，无翅，有棱，沿棱具乳突，花萼裂片钻形，中脉突起并下延到萼筒的棱；花冠筒漏斗状，蓝色到蓝紫色，花冠裂片卵形；褶卵圆形，边缘齿状。

特产中国云南省西北部。生长在向阳的山坡草地与灌丛边缘。海拔 2000~2800 m。花果期 9~11 月。

Annual; 10-30 cm high, much branched from base and above; main stem erect to ascending, woody at base, angulate, branches spreading; stem leaves moderately spaced, blades broad ovate to cordate, margins serrulate; flowers solitary and terminal; calyx tube obconic, without wing, angulate, calyx lobes subulate, midveins decurrent into angles of calyx tube; corolla tube funnelform, blue to blue purple; corolla lobes ovate; plicae ovate, margins denticulate.
Endemic to northwestern Yunnan provinces of China. On sunny grassy slopes and bush margins. Altitude: 2000-2800 m. Flowering and fruiting: September-November.

A–C. 中国云南省丽江市 Lijiang, Yunnan, China. 20.09.1990

C

狭蕊

Metagentiana primuliflora

一年生草本，高 5~20 cm，斜升或平铺，茎纤细有棱；茎生叶稀疏，叶片阔卵形到心形，边缘细锯齿状；花单生花枝顶端；花萼筒管状，无翅，有棱，花萼裂片钻形，中脉突起并下延成萼筒的棱；花冠高脚杯状，花冠筒纤细，檐部陡然扩大，亮蓝色到蓝紫色，喉部有绿色斑块，花冠裂片卵形；褶卵卵圆形，边缘啮齿状。

特产中国云南省中部到西北部及四川省西南部。生长在向阳的山坡草地与灌丛边缘。海拔 2000~3000 m。花果期 9~10 月。

Annual; 5-20 cm high, ascending to spreading, slender and angulate; stem leaves widely spaced, blades broad ovate to cordate, margins serrulate; flowers solitary and terminal; calyx tube tubular, without wing, angulate, calyx lobes subulate, midveins decurrent into angles of calyx tube; corolla salverform, corolla tube fine, abruptly expanded to limb, bright blue to blue purple; with green spots at throat, corolla lobes ovate; plicae ovate to round, margins erose.
Endemic to central to northwestern Yunnan and southwestern Sichuan provinces of China. On sunny grassy slopes and bush margins. Altitude: 2000-3000 m. Flowering and fruiting: September-October.

A-E. 中国云南省宾川县
Binchuan, Yunnan, China. 05.10.1999

C

D

E

翼萼狭蕊

Metagentiana pterocalyx

一年生草本，高 15~30 cm，直立、斜升或平铺，茎四棱，分枝较少；茎生叶稀疏，叶片阔卵形到心形，沿叶脉被柔毛，边缘细锯齿状，有缘毛；花单生花枝顶端；花萼筒倒锥形，具宽翅，花萼裂片披针形，中脉突起并下延成萼筒的翅，被柔毛；花冠筒管状，亮蓝色到蓝紫色，花冠裂片卵形；褶卵卵圆形，边缘啮齿状。

特产中国云南省中部到北部、四川省西南部及贵州西部。生长在向阳的山坡草地与灌丛边缘。海拔 1800~3000 m。花果期 9~11 月。

Annual; 15-30 cm high, erect, ascending to spreading, 4-angled, less branched; stem leaves widely spaced, blades broad ovate to cordate, pubescent along veins, margins serrulate and ciliolate; flowers solitary and terminal; calyx tube obconic, with wide wings, calyx lobes lanceolate, midveins decurrent into wings of calyx tube; corolla tubular, bright blue to blue purple; corolla lobes ovate; plicae ovate to round, margins erose. Endemic to central to northern Yunnan, southwestern Sichuan and western Guizhou provinces of China. On sunny grassy slopes and bush margins. Altitude: 1800-3000 m. Flowering and fruiting: September-November.

A

B

A. B. E. 中国云南省丽江市 Lijiang, Yunnan, China. 01.10.1998

C. 中国云南省丽江市 Lijiang, Yunnan, China. 09.10.1999

D. 中国云南省丽江市 Lijiang, Yunnan, China. 02.10.1992

C

D

E

红花狭蕊

Metagentiana rhodantha

多年生草本，有短根状茎，高 15~50 cm，直立，分枝开展；茎生叶较稀疏，叶片卵形到卵状三角形，下表面沿叶脉被柔毛，边缘细锯齿状，具缘毛；花单生花枝顶端；花萼筒倒锥形，具窄翅，花萼裂片线状披针形，中脉突起并下延成萼筒的翅，被刚毛；花冠筒管状，紫红色到粉红色，花冠裂片卵形；褶卵卵状三角形，边缘具长流苏。

特产中国中部到西南部多省较大地区。生长在向阳的山坡草地、森林与灌丛边缘。海拔 800~2000 m。花果期 9~12 月。

Perennial with short rhizome; 15-50 cm high, erect, branches spreading; stem leaves widely spaced, blades ovate to ovate-triangular, abaxially pubescent along veins, margins serrulate and ciliolate; flowers solitary and terminal; calyx tube obconic, with narrow wings, calyx lobes linear-lanceolate, midveins decurrent into wings of calyx tube, hispid; corolla tubular, reddish purple to pink; corolla lobes ovate; plicae ovate-triangular, margins with long fringes.

Endemic to a wide area from central to southwestern provinces of China. On sunny grassy slopes, forest and bush margins. Altitude: 800-2000 m. Flowering and fruiting: September-December.

A

B

A-C. 中国云南省大理白族自治州 Lijiang, Yunnan, China. 08.10.1999

D. 中国云南省昆明市 Kunming, Yunnan, China. 02.10.1999

C

D

中国四川省西部山区的高山草甸景观，孕育了众多的龙胆及其近缘属植物
Landscape of alpine meadows in western Sichuan of China, where extreme high diversity of gentians occurs. 04.07.2021

4.1.5
华龙胆属
Sinogentiana

一年生（严格讲为越年或跨年生）草本；茎直立或斜升；花黄白色到亮黄色；花冠筒较长，超过花冠裂片 4 倍以上；雄蕊先端弯曲。

本属 2 种，特产于中国横断山及其外围地区。

Annual (strictly speaking biannual); stems erect or ascending; flowers yellowish white to bright yellow; corolla tube long, more than 4 times longer than the lobes; anthers apex recurved.

2 species, endemic to the Hengduanshan Mountains and adjacent regions in China.

毛脉华龙胆

Sinogentiana souliei

一年生草本，高 10~30 cm，主枝直立，分枝平展，有棱角，沿棱角具柔毛；茎生叶稀疏，较节间短，无柄，叶片卵形到卵状披针形，沿叶脉生柔毛；花单生分枝顶端；花萼筒钟形，具翅，花萼裂片披针形，中脉突起并下延成萼筒的翅，沿翅密生柔毛；花冠筒管状，黄色，有绿色条纹，沿脉密生柔毛，花冠裂片卵形；褶圆形，边缘啮齿状。

特产中国云南省西北部和四川西部。生长在山坡草丛中与灌丛边缘。海拔 2900~3500 m。花果期 9~11 月。

Annual; 10-30 cm high, main stems erect, branches spreading, angulate, pubescent along the angles; stem leaves widely spaced, sessile, shorter than internodes, blades ovate to ovate-lanceolate, pubescent along veins; flowers solitary and terminal; calyx tube campanulate, calyx lobes lanceolate, midveins keeled and decurrent into wings of calyx tube, densely pubescent; corolla tube tubular, yellow, with green stripes, pubescent along veins; Corolla lobes ovate; plicae rounded, margins erose.
Endemic to northwestern Yunnan and western Sichuan provinces of China. On grassland scrub slopes and bush margins. Altitude: 2900-3500 m. Flowering and fruiting: September-November.

A

B

A. D. 中国四川省雅江县 Yajiang, Sichuan, China. 13.08.2002

B. 中国云南省丽江市 Lijiang, Yunnan, China. 20.09.1990

C. 中国云南省丽江市 Lijiang, Yunnan, China. 03.10.1992

C

D

华龙胆

Sinogentiana striata

一年生草本，高 5~20 cm，茎直立或斜升，有棱角；茎生叶稀疏，无柄，叶片卵状披针形，仅下表面沿中脉生柔毛；花单生分枝顶端；花萼筒管状，具翅，花萼裂片钻形，中脉突起并下延成萼筒的翅，无毛；花冠筒管状，黄白色到黄色，有绿色条纹，无毛，花冠裂片卵形，具尾尖；褶圆形，偏斜，边缘细齿状。

特产中国青海东部、宁夏南部、甘肃中部到南部及四川西北部。生长在高山山坡草丛中与灌丛边缘。海拔 2200~3500 m。花果期 8~10 月。

Annual; 5-20 cm high, stems erect or ascending, angulate; stem leaves widely spaced, sessile, blades ovate-lanceolate, pubescent only abaxially along mid veins; flowers solitary and terminal; calyx tube tubular, winged, calyx lobes subulate, midveins keeled and decurrent into wings of calyx tube, glabrous; corolla tube tubular, yellowish white to yellow, with green stripes, glabrous; corolla lobes ovate, caudate; plicae rounded, oblique, margins denticulate.

Endemic to eastern Qinghai, southern Ningxia, central to southern Gansu and northwestern Sichuan provinces of China. On grassland scrub slopes and bush margins. Altitude: 2200-3500 m. Flowering and fruiting: August-October.

A-B. 中国青海省大通回族土族自治县 Datong, Qinghai, China. 01.09.1992

C-D. 中国四川九寨沟 Jiuzhaigou, Sichuan, China. 14.09.1994

E. 中国四川省阿坝藏族羌族自治州 Aba, Sichuan, China. 11.09.2023

F. 中国甘肃省宕昌县 Tanchang, Gansu, China. 05.09.2023

C

D

E

F

中国浙江省东白山，双蝴蝶属植物的典型生境之一

One of the typical habitats of *Tripterospermum* plants, Dongbaishan Mt., Zhejiang. 18.10.2022

4.1.6
双蝴蝶属
Tripterospermum

多年生缠绕草质藤本；花萼筒 5 条维管束；花冠管状或阔管状，有褶；雄蕊 5 枚，异型不等长，先端弯曲；蜜腺盘状围绕子房基部；种子具翅，三棱形或盘状。

本属约 25 种，分布于东亚及东南亚国家，以中国最为丰富。属下分类较为混乱，种间界限也较为含混，有待深入研究。

Perennial; stems twining or prostrate and creeping; calyx tube with 5 vascular bundles; corolla tubular or broadly tubular, with plicae; stamens unequal, recurved; nectaries forming a collar-like disc around gynophore; seeds winged, triquetrous to compressed.

About 25 species, spread in eastern and southeastern Asian countries, mostly in China. Infrageneric taxonomy of the genus is not well understood. Species delimitations are quite controversial. Further taxonomic investigation badly needed.

双蝴蝶

Tripterospermum chinense

多年生草质藤本，茎缠绕，光滑；基生叶2对，略肉质，无柄，椭圆形；茎生叶近无柄，叶片卵状披针形；花单生或2~4成聚伞花序，腋生；花萼筒倒锥形，花萼裂片钻形到线形，直立到平展；花冠白色、淡蓝到淡紫色，狭钟状，花冠裂片卵状三角形，褶半圆形，边缘整齐或波状。

散布于中国东部、中部到南部的山区，见于浙江、江苏、安徽、江西、福建和广西等地。生长在山坡疏林中及边缘和灌丛中。海拔 200~1500 m。花果期 9~11 月。

Perennial, stems twisting, glabrous; basal leaves 2 pairs, fleshy, sessile, blades elliptic; stem leaves subsessile, blades ovate-lanceolate; flowers solitary or 2-4 in cyme, axillary; calyx tubes obconic, calyx lobes subulate to linear, erect to spreading; corolla white, pale blue to pale purple, narrowly campanulate, lobes ovate-triangular, plicae semiorbicular, margins entire to crenulate. Scattered in the eastern, central and southern China, in Zhejiang, Jiangsu, Anhui, Jiangxi, Fujian and Guangxi provinces of China. In sparse forest, forest margin and bushes. Altitude: 200-1500 m. Flowering & fruiting: September-November.

A

B

A–E. 中国浙江省黄茅尖
Huangmaojian Mt., Zhejiang, China. 17.10.2022

F. 中国浙江省东白山
Dongbaishan Mt., Zhejiang, China. 17.10.2022

C
D
E
F

峨眉双蝴蝶

Tripterospermum cordatum

多年生草质藤本，茎平卧或缠绕，光滑；花期无基生叶；茎生叶具柄，叶片心形、卵形或卵状披针形，近革质；花多单生或 2~4 成聚伞花序，腋生；花萼筒钟形，具翅，花萼裂片线状披针形，直立到开展；花冠淡粉到淡紫色，狭钟状，花冠裂片卵状三角形，褶半圆形，边缘整齐或圆齿状。

散布于中国中部到西南部的山区，见于陕西、湖北、湖南、四川、云南和贵州等地。生长在山谷疏林中及边缘和灌丛中。海拔 500~2500 m。花果期 9~11 月。

Perennial, stems prostrate or twisting, glabrous; Basal leaves absent at anthesis; stem leaves petiolate, blades cordate, ovate or ovate-lanceolate; flowers solitary or 2-4 in cyme, axillary; calyx tubes campanulate, winged, calyx lobes linear-lanceolate, erect to spreading; corolla pale pink to pale purple, narrowly campanulate, lobes ovate-triangular, plicae semiorbicular, margins entire to crenulate.

Scattered in central to southwestern China, in Shaanxi, Hubei, Hunan, Sichuan, Yunnan and Guizhou provinces of China. In sparse forest, forest margin and bushes. Altitude: 500-2500 m. Flowering & fruiting: September - November.

A

B

A-B. 中国四川省峨眉山
Emeishan Mt., Sichuan, China. 24.08.2006

C. 中国四川省贡嘎山
Gonggashan Mt., Sichuan, China. 04.10.1993

D-E. 中国四川省贡嘎山
Gonggashan Mt., Sichuan, China. 10.08.2004

C

D

E

湖北双蝴蝶

Tripterospermum discoideum

多年生草质藤本，茎缠绕，光滑；花期无基生叶；茎生叶具短柄，叶片披针形或卵状披针形；花多单生或2~4成聚伞花序，腋生；花萼筒钟形，具棱或狭翅，花萼裂片线状披针形；花冠淡蓝色到淡紫色，狭钟状，花冠裂片卵状三角形，褶半圆形，边缘整齐或圆齿状。

散布于中国中部的陕西、甘肃、湖北、和四川等地。生长在山谷疏林中及边缘和灌丛中。海拔 600~2000 m。花果期 9~11 月。

Perennial, stems prostrate or twisting, glabrous; basal leaves absent at anthesis; stem leaves with short petiole, blades lanceolate or ovate-lanceolate; flowers solitary or 2-4 in cyme, axillary; calyx tubes campanulate, angulate or narrowly winged, calyx lobes linear-lanceolate; corolla pale blue to pale purple, narrowly campanulate, lobes ovate-triangular, plicae semiorbicular, margins entire to crenulate.

Scattered in central China, in Shaanxi, Gansu, Hubei and Sichuan provinces of China. In sparse forest, forest margin and bushes. Altitude: 600-2000 m. Flowering & fruiting: September-November.

A

B

A-B. 中国陕西省宁陕县 Ningshan, Shaanxi, China. 24.08.2004

C. 中国陕西省岚皋县 Langao, Shaanxi, China. 07.08.2013

D. 中国陕西省留坝县 Liuba, Shaanxi, China. 03.09.2023

C

D

细茎双蝴蝶

Tripterospermum filicaule

多年生草质藤本，茎缠绕，光滑；茎生叶具柄，叶片卵形或卵状披针形；花单生或2~3朵成聚伞花序，腋生；花萼筒钟形，具翅，花萼裂片钻形到线状披针形；花冠蓝色、紫色到粉红色，狭钟状，花冠裂片卵状三角形，褶卵形，边缘整齐或微波状。

散布于中国中部到南部及西南部。生长在山谷疏林中及边缘。海拔 500~2500 m。花果期 9~12 月。

Perennial, stems twisting, glabrous; stem leaves with petiole, blades ovate or ovate-lanceolate; flowers solitary or 2-3 in cyme, axillary; calyx tubes campanulate, winged, calyx lobes subulate to linear-lanceolate; corolla blue, purple to pink, narrowly campanulate, lobes ovate-triangular, plicae ovate, margins entire to slightly undulate. Scattered in central to southern and southwestern provinces of China. In sparse forest, forest margin. Altitude: 500-2500 m. Flowering & fruiting: September-December.

A

B

A–B. 中国云南省龙陵县 Longling, Yunnan, China. 27.08.2001
C. 中国安徽黄山 Huangshan Mt., Anhui, China. 08.10.2010

C

香港双蝴蝶

Tripterospermum nienkui

多年生草质藤本，茎缠绕，光滑；茎生叶具短柄，叶片披针形、卵形或卵状披针形；花多单生或 2~5 成聚伞花序，腋生；花萼筒钟形，具翅，花萼裂片钻形到线状披针形，平展；花冠淡蓝色到淡紫色，钟状，花冠裂片卵状三角形，褶卵形，边缘整齐或啮齿状。

散布于中国南部的福建、湖南、广东和广西等省（区）及越南。生长在山谷疏林中及边缘。海拔 500~2000 m。花果期 9~12 月。

Perennial, stems twisting, glabrous; stem leaves with short petiole, blades lanceolate, ovate or ovate-lanceolate; flowers solitary or 2-5 in cyme, axillary; calyx tubes campanulate, winged, calyx lobes subulate to linear-lanceolate; corolla pale blue to pale purple, campanulate, lobes ovate-triangular, plicae ovate, margins entire to erose. Scattered in southern China, in Fujiang, Hunan, Guangdong, and Guangxi provinces of China, and Vietnam. In sparse forest, forest margin. Altitude: 500-2000 m. Flowering & fruiting: September-December.

A-D. 中国广西省大明山
Damingshan Mt., Guangxi, China. 09.09.2001

C

D

尼泊尔双蝴蝶

Tripterospermum volubile

多年生草质藤本，茎缠绕，光滑；茎生叶具柄，叶片卵形或卵状披针形；花多单生或偶尔 2 朵腋生；花萼筒钟形，具翅，花萼裂片钻形；花冠白色到黄白色，钟状，花冠裂片卵状三角形，褶卵形，边缘整齐或波状。

散布于喜马拉雅东部及邻近地区，包括中国西藏和云南等省。生长在山谷疏林中及边缘。海拔 1500~3000 m。花果期 8~10 月。

Perennial, stems twisting, glabrous; stem leaves with petiole, blades ovate or ovate-lanceolate; flowers often solitary or occasionally 2 axillary; calyx tubes campanulate, winged, calyx lobes subulate; corolla white to yellowish white, campanulate, lobes ovate-triangular, plicae ovate, margins entire to undulate.

Scattered in eastern Himalayan regions, including Xizang and Yunnan provinces of China. In sparse forest, forest margin. Altitude: 1500-3000 m. Flowering & fruiting: August-October.

A

B

A-B. 中国云南省贡山独龙族怒族自治县
Gongshan, Yunnan, China. 04.08.2000

C. 中国云南省高黎贡山的山水植被
Wild nature in Gaoligongshan Mt., Yunnan, China. 17.08.2001

C

4.2 獐牙菜亚族

Subtribe *Swertiinae*

粗壮獐牙菜 *Swertia hookeri*

中国西藏自治区错那市 Cuona, Xizang, China. 14.08.2006

4.2.1
巴脱泥属
Bartonia

半寄生（共生）小草本，叶鳞片状，完全或部分依靠菌根获取养分（菌根异养）；花四数。

本属 3 种，特产于美国东部及邻近的加拿大。

Semiparasitic diminutive herb. Leaves scale-like, wholly or partially dependent on mycorrhizal fungi (mycotrophic). Flowers 4-merus.

3 species, endemic to eastern United States of America and adjacent Canada.

春巴脱泥

Bartonia verna

一年生半寄生小草本，高 5~10 cm，直立；茎生叶鳞片状，稀疏；花单生或伞房花序，顶生；花萼裂片 4 枚，鳞片状，三角形；花冠裂片 4 枚，白色或黄白色倒卵形到匙形。

特产美国东部及加拿大。生长在林下湿地。花果期 12 月至来年 2 月。

Annual diminutive herb, partially mycoheterotrophic; 5-10 cm high; stem leaves scale-like, widely spaced; flowers solitary or in cyme and terminal; calyx lobes 4, scale-like, triangular; corolla lobes 4, white to yellowish white obovate to spathulate. Endemic to eastern USA and Canada. On wetlands in forest. Flowering and fruiting: December-February.

A

B

A-B. 美国佛罗里达州奥西奥拉市 摄影 /Photo: Peter Connolly
Osceola County, Florida, USA. 18.01.2021

中国四川省康定市亚拉雪山

Yalaxueshan Mt., Kangding, Sichuan, China. 13.09.2023

4.2.2 喉毛花属 *Comastoma*

一年生或多年生草本；茎从基部分枝；花单生或聚伞花序；花四或五基数；花萼裂至基部，无萼内膜；花冠裂片间无褶，花冠裂片基部附生流苏状附属物（副冠），常呈 1~2 束，开花时封盖花冠筒口部；蜜腺着生于花冠筒基部；雄蕊着生花冠筒上；种子光滑。

本属约 15 种，广泛分布于亚洲、欧洲北美温带地区高山地带。

Annual or perennial herb; stems branched from base; flowers solitary or arranged in cyme, 4- or 5-merus; calyx divided nearly to base, without intracalycular membrane; corolla without plicae; at the base of each corolla lobe bears fringed scales with white fimbriae covering the throat of corolla tube at anthesis; nectaries at base of corolla tube; stamens inserted on corolla tube; seeds smooth.

About 15 species, widespread in alpine areas of temperate Asia, Europe and North America.

蓝钟喉毛花

Comastoma cyananthiflorum

多年生草本，有短根状茎，高 5~10 cm；茎自基部分枝，数枚丛生，斜升；叶无柄或有短柄，叶片倒卵状匙形；花单生分枝顶端；花萼裂至基部，萼片不等，卵状披针形；花冠蓝色，高脚杯状，喉部深蓝紫色，花冠裂片开张，长圆形，每裂片基部具 2 束白色副冠。

分布于中国西藏、青海、四川和云南相邻接地区。生长在高山草甸、灌丛与森林边缘。海拔 3200~4500 m。花果期 6~9 月。

Perennial with short rhizomes, 5-10 cm high; stems branched from base to several clumps, ascending; stem leaves sessile or with short petioles, blade obovate-spathulate; flowers solitary at terminals; calyx divided to base, calyx lobes unequal, ovate-lanceolate; corolla salverform, blue, throat dark blue-purple, corolla lobes oblong, spreading, each lobe with 2 white fimbriate scales at base.

Scattered in high mountains in the area connecting Xizang, Qinghai, Sichuan and Yunnan provinces of China. In alpine meadows, bushes and forest margins. Altitude: 3200-4500 m. Flowering & fruiting: June-September.

A

A–C. 中国云南省丽江市玉龙雪山
Lijiang, Yunnan, China. 03.10.1992

D. 中国西藏自治区错那市
Cuona, Xizang, China. 14.08.2006

C

D

镰萼喉毛花

Comastoma falcatum

一年生草本，高 5~20 cm；茎自基部多分枝，分枝斜升；叶多数基生，叶片长圆状匙形，茎生叶近无柄，叶片卵形；花单生分枝顶端；花萼裂至基部，萼片不等，卵状披针形，弯曲成镰刀状，边缘波状；花冠蓝色到蓝紫色，高脚杯状，喉部深色，花冠裂片开张，长圆形，每裂片基部具 2 束白色流苏状副冠。

较广泛分布于中国北部及西北部及相邻的蒙古、俄罗斯、尼泊尔和印度等地。生长在高山草甸、高山灌丛与草地。海拔 2500~4500 m。花果期 7~9 月。

Annual, 5-20 cm high; stems much branched from base, ascending; leaves mostly basal, blade oblong-spathulate, stem leaves subsessile, ovate; flowers solitary at terminals; calyx divided to base, calyx lobes ovate-lanceolate, unequal, falcate-curved, margin undulate; corolla salverform, blue to blue-purple, throat darker, corolla lobes extending, oblong, spreading, each lobe with 2 white fimbriate scales at base.

Widespread in northern and northwestern China and adjacent Mongolia, Russia, Nepal and India. In alpine meadows, bushes and grassland. Altitude: 2500-4500 m. Flowering & fruiting: July-September.

A. 中国新疆维吾尔自治区奎屯市
Kuitun, Xinjiang, China. 12.08.1995

B. 中国西藏自治区安多县 Anduo, Xizang, China. 08.09.1992

C. 中国青海省格尔木市
Geermu, Qinghai, China. 04.08.2006

长梗喉毛花

Comastoma pedunculatum

一年生草本，高 5~15 cm；茎自基部多分枝，分枝斜升；叶稀疏，多数基生，叶片长圆状匙形，茎生叶 1-2 对，无柄，叶片卵状长圆形；花单生分枝顶端，花梗细长；花萼裂至基部，萼片近整齐，披针形，边缘整齐；花冠蓝色到蓝紫色，筒状，喉部深色，花冠裂片直伸或开张，长圆形，每裂片基部具 1 束白色流苏状副冠。

较广泛分布于中国青藏高原及邻近地区。生长在高山草甸和河滩等向阳处。海拔 3500~4500 m。花果期 7~10 月。

Annual, 5-15 cm high; stems much branched from base, ascending; few leaves, mostly basal, blade oblong-spathulate, stem leaves 1-2 pairs, sessile, ovate-oblong; flowers solitary at terminals with long slender peduncles; calyx divided to base, calyx lobes lanceolate, subequal, margins entire; corolla tubular, blue to blue-purple, throat darker, corolla lobes oblong, spreading, each lobe with 1 white fimbriate scale at base.

Scattered in the Qinghai-Xizang Plateau of China and adjacent areas. In sunny places of alpine meadows and stream side. Altitude: 3500-4500 m. Flowering & fruiting: July-October.

A. 中国西藏自治区八宿县 Basu, Xizang, China. 08.10.2003

B-C. 中国西藏自治区左贡县

Zuogong, Xizang, China. 10.10.2003

D. 中国四川省道孚县 Daofu, Sichuan, China. 12.09.2023

E. 中国青海省河南蒙古族自治县 Henan, Qinghai, China. 09.09.2023

多枝喉毛花

Comastoma polycladum

一年生草本，高 5~15 cm；茎自基部及以上多分枝，分枝直立；叶稀疏，叶片椭圆形到椭圆状披针形；花单生分枝顶端，花梗细长；花萼裂至基部，萼片近整齐，披针形，边缘黑紫色，皱波状；花冠蓝色到蓝紫色，筒状，喉部深色，花冠裂片直伸或开张，长圆形，每裂片基部具 2 束白色流苏状副冠。

散布于中国青海、甘肃、山西和内蒙古等地区。

生长在山坡草地、河滩等向阳处。海拔 2000~4500 m。花果期 7~9 月。

Annual, 5-15 cm high; stems much branched from base and above, erect; leaves sparsely spaced, blade elliptic to elliptic-lanceolate; flowers solitary at terminals with long and slender peduncles; calyx divided to base, calyx lobes lanceolate, subequal, margins dark purple, undulate; corolla tubular, blue to blue-purple, throat darker, corolla lobes oblong, spreading, each lobe with 1 white fimbriate scale at base. Scattered in the Qinghai-Xizang Plateau of China and adjacent areas. In sunny places of alpine meadows and stream side. Altitude: 2000-4500 m. Flowering & fruiting: July-October.

中国甘肃省夏河县 Xiahe, Gansu, China. 09.10.1990

喉毛花

Comastoma pulmonarium

一年生草本，高 5~30 cm；茎单一或基部以上分枝，直立；叶较稀疏，多数茎生，无柄，叶片卵状披针形；花单生或聚伞花序顶生；花萼长约花冠 1/4，裂至基部，萼片近整齐，披针形，边缘整齐；花冠白色到淡蓝色，筒状，喉部深色，花冠裂片直伸或开张，长圆状卵形，每裂片基部具 1 束白色流苏状副冠。

较广泛分布于中国青藏高原及邻近的中国北部、西北部及日本、俄罗斯等地区。形态变异较大。生长在山坡草地和河滩等向阳处。海拔 3000~4500 m。花果期 7~11 月。

Annual, 5-30 cm high; stems simple or branched by upper part, erect; leaves sparsely spaced, sessile, blade ovate-oblong; flowers solitary or in cyme at terminals; calyx about ¼ of the length of corolla, divided to base, calyx lobes lanceolate, equal, margins entire; corolla tubular, white to pale blue, throat darker, corolla lobes oblong-ovate, spreading, each lobe with 1 white fimbriate scale at base.

Widespread in the Qinghai-Xizang Plateau of China and adjacent areas in northern and northwestern China, Japan and Russia. In sunny places of mountain slopes and stream side. Altitude: 3000-4500 m. Flowering & fruiting: July-November.

A–C. 中国四川省红原县
Hongyuan, Sichuan, China. 28.07.2004

D. 中国四川省康定市
Kangding, Sichuan, China. 13.09.2023

E. 中国四川省阿坝藏族羌族自治州
Aba, Sichuan, China. 11.09.2023

C

D

E

纤枝喉毛花

Comastoma stellariifolium

多年生草本，基部分枝，有短根状茎及营养枝，高 10~15 cm，极纤细；直立或斜升；叶无柄，4~6 对，叶片倒卵状披针形；花单生分枝顶端，花梗纤细；花萼裂至基部，萼片近相等，披针形；花冠蓝色，高脚杯状，花冠裂片开张，长圆状卵形，每裂片基部具 1 束白色副冠。

特产于喜马拉雅山东部地区。生长在高山灌丛中及边缘。海拔 3200~4500 m。花果期 8~10 月。

Perennial, branched at base with short rhizomes and vegetative shoots, 10-15 cm high, slender, erect or ascending; leaves sessile, 4-6 pairs, blade obovate-lanceolate; flowers solitary at terminals, pedicel slender; calyx divided to base, calyx lobes subequal, lanceolate; corolla salverform, blue, corolla lobes spreading, oblong-ovate, each lobe with 1 white fimbriate scales at base.
Endemic to the eastern Himalaya mountains. In alpine bushes and bush margins. Altitude: 3200-4500 m. Flowering & fruiting: August-October.

A-D. 中国西藏自治区墨脱县
Motuo, Xizang, China. 24.09.2003
E. 中国云南省大理白族自治州
Dali, Yunnan, China. 23.09.2023

C

D

E

高杯喉毛花

Comastoma traillianum

一年生草本，高 5~30 cm；茎自基部及以上多分枝，分枝直立或铺散，紫色或紫红色；茎生叶无柄，叶片卵形到长圆状卵形；聚伞花序顶生和腋生；花萼裂至基部，萼片不整齐，卵形到卵状披针形，边缘整齐；花冠蓝色，高脚杯状，花冠裂片开张，长圆形，每裂片基部具 2 束白色流苏状副冠。

特产于中国云南西北部和四川南西南部。生长在高山草甸及灌丛边缘等向阳处。海拔 3000~4000 m。花果期 8~10 月。

Annual, 5-30 cm high; stems much branched from base and above, erect and ascending, reddish purple; leaves mostly on stems, sessile, blade ovate to oblong-ovate; flowers in cymes, terminal and axillary; calyx divided to base, calyx lobes ovate to ovate-lanceolate, unequal, margins entire; corolla salverform, blue, corolla lobes oblong, spreading, each lobe with 2 white fimbriate scale at base.
Endemic to northwestern Yunnan and southwestern Sichuan provinces of China. In sunny places of alpine meadows and bush margins. Altitude: 3000-4000 m. Flowering & fruiting: August-October.

A. E. F. 中国云南省香格里拉市
Xianggelila, Yunnan, China. 19.09.2023
B. 中国云南省丽江市 Lijiang, Yunnan, China. 11.10.1999
C. 中国云南省香格里拉市
Xianggelila, Yunnan, China. 28.09.1999
D. 中国西藏自治区左贡县
Zuogong, Xizang, China. 10.08.2002

D

E

F

美丽轮叶菜 *Frasera speciosa*
美国科罗拉多州 Colorado, USA. 18.07.1992

4.2.3
轮叶菜属
Frasera

多年生或一次性结果多年生草本；叶多轮生；通常大型圆锥状聚伞花序；花常 4 基数，偶尔 5 基数，花萼和花冠深裂至基部；每个花冠裂片中部具 1 或 2 枚圆形腺窝，被环形流苏覆盖。

本属约 15 种，特产于北美，尤以北美西部落基山地区种类最为丰富。

Perennial, mostly monocarpic; leaves mostly whorled; large terminal paniculiform cymes; flowers 4- or 5-merus; calyx and corolla deeply divided to nearly base; each corolla lobes 1 or 2 large glands below the middle, the gland ± completely surrounded by a usually fimbriate membrane.

About 15 species, endemic to North America, mostly western North America.

美丽轮叶莆

Monument Plant

Frasera speciosa

一次性结果多年生（5~15 年）草本，花枝高 30~200 cm，粗壮，直立；莲座叶发达，叶无柄，叶片长圆状披针形，茎生叶无柄，3~4 叶轮生，叶片披针形；大型圆锥状聚伞花序，有 20~200 朵花；花 4 基数，花萼裂至基部，萼片线状披针形；花冠黄绿色，具紫色斑点，裂至基部，花冠裂片开张，长圆状卵形，每裂片基部具 2 枚圆形腺窝，被流苏覆盖。

特产于北美西部落基山区。生长在高山草地及林地边缘。海拔 1500~2800 m。花果期 7~8 月。

Monocarpic perennial (5-15 years), flowering stem 30-200 cm high, stout, erect; rosette leaves well developed, sessile, blade oblong-lanceolate; stem leaves sessile, 3-4 whorled, lanceolate; gigantic panicle with 20-200 flowers; flowers 4-merus; calyx divided to base, calyx lobes linear-lanceolate; corolla yellowish green with purple spots, corolla lobes spreading, oblong-ovate, each lobe with 2 nectar pads covered by fringes.

Native to Western N America mostly in the Rocky Mountains. In alpine grassland and margins of woodland. Altitude: 1500-2800 m. Flowering & fruiting: July-August.

A

B

A–B. 美国科罗拉多落基山
Rocky Mt., Colorado, USA. 18.07.1992

4.2.4 假龙胆属 *Gentianella*

黑边假龙胆 *Gentianella azurea*；
中国青海省河南蒙古族自治县 Henan, Qinghai, China. 09.09.2023

一年生或多年生草本；花 4 或 5 基数；花萼几近全裂，无萼内膜；花冠管状或漏斗状，无褶；雄蕊 4 或 5 枚，着生于花冠筒上；蜜腺小，着生于花冠筒；种子细小，光滑无翅。

本属约 250 种，广泛分布于除非洲外的全球南北温带地区，以南美洲种类最为丰富。

Annual or perennial herbs; flowers 4- or 5-merus; calyx divided to base without intracalycular membrane; corolla tubular or funnelform, without plicae; stamens 4 or 5 attached to corolla tube; nectaries small and attached to corolla tube; seeds small, smooth and wingless.

About 250 species, widespread globally in both southern and northern temperate areas except for Africa, highly diversified in South America.

紫红假龙胆

Gentianella arenaria

一年生矮小草本，高 2~5 cm；自基部多分枝，分枝斜升或铺散，全株紫红色；叶多数基生，具短柄，叶片匙形；花单生分枝顶端，4 基数；花萼裂至基部，裂片整齐，外弯，匙形；花冠淡蓝紫色，管状，花冠裂片长圆形，每裂片基部着生蜜腺。

特产于中国西北部的甘肃、青海、新疆和西藏。生长在高海拔沙质和砾石质河滩等空旷地。海拔 3500~5000 m。花果期 7~9 月。

Annual diminutive herb, 2-5 cm high; stems much branched from base, prostrate and ascending, reddish purple; leaves mostly basal, shortly petiolate, blade spathulate; flowers solitary on terminal; flowers 4-merus; calyx divided to base, calyx lobes spathulate, recurved, equal, entire; corolla tubular, pale blue-purple, corolla lobes oblong, each lobe with 1 nectary at base.

Endemic to northwestern China in Gansu, Qinghai, Xinjiang and Xizang. In sandy and stony open places of stream side. Altitude: 3500-5000 m. Flowering & fruiting: July-September.

A. 中国新疆维吾尔自治区阿克苏市
Akesu, Xinjiang, China. 22.08.1995

B-C. 中国青海省沱沱河 Tuotuohe, Qinghai, China. 05.08.2006

粗糙假龙胆

Gentianella aspera

一年生草本，高 5~30 cm；自基部多分枝；茎生叶卵形到卵状披针形，中脉边缘具刚毛；聚伞花序；花 4 或 5 基数，花萼裂片不整齐，卵状披针形；花冠粉色到紫红色，花冠裂片卵形。

特产于欧洲阿尔卑斯山中部到东部地区。生长在山地草丛中。海拔 1500~2500 m。花果期 7~9 月。

Annual, 5-30 cm; stems branched from base; stem leaves ovate to ovate-lanceolate, hirsute on midveins and margins; flowers in cymes; flowers 4- or 5-merus; calyx unequal, ovate-lanceolate; corolla pink to purple, lobes ovate.

Endemic to central to eastern Alps in Europe. In alpine grassland. Altitude: 1500-2500 m. Flowering & fruiting: July-September.

A–B. 德国巴伐利亚鹰巢峰

Eagle's Nest, Berchtesgaden, Germany. 19.08.2005

黑边假龙胆

Gentianella azurea

一年生草本，高 5~20 cm；自基部多分枝；茎生叶椭圆形到长圆状披针形；聚伞花序；花 5 基数，花萼裂片整齐，卵状披针形，边缘黑色；花冠蓝色，花冠裂片卵形。

分布于中国西部及西北部及相邻的蒙古和俄罗斯。生长在山地草丛中。海拔 2000~3500 m。花果期 7~9 月。

Annual, 5-20 cm; stems branched from base; stem leaves elliptic to oblong-lanceolate; flowers in cymes; flowers 5-merus; calyx lobes equal, ovate-lanceolate, margins black; corolla blue, corolla lobes ovate.

Scattered in western and northwestern China and adjacent Mongolia and Russia. In alpine grassland. Altitude: 2000-3500 m. Flowering & fruiting: July-September.

A-B. 中国新疆维吾尔自治区奎屯市
Kuitun, Xinjiang, China. 12.08.1995

C. 中国四川省阿坝藏族羌族自治州
Aba, Sichuan, China. 11.09.2023

D-E. 中国青海省河南蒙古族自治县
Henan, Qinghai, China. 09.09.2023

D

E

保加利亚假龙胆

Gentianella bulgarica

一年生草本，高 5~20 cm；枝纤细斜升；茎生叶长圆状披针形；聚伞花序；花 5 基数，花萼裂片近整齐，线状披针形，边缘外卷；花冠淡紫色，花冠裂片卵形。

分布于欧洲喀尔巴阡山山南部到巴尔干半岛中部。生长在山地草丛中。海拔 1000~2000 m。花果期 7~9 月。

Annual, 5-20 cm; stems slender, procumbent; stem leaves oblong-lanceolate; flowers in cymes; flowers 5-merus; calyx lobes subequal, linear-lanceolate, margins revolute; corolla pale purple, corolla lobes ovate.

Scattered in the southern Carpathians to central Balkan Pennisula. In alpine grassland. Altitude: 1000-2000 m. Flowering & fruiting: July-September.

A–B. 保加利亚贝格里克 Beglika, Bulgaria. 22.07.1993

田野假龙胆

Gentianella campestris

一年生草本，高 5~20 cm；枝纤细，直立到斜升；茎生叶卵形到卵状披针形；聚伞花序；花 4 基数，花萼裂片异型，外侧 2 片较大，卵形或卵状披针形，包裹内侧 2 片，较小，线状披针形，内侧萼片有时退化消失；花冠粉紫色，花冠裂片倒卵形或长圆状卵形。

广布于欧洲中部到北部各地。生长在山地草丛中。海拔 1000~2500 m。花果期 7~9 月。

Annual, 5-20 cm; stems slender, erect to ascending; stem leaves ovate to ovate-lanceolate; flowers in cymes; flowers 4-merus; calyx lobes very unequal, outer 2 larger ovate to ovate-lanceolate, enfolding the inner 2, smaller, linear-lanceolate, reduced or sometimes missing; corolla pink- purple, corolla lobes obovate or oblong-ovate.

Widespread in central to northern Europe. In alpine grassland. Altitude: 1000-2500 m. Flowering & fruiting: July-September.

A–B. 瑞士卢塞恩瑞吉山
Rigi Mt., Luzern, Switzerland. 16.07.2002

C. 法国奥维尼中央山脉 Massif central, Auvergne, France. 01.08.2005

密花假龙胆

Gentianella gentianoides

一年生草本，高 5~8 cm；自基部以上多分枝，密被紫色乳突；茎生叶卵形到椭圆形；聚伞花序，花密集；花 5 基数，花萼裂片整齐，狭椭圆形；花冠淡蓝到紫红色，花冠裂片长圆状卵形，先端具小尖头。

分布于中国西南部的云南和四川。生长在山地草丛中。海拔 2700~3500 m。花果期 9~11 月。

Annual, 5-8 cm; stems branched from base and above, densely papillate; stem leaves ovate to elliptic; flowers in cymes, densely spaced; flowers 5-merus; calyx lobes equal, narrowly elliptic; corolla pale blue to pink-purple, corolla lobes oblong-ovate, apex mucronate.

Scattered in Yunnan and Sichuan provinces of southwestern China. In alpine grassland. Altitude: 2700-3500 m. Flowering & fruiting: September-November.

A–B. 中国云南省丽江市 Lijiang, Yunnan, China. 09.10.1999

C. 中国云南省丽江市 Lijiang, Yunnan, China. 09.10.1999

德国假龙胆

Gentianella germanica

一年生草本，高 5~20 cm；枝纤细，直立或斜升；茎生叶卵状披针形；聚伞花序；花 5 基数，花萼裂片近整齐，线状披针形，边缘外卷；花冠粉紫色，花冠裂片长圆状卵形。

分布于欧洲中部到西部。生长在山地草丛中。海拔 1500~2000 m。花果期 7~9 月。

本种与粗糙假龙胆和保加利亚假龙胆都比较接近，区别在于叶无毛。

Annual, 5-20 cm; stems slender, erect or procumbent; stem leaves ovate-lanceolate; flowers in cymes; flowers 5-merus; calyx lobes subequal, linear-lanceolate, margins revolute; corolla pale pink-purple, corolla lobes ovate.

Scattered in central to western Europe. In alpine grassland. Altitude: 1500-2000 m. Flowering & fruiting: July-September.

Very similar to *Gentianella aspera* and *Gentianella bulgarica*, differences mainly due to glabrous leaves.

A–B. 瑞士汝拉洲波杭退

Porrentruy, Jura, Switzerland. 30.09.2006

矮假龙胆

Gentianella pygmaea

一年生矮小草本，高 2~3 cm；自基部多分枝，分枝铺散或斜升；叶较少，具短柄，叶片匙形到倒卵形；花单生分枝顶端，4 基数；花萼裂至基部，裂片整齐，直立，倒卵形；花冠白色到黄绿色，花冠裂片卵形。

散布于中国西部的青海、四川、新疆和西藏及相邻的中亚和南亚部分国家（印度、吉尔吉斯斯坦、塔吉克斯坦）。生长在高海拔沙质和砾石质河滩等空旷地。海拔 3500~5000 m。花果期 7~9 月。

Annual diminutive herb, 2-3 cm high; stems much branched from base, prostrate and ascending; leaves few, shortly petiolate, blade spathulate to obovate; flowers solitary on terminal; flowers 4-merus; calyx divided to base, calyx lobes obovate, equal, entire; corolla white to yellowish green, corolla lobes obovate.
Scattered in northwestern China in Qinghai, Sichuan, Xinjiang and Xizang, and adjacent countries in central and southern Asia (India, Kyrgyzstan, Sikkim, Tajikistan). In sandy and stony open places of stream side. Altitude: 3500-5000 m. Flowering & fruiting: July-September.

A–B. 中国青海沱沱河 Tuotuohe, Qinghai, China. 05.08.2006
C. 中国西藏自治区安多县
Anduo, Xizang, China. 08.09.1992

新疆假龙胆

Gentianella turkestanorum

一年生草本，高 10~40 cm；直立；茎生叶无柄，卵状披针形到卵形；聚伞花序顶生和腋生；花 5 基数，花萼裂片近整齐，线形到狭椭圆形；花冠淡蓝色到蓝紫色，花冠裂片长圆状卵形或长圆状椭圆形，先端具小尖头。

分布于中国新疆及邻近中亚国家哈萨克斯坦、吉尔吉斯斯坦、蒙古和俄罗斯。生长在潮湿的溪边草地、沙质草地和灌丛边缘。海拔 1500~2500 m。花果期 7~9 月。

Annual, 10-40 cm; stems erect; stem leaves sessile, ovate-lanceolate to ovate; flowers in cymes, terminal and axillary; flowers 5-merus; calyx lobes subequal, linear to narrowly elliptic; corolla pale blue to blue-purple, corolla lobes oblong-ovate or oblong-elliptic, apex mucronate. Scattered in Xinjiang of China and adjacent central Asian countries Kazakhstan, Tajikistan, Mongolia, and Russia. In wet grassland, streamside, and sandy meadows, bush margins. Altitude: 1500-2500 m. Flowering & fruiting: July-September.

A-B. 中国新疆维吾尔自治区巩乃斯
Gongnaisi, Xinjiang, China. 07.09.2004

C. 中国新疆维吾尔自治区奎屯市 Kuidun, Xinjiang, China. 12.08.1995

4.2.5
扁蕾属
Gentianopsis

湿生扁蕾 *Gentianopsis paludosa*；
中国甘肃省宕昌县 Tanchang, Gansu, China. 05.09.2023

1~2 年生草本；花单生茎顶，4 基数；花在蕾期两侧扁压，近 4 棱形；萼片 4 枚，异形，外侧 2 枚萼片狭而长，内侧两枚宽而短；具不连续三角形萼内膜；花冠筒状或漏斗状，4 裂片，无褶；蜜腺 4，着生花冠筒基部。

本属约 24 种，分布于北半球温带和亚热带高山地区。

Annual or biennial; flowers solitary on terminal of stems; flowers 4-merus; flower buds subquadriangular, flattened; calyx lobes 4, dimorphic, 2 outer ones narrower and longer, whereas 2 inner ones shorter and wider; intracalycular membrane discontinuous, triangular; corolla tubular or funnelform, without plicae, 4 lobes; 4 nectaries attached to the base pf corolla tube.

About 24 species, widespread in temperate and alpine subtropical area of the northern hemisphere.

扁蕾（原变种）

Gentianopsis barbata var. *barbata*

一年生草本，高 10~40 cm；直立，基部多分枝；茎生叶无柄，线形至狭披针形；花单生分枝顶端；花萼筒管状，花萼裂片三角状披针形，长超过花冠筒；花冠蓝色，花冠筒管状，花冠裂片长圆状椭圆形，先端具不规则细齿，两侧基部具少数短流苏。

广泛散布于中国北部、西部及相邻近的中亚和东亚部分国家。生长在较潮湿向阳的山坡、草地和灌丛边缘。海拔 1000~3500 m。花果期 7~9 月。

Annual, 10-40 cm, erect, much branched from base; stem leaves sessile, linear to narrowly lanceolate; flowers solitary at terminals of branches; calyx tube tubular, calyx lobes triangular-lanceolate, as long as or longer than corolla tube; corolla blue, corolla tube tubular, corolla lobes oblong-elliptic, apex irregularly denticulate, lateral margins shortly fringed at base.

Widespread in northern and western China and adjacent central and eastern Asian countries. In wet slopes, grassland, streamside, meadows, and bush margins. Altitude: 1000-3500 m. Flowering & fruiting: July-September.

A

B

A-B. 中国青海省都兰县 Dulan, Qinghai, China. 03.08.2006

C. 中国新疆维吾尔自治区喀纳斯 Hanasi, Xinjiang, China. 08.08.1995

C

细萼扁蕾（变种）

Gentianopsis barbata var. *stenocalyx*

与原变种的区别是茎和花梗纤细，叶和花萼裂片窄细。散布于中国西南部局部区域。

This variety is characterized with much slenderer stems and peduncles, narrower and longer leaves and calyx lobes. Scattered in restricted areas in southwestern China.

中国云南省丽江市 Lijiang, Yunnan, China. 24.09.1990

黄白扁蕾（变种）

Gentianopsis barbata var. *albiflavida*

与原变种的区别是茎和花梗纤细，叶和花萼裂片窄细；花冠黄白色或黄色。散布于中国西北部局部区域。

This variety is characterized with much slenderer stems and peduncles, narrower and longer leaves and calyx lobes; corolla yellowish white or yellow. Scattered in restricted areas in northwestern China.

中国新疆维吾尔自治区奎屯市 Kuitun, Xinjiang, China.
12.08.1995

欧洲扁蕾

Gentianopsis ciliata

一年生草本，高 5~30 cm；直立，单一或少分枝；茎生叶无柄，线形至狭披针形；花单生分枝顶端；花萼筒管状，花萼裂片狭披针形，长约花冠的一半；花冠蓝色，花冠筒管状，花冠裂片长圆状倒卵形，先端具钝，两侧基部具较流苏。

广布于欧洲大陆大部分地区。生长在山坡草地和林地边缘。海拔 1000~2500 m。花果期 6~8 月。

Annual, 5-30 cm, erect, simple or few branches; stem leaves sessile, linear to narrowly lanceolate; flowers solitary at terminals of branches; calyx tube tubular, calyx lobes narrowly lanceolate, about ½ as long as corolla; corolla blue, corolla tube tubular, corolla lobes oblong-obovate, apex obtuse, lateral margins fringed at base.

Widespread in continental Europe. In slope grassland, and woodland margins. Altitude: 1000-2500 m. Flowering & fruiting: June-August.

A-B. 瑞士卢塞恩州瑞吉山

Rigi Mt., Luzern, Switzerland. 20.09.2004

C. 瑞士伯尔尼州莎斯哈尔山

Chasseral, Bern, Switzerland. 16.09.2004

C

回旋扁蕾

Gentianopsis contorta

一年生草本，高 5~30 cm；直立，单一或上部少分枝；茎生叶无柄，椭圆形至卵状椭圆形；花单生分枝顶端；花萼筒管状，花萼裂片三角形，长达花冠的 2/3；花冠蓝紫色，花冠筒管状，花冠裂片椭圆状倒卵形，先端钝，两侧整齐，无流苏。

零星散布于中国西部及尼泊尔、日本等地。生长在山坡草地和林地边缘。海拔 2000~3500 m。花果期 8~10 月。

Annual, 5-30 cm, erect, simple or few branches in upper part; stem leaves sessile, elliptic to ovate-elliptic; flowers solitary at terminals of branches; calyx tube tubular, calyx lobes triangular, about 2/3 as long as corolla; corolla blue-purple, corolla tube tubular, corolla lobes elliptic-obovate, apex obtuse, lateral margins entire, not fringed.
Scattered in western China, Nepal, and Japan. In slope grassland, and woodland margins. Altitude: 2000-3500 m. Flowering & fruiting: August-October.

A-B. 中国云南省香格里拉市
Xianggelila, Yunnan, China. 21.09.1999
C. 中国云南省德钦县 Deqin, Yunnan, China. 26.09.1999

C

大花扁蕾

Gentianopsis grandis

一年生草本，高 20~50 cm；直立，粗壮单一，基部以上多分枝；茎生叶无柄，狭披针形；花大，长 5~10 cm，单生分枝顶端；花萼筒管状，花萼裂片三角状披针形，长达花冠的 2/3；花冠蓝色，花冠漏斗状，花冠裂片长圆状椭圆形，先端具不规则粗齿，两侧基部具流苏。

特产于中国西南部的云南和四川。生长在较潮湿向阳的山坡草地和林地边缘。海拔 2000~3500 m。花果期 7~10 月。

Annual, 20-40 cm, erect, stout and simple, branched above base; stem leaves sessile narrowly lanceolate; flowers gigantic, 5-10 cm long, solitary at terminals of branches; calyx tube tubular, calyx lobes triangular-lanceolate, 2/3 as long as corolla; corolla blue, funnelform, corolla lobes oblong-elliptic, apex irregularly crenulate, lateral margins fringed at base.

Endemic to southwestern China in Yunnan and Sichuan. In wet slopes, grassland, and woodland margins. Altitude: 2000-3500 m. Flowering & fruiting: July-October.

A

B

A-C. 中国云南省丽江市 Lijiang, Yunnan, China. 30.09.1992

黄花扁蕾

Gentianopsis lutea

一年生草本，高 5~30 cm；直立或斜升，单一或上部少分枝；茎生叶无柄，椭圆形至椭圆状披针形；花单生分枝顶端；花萼筒管状，花萼裂片三角状披针形，长达花冠的 2/3；花冠淡黄色，花冠筒管状钟形，花冠裂片椭圆形，先端钝，两侧基部具短流苏。

零星散布于中国西南部云南和四川。生长在山坡草地和林地边缘。海拔 2000~3000 m。花果期 8~10 月。

Annual, 5-30 cm, erect or ascending, simple or few branches in upper part; stem leaves sessile, elliptic to elliptic-lanceolate; flowers solitary at terminals of branches; calyx tube, calyx lobes triangular-lanceolate, about 2/3 as long as corolla; corolla pale yellow, corolla tube tubular tubular-campanulate, corolla lobes elliptic, apex obtuse, lateral margins fringed at base.

Scattered in southwestern China, in Yunnan and Sichuan. In slope grassland, and woodland margins. Altitude: 2000-3000 m. Flowering & fruiting: August-October.

A. 中国四川省泸定县 Luding, Sichuan, China. 27.09.1994

B. 中国云南省昆明市 Kunming, Yunnan, China. 01.09.1990

高原扁蕾（变种）

Gentianopsis paludosa var. *alpina*

本变种与原变种的区别是花白色或黄白色。

零星散布于中国甘肃、青海和西藏。生长在较潮湿向阳的山坡草地、溪边和林地边缘。海拔 1500~3500 m。花果期 7~10 月。

This variety can be distinguished from the var. paludosa in having white or yellowish white flowers.

Occasionally seen in Gansu, Qinghai and Xizang of China. In wet slopes, grassland, streamside, meadows, and woodland margins. Altitude: 1500-3500 m. Flowering & fruiting: July-October.

A–B. 中国甘肃省玛曲县 Maqu, Gansu, China. 27.07.1990

卵叶扁蕾（变种）

Gentianopsis paludosa var. *ovatodeltoidea*

本变种与原变种的区别在于本变种的茎生叶卵状或三角状披针形，茎的中上部多分枝。

零星散布于中国中部到北部。生长在较潮湿向阳的山坡草地、溪边和林地边缘。海拔 1200~3000 m。花果期 7~10 月。

This variety can be distinguished from the var. paludosa in having stem leaves ovate-lanceolate or triangular-lanceolate, much branched from middle and above.

Scattered in central to northern China. In wet slopes, grassland, streamside, meadows, and woodland margins. Altitude: 1200-3000 m. Flowering & fruiting: July-October.

A-B. 中国甘肃省宕昌县 Tanchang, Gansu, China. 25.08.1990

C. 卵叶扁蕾 *Gentianopsis paludosa* var. *ovatodeltoidea* + 椭圆叶花锚 *Halenia elliptica*
中国甘肃省宕昌县 Tanchang, Gansu, China. 05.09.2023

湿生扁蕾（原变种）

Gentianopsis paludosa var. *paludosa*

一年生草本，高 5~40 cm；直立，单一或基部分枝；茎生叶无柄，椭圆状披针形；花单生分枝顶端；花萼筒管状，花萼裂片三角状披针形，长约花冠的 1/2；花冠蓝色，花冠筒管状，花冠裂片长圆状椭圆形，先端具不规则细齿，两侧基部具少数流苏。

广泛散布于中国北部和西部及相邻近的喜马拉雅地区部分国家。生长在较潮湿向阳的山坡草地、溪边和林地边缘，尤其是扰动过的草地。海拔 1500~3500 m。花果期 7~10 月。

Annual, 5-40 cm, erect, simple or branched from base; stem leaves sessile, elliptic-lanceolate; flowers solitary at terminals of branches; calyx tube tubular, calyx lobes triangular-lanceolate, about ½ as long as corolla; corolla blue, corolla tube tubular, corolla lobes oblong-elliptic, apex irregularly denticulate, lateral margins fringed at base.

Widespread in northern and western China and adjacent Himalayan countries. In wet slopes, grassland, streamside, meadows, and woodland margins. Altitude: 1500-3500 m. Flowering & fruiting: July-October.

A

B

A–B. 中国四川省康定市 Kangding, Sichuan, China. 05.07.2021
C. 中国四川省道孚县 Daofu, Sichuan, China. 03.07.2021
D. 中国四川省红原县 Hongyuan, Sichuan, China. 28.07.2004
E. 中国甘肃省宕昌县 Tanchang, Gansu, China. 05.09.2023

C

D

E

椭圆叶花锚 *Halenia elliptica*
中国云南省丽江市 Lijiang, Yunnan, China. 20.09.1990

4.2.6
花锚属
Halenia

一年或多年生草本；单叶对生，全缘；松散的圆锥花序；花 4 基数；花萼深裂，萼片 4 枚；花冠钟形，4 深裂，花冠裂片基部具长距，内生蜜腺；雄蕊 4，着生于花冠筒，于花冠裂片互生。

本属约 80 种，分布于亚洲和美洲，以北美西南部和南美西北部种类最为丰富。也有学者认为南美的许多种类应该合并，总种数约 40 种。

Annual or perennial; leaves simple, opposite; lax panicles of cymes; flowers 4-merus; calyx lobed 4 nearly to the base; corolla campanulate with 4 lobes, each lobe with a spur near its base, nectary inside the spur; stamens 4, attached to corolla tube, alternative with corolla lobes.

About 80 species, distributed in Asia, North and South America, mostly in southwestern North America and northwestern South America. Some researchers suggest merging of many species from South America, so the total species number of the genus reduces to -40 species.

花锚

Halenia corniculata

一年生草本，高 20~60 cm，直立，茎四棱形；茎生叶无柄，椭圆状披针形至卵形；聚伞花序顶生和腋生；花萼裂片三角状披针形；花冠黄色，花冠筒钟状，花冠裂片椭圆形或卵形，距开展。

较广布于东北亚和北美温带地区，在中国华北和东北常见。生长在山坡草地和林地边缘。海拔 200~1500 m。花果期 7~9 月。

Annual, 20-60 cm, erect, stem quadrangular; stem leaves sessile, elliptic-lanceolate to ovate; cymes terminal and axillary; calyx lobes triangular-lanceolate; corolla yellow, corolla tube campanulate, corolla lobes elliptic or ovate, spurs spreading.

Widely scattered in temperate areas in Northeast Asia and North America; commonly seen in northern and northeastern China. In slope grassland, and woodland margins. Altitude: 200-1500 m. Flowering & fruiting: July-September.

A–B. 中国吉林省长白山

Changbaishan Mt., Jilin, China. 19.08.2000

椭圆叶花锚

Halenia elliptica

一年生草本，高 10~60 cm，直立，茎四棱形，光滑，中上部多分枝；茎生叶无柄，椭圆状披针形至卵状披针形；聚伞花序顶生和腋生；花萼裂片卵形或椭圆形；花冠蓝色到蓝紫色，偶有白花变异，花冠筒钟状，花冠裂片椭圆形或卵形，距开展。

较广布于中国北部、中部到西部及喜马拉雅地区国家。生长在山坡草地和林地边缘。海拔 800~3500 m。花果期 7~9 月。

Annual, 10-60 cm, erect, stem quadrangular, glabrous, much divided from middle and above; stem leaves sessile, elliptic-lanceolate to ovate-lanceolate; cymes terminal and axillary; calyx lobes ovate or elliptic; corolla blue to blue-purple, occasionally white corolla tube campanulate, corolla lobes elliptic or ovate, spurs spreading.

Widely spread in central, northern and western China, and the Himalayan countries. In slope grassland, and woodland margins. Altitude: 800-3500 m. Flowering & fruiting: July-September.

A–B. 椭圆叶花锚 *Halenia elliptica*
中国云南省香格里拉市
Xianggelila, Yunnan, China. 18.09.2023

C

D

C. 椭圆叶花锚 *Halenia elliptica*；中国湖北省巴东县 Badong, Hubei, China. 15.08.2004

D. 白花型椭圆叶花锚 *Halenia elliptica*；中国四川省雅江县 Yajiang, Sichuan, China. 14.08.2002

E. 白花型椭圆叶花锚 *Halenia elliptica*；中国云南省宁蒗彝族自治县 Ninglang, Yunnan, China. 17.09.2023

F. 紫花型椭圆叶花锚 *Halenia elliptica*；中国四川省盐源县 Yanyuan, Sichuan, China. 16.09.2023

4.2.7
口药花属
Jaeschkea

一年生草本；茎铺散或直立；茎生叶对生；聚伞花序；花 4~5 基数；花萼深裂； 花冠筒状，裂至中部，花冠筒基部有腺体，花冠裂片覆瓦状排列，无褶；雄蕊着生于花冠裂片弯缺处下方，于花冠裂片互生。本属 3 种，特产喜马拉雅山西部。

Annual; stems prostrate or erect; stem leaves opposite; flowers in cymes, 4-5-merous; calyx deeply divided to base; corolla tube tubular, divided to middle; nectaries at base of corolla tube, corolla lobes imbricate, plicae absent; anthers inserted just below corolla sinus and alternative with corolla lobes. 3 species, endemic to the western Himalayas.

小籽口药花

Jaeschkea microsperma

植株矮小，高 1~5 cm，茎铺散斜升，紫红色；花单生分枝顶端，花萼长约花冠的一半，花冠淡蓝色，花药黄色。特产喜马拉雅山西部。生长在高山砾石滩河床和砾石山坡。海拔 4300~4500 m。花果期 7~9 月。

Plant small, 1-5 cm high, stems prostrate, reddish purple; flowers solitary at branch terminals; calyx about ½ as long as corolla; corolla pale blue; anthers yellow. Endemic to western Himalayas in India, Sikkim and Xizang of China. In stony river beds or slopes. Altitude: 4300-4500 m. Flowering & fruiting: July-September.

A–B. 中国西藏自治区聂拉木 摄影 /Photo：刘彬彬
Nielamu, Xizang, China. 26.07.2013

中国福建省黄冈山森林景观，匙叶草的典型生境

Forest, typical inhabitat of *Latouchea fokiensis*, in Huanggangshan Mt., Fujian, China. 21.07.2000

4.2.8
匙叶草属
Latouchea

多年生草本，有根状茎；叶大部分基生，平铺地面呈莲座状，匙形；轮生聚伞花序；花 4 基数；花萼深裂，萼片 4 枚；花冠钟形，4 裂，无褶；雄蕊 4，着生于花冠裂片弯缺处，于花冠裂片互生。

本属仅 1 种，特产中国南部南岭山脉。

Perennial, rhizomatous; leaves mostly basal as a rosette, spathulate; flowers 3-4 in whorled cymes; flowers 4-merus; calyx lobed 4 nearly to the base; corolla campanulate with 4 lobes, without plicae; stamens 4, inserted at sinus between corolla lobes, alternative with corolla lobes.

1 species, endemic to southern China in the Nanling Range.

匙叶草

Latouchea fokiensis

多年生草本，高 10~20 cm，光滑，茎（花葶）单一，不分枝，有根状茎；叶基生，平铺地面呈莲座状，匙形，叶脉显著；聚伞花序轮生，每节有 2 紫色苞片；每轮有 3~8 朵花，每枚花下有 2 枚小苞片；花 4 数；花萼深裂，4 裂片线状披针形；花冠淡蓝色，花冠筒钟状，花冠裂片卵状三角形。

特产中国南部，沿南岭分布。生长在山地未被扰动的针阔叶林下，土层较薄的岩石上潮湿处。海拔 1100~1500 m。花果期 3~5 月。

Annual, 10-20 cm, erect, glabrous, simple without branch, with rhizomes; leaves basal, prostrate on ground as rosette, blade spathulate, veins conspicuous; cymes with 3-8 flowers whorled, each whorl with 2 purple bracts; flowers 4-merus; calyx divided deeply, 4 lobes linear-lanceolate; corolla pale blue, campanulate, corolla lobes ovate-triangular.

Endemic to southern China, seen only along the Nanling Range. Under undisturbed forest of mixed needle-leaf and broad-leaf trees, on wet rocks with thin soils. Altitude: 1100-1500 m. Flowering & fruiting: March-May.

A–C. 中国福建省黄冈山

Huanggangshan Mt., Fujian, China. 12.04.2001

C

中国四川省阿坝藏族羌族自治州 Aba, Sichuan, China; 11.09.2023

4.2.9
辐花属
Lomatogoniopsis

一年生草本；花辐状，5 基数；花萼深裂至基部；花冠深裂，冠筒甚短，裂片蕾期旋转排列，无褶，裂片基部对生薄片状附属物覆盖雌雄蕊；雄蕊 5 枚，着生于花冠筒，与花冠裂片互生；柱头二裂，自雌蕊顶端沿心皮的缝合线下延。

本属约 3 种，特产中国青藏高原。

Annual; flowers 5-merus, rotate; calyx divided to base; corolla deeply divided, with short tube, corolla lobes rotate without plicae, appendaged with one linear membrane covering pistil and stamens; stamens 5 attached to corolla tube, alternative with corolla lobes; stigmata 2-lobed, decurrent along apical part of ovary.

About 3 species, endemic to the Qinghai-Xizang Plateau of China.

辐花

Lomatogoniopsis alpina

一年生草本，高 5~10 cm；茎紫色，自基部多分枝，铺散或斜升；茎生叶无柄，卵形；聚伞花序，顶生和腋生；花 5 基数，花萼深裂至基部，裂片整齐，狭椭圆形；花冠淡蓝色，深裂，花冠裂片长圆状卵形，每个裂片基部着生条状披针形附属物，附属物直立，覆盖雌雄蕊。

特产于中国西藏、青海和四川邻接地区。生长在高山草地和灌丛边缘。海拔 3500~4500 m。花果期 8~10 月。

Annual, 5-10 cm; stems purple, branched from base, prostrate and ascending; stem leaves sessile, ovate; flowers in cymes terminal and axillary; flowers 5-merus; calyx divided to base, lobes equal, narrowly elliptic; corolla deeply divided, pale blue, corolla lobes oblong-ovate, each lobe with 1 appendage, corolla lobe appendages linear-lanceolate, erect, covering pistil and stamens.
Endemic to the high plateau area connecting Xizang, Sichuan and Qinghai in China. In alpine meadows and bush margins. Altitude: 3500-4500 m. Flowering & fruiting: August-October.

A

B

A-D. 中国四川省理塘县 Litang, Sichuan, China. 14.08.2002
E. 中国四川省道孚县 Daofu, Sichuan, China. 13.09.2023

C
D
E

合萼肋柱花 *Lomatogonium gamosepalum*

中国甘肃省碌曲县 Luqu, Gansu, China. 09.09.2023

4.2.10
肋柱花属
Lomatogonium

一年生或多年生草本；茎铺散、斜升或直立；花多 5 基数；花萼和花冠深裂至基部，裂片辐射状排列，无褶；花冠裂片基部生 2 蜜腺；雄蕊 5 枚，着生于花冠筒基部，与花冠裂片互生，花药多蓝色；无花柱，柱头沿两侧心皮的缝合线下延。

本属约 20 种，散布于亚欧和北美，以亚洲种类最为丰富。

Annual or perennial; stems prostrate, ascending or erect; flowers 5-merus; calyx and corolla divided to base, lobes rotate, without plicae; each corolla lobe with 2 nectary; stamens 5 attached to base of corolla tube, alternative with corolla lobes, anthers mostly blue; styles absent; stigmata lobes decurrent along carpel sutures.

About 20 species, spread in Eurasia and North America, mostly in Asia.

美丽肋柱花

Lomatogonium bellum

一年生草本，高 5~30 cm；茎紫色，直立，基部单一，中上部分枝；茎生叶无柄，卵形到卵状披针形，叶片基部圆钝，半抱茎；聚伞花序圆锥状或总状；花 5 基数，花萼深裂至基部，裂片整齐，披针形；花冠淡蓝到蓝色，深裂，花冠裂片椭圆状卵形，脉纹深蓝色，每个裂片基部两侧有腺窝。

散布于中国中部、西部到西南部。生长在高山草地和灌丛边缘。海拔 1500~3000 m。花果期 8~10 月。

Annual, 5-30 cm; stems purple, erect, simple at base and branched in middle and upper part; stem leaves sessile, ovate to ovate-lanceolate, leaf base rounded, subamplexicaul; cymes in panicle or racemelike; flowers 5-merus; calyx divided to base, lobes equal, lanceolate; corolla deeply divided, pale blue to blue, corolla lobes elliptic-ovate, veins dark blue, each lobe with 2 nectaries at base.

Scattered in central to western and southwestern China. In alpine meadows and bush margins. Altitude: 1500-3000 m. Flowering & fruiting: August-October.

A

B

A–C. F. 中国甘肃省迭部县扎尕那
Zagana, Gansu, China. 18.10.2018

D–E. 中国陕西省太白山
Taibaishan Mt., Shaanxi, China. 02.09.2023

C
D
E
F

成为旅游新热点前的中国甘肃省迭部县扎尕那
Before becoming a tourism hotspot, Zagana, Gansu, China. 18.10.2018

短药肋柱花

Lomatogonium brachyantherum

一年生草本，高 5~15 cm；茎紫红色，茎从基部起多分枝，分枝纤细，铺散或斜升；基生叶较大，匙形，茎生叶无柄，椭圆形到椭圆状披针形；花单生分枝顶端，花梗纤细，开散；花萼裂片椭圆状披针形；花冠淡蓝色到蓝色，深裂，花冠裂片椭圆形，脉纹深蓝色，每个裂片基部两侧有腺窝；花药小，黄色。

广布于中国西部的西藏、青海和甘肃等省及喜马拉雅和南亚部分国家。生长在河滩、沼泽边、高山草地和灌丛边缘。海拔 2500~4000 m。花果期 8~10 月。

Annual, 5-15 cm; stems reddish purple, much branched from base, branches slender, prostrate or ascending; basal leaves larger, spathulate, stem leaves sessile, elliptic to elliptic-lanceolate; flowers solitary on terminals of branches, pedicels slender, spreading; calyx lobes elliptic-lanceolate; corolla deeply divided, pale blue to blue, corolla lobes elliptic, veins dark blue, each lobe with 2 nectaries at base; Anthers small, yellow.

Widespread in Xizang, Qinghai, and Gansu provinces of China, and adjacent Himalayan and southern Asian countries. Beside streams and bogs, in alpine meadows. Altitude: 2500-4000 m. Flowering & fruiting: August-October.

A

B

C

A. C. 中国西藏自治区康马县
Kangma, Xizang, China. 10.08.2006

B. 中国青海省拉脊山
Lajishan Mt., Qinghai, China. 03.09.1992

肋柱花

Lomatogonium carinthiacum

一年生草本，高 5~20 cm；茎紫色，直立到斜升，基部开始分枝；基生叶有短柄，茎生叶无柄，椭圆形、卵形到卵状披针形；聚伞花序顶生，花少，松散；花 5 基数，花萼深裂至基部，裂片整齐，椭圆形到卵状披针形；花冠淡蓝到蓝色，深裂，花冠裂片椭圆形或卵状椭圆形，脉纹深蓝色，每个裂片基部两侧有腺窝。

广布于亚欧大陆温带高山地区。生长在高山草地、草甸和灌丛边缘。海拔 1500~4000 m。花果期 8~10 月。

Annual, 5-20 cm; stems purple, erect to ascending, branched from base; basal leaves short petiolate, stem leaves sessile, elliptic, ovate to ovate-lanceolate; cymes terminal sparse; flowers 5-merus; calyx divided to base, lobes equal, elliptic to ovate-lanceolate; corolla deeply divided, pale blue to blue, corolla lobes elliptic or ovate-elliptic, veins dark blue, each lobe with 2 nectaries at base.

Widespread in Eurasian temperate and alpine areas; In alpine grassland, meadows and bush margins. Altitude: 1500-4000 m. Flowering & fruiting: August-October.

A. 中国四川省小金县 Xiaojin, Sichuan, China. 23.09.1994

B. 中国青海省拉脊山 Lajishan Mt., Qinghai, China. 20.09.1992

C. 中国新疆维吾尔自治区奎屯市 Kuitun, Xinjiang, China. 12.08.1995

D. 中国陕西省太白山 Taibaishan Mt., Shaanxi, China. 02.09.2023

云南肋柱花

Lomatogonium forrestii

一年生草本，高 10~30 cm；茎紫红色，直立，基部起分枝；茎生叶无柄，狭椭圆形到线形；聚伞花序排列成圆锥状，开散；花 5 基数，花萼深裂至基部，裂片整齐，线状披针形；花冠淡蓝色，深裂，花冠裂片椭圆形，脉纹深蓝色，每个裂片基部两侧有腺窝。

特产于中国云南西北部。生长在高山草地和灌丛边缘。海拔 2500~3000 m。花果期 8~10 月。

Annual, 10-30 cm; stems reddish purple, erect, branched from base; stem leaves sessile, narrowly elliptic to linear; cymes in loose panicle; flowers 5-merus; calyx divided to base, lobes equal, linear-lanceolate; corolla deeply divided, pale blue, corolla lobes elliptic, veins dark blue, each lobe with 2 nectaries at base.

Endemic to northwestern Yunnan of China. In alpine grassland and bush margins. Altitude: 2500-3000 m. Flowering & fruiting: August-October.

A-B. 中国云南省德钦县 Deqin, Yunnan, China. 29.09.1992

合萼肋柱花

Lomatogonium gamosepalum

一年生草本，高 10~20 cm；茎紫红色，茎从基部起多分枝，铺散或斜升；茎生叶无柄，卵形、倒卵形或椭圆形；聚伞花序顶生，松散；花 5 基数，偶尔 4 基数，花萼深裂，萼筒明显，花萼裂片稍不整齐，狭卵形；花冠淡蓝色到蓝色，深裂，花冠裂片卵形，脉纹深蓝色，每个裂片基部两侧有腺窝。

散布于中国西部的西藏、青海、四川和甘肃等省（区）。生长在河滩、高山草地和灌丛边缘等开阔地。海拔 2500~4000 m。花果期 8~10 月。

Annual, 10-20 cm; stems reddish purple, much branched from base, prostrate or ascending; stem leaves sessile, ovate, obovate to elliptic; cymes terminal, loose; flowers 5-merus, occasionally 4-merus; calyx deeply divided, calyx tube conspicuous, calyx lobes subequal narrowly ovate; corolla deeply divided, pale blue to blue, corolla lobes ovate, veins dark blue, each lobe with 2 nectaries at base.

Native to western China in Xizang, Qinghai, Sichuan and Gansu provinces. In open places of alpine river banks, grassland, and bush margins. Altitude: 2500-4000 m. Flowering & fruiting: August-October.

A

B

C

A. 中国四川省甘孜藏族自治州
Ganzi, Sichuan, China. 29.09.1993

B. 中国云南省香格里拉市
Xianggelila, Yunnan, China. 28.09.1999

C. 中国甘肃省碌曲县 Luqu, Gansu, China. 30.08.1994

D–G. 中国甘肃省卓尼县 Zhuoni, Gansu, China. 08.09.2023

D
E
F
G

大花肋柱花

Lomatogonium macranthum

与美丽肋柱花（*Lomatogonium bellum*）很相似，主要区别在于本种叶片基部钝，不抱茎，伞房花序，花数量较少；花萼裂片更窄，线形到线状披针形；而前者的叶片基部圆形到心形，半抱茎，伞房花序排列成总状或圆锥状，花数量较多；花萼裂片较宽，卵状披针形。

散布于中国西部的甘肃、青海、四川、西藏和喜马拉雅山地区。生长在高山草地和灌丛边缘。海拔 2500~3500 m。花果期 8~10 月。

Highly resemble to *Lomatogonium bellum*, while *Lomatogonium macranthum*: leaf bases obtuse, not amplexicaul, cymes with fewer flowers, calyx lobes narrower, linear to linear-lanceolate; whereas Lomatogonium bellum: leaf bases rounded to cordate, subamplexicaul, cymes arranged in raceme or panicle, with more flowers, calyx lobes wider, ovate-lanceolate.

Scattered in western China and the Himalayan areas. In alpine meadows and bush margins. Altitude: 2500-3500 m. Flowering & fruiting: August-October.

A

B

C

A-B. 大花肋柱花 *Lomatogonium macranthum*
中国四川省甘孜藏族自治州
Ganzi, Sichuan, China. 29.09.1993

C. 大花肋柱花 *Lomatogonium macranthum*
中国青海省久治县
Jiuzhi, Qinghai, China. 11.09.2023

D-E. 大花肋柱花 *Lomatogonium macranthum*
中国青海省河南蒙古族自治县
Henan, Qinghai, China. 09.09.2023

F-G. 雄性不育的重瓣大花肋柱花 *Lomatogonium macranthum* 个体突变
中国四川省阿坝藏族羌族自治州
Male sterile and double corolla mutant, Aba, Sichuan, China. 09.09.2023

D
E
F
G

圆叶肋柱花

Lomatogonium oreocharis

多年生草本，高 10~20 cm，根状茎多分枝，形成较多营养枝和花枝；花枝紫色，直立或斜升；茎生叶具柄，叶片圆形到倒卵形；花单生或数朵呈顶生聚伞花序；花 5 基数，花萼深裂，萼筒钟形，花萼裂片稍不整齐，倒卵形或匙形；花冠蓝色到蓝紫色，深裂，花冠裂片倒卵形，脉纹深蓝色，每个裂片基部两侧有腺窝。

特产于中国西藏和云南。生长于高山森林和灌丛中。海拔 3000~4500 m。花果期 8~10 月。

Perennial, 10-20 cm, rhizomes well developed and much divided to vegetative and flowering stems; flowering stems purple, erect or ascending; stem leaves petiolate, blade orbicular to obovate; flowers solitary or few in cymes terminal; flowers 5-merus; calyx deeply divided, calyx tube campanulate, calyx lobes subequal, obovate to spathulate; corolla deeply divided, blue to blue-purple, corolla lobes obovate, veins dark blue, each lobe with 2 nectaries at base.

Endemic to Xizang and Yunnan of China. In alpine forest and bushes. Altitude: 3000-4500 m. Flowering & fruiting: August-October.

A

B

A–C. 中国云南省丽江市 Lijiang, Yunnan, China. 11.10.1999

C

宿根肋柱花

Lomatogonium perenne

多年生草本，高 10~20 cm，根状茎多分枝，形成莲座状营养枝和花枝；花枝黑紫色，直立；莲座叶和基部茎生叶具柄，叶片匙形；花单生或数朵呈顶生聚伞花序；花 5 基数，花萼深裂，花萼裂片狭椭圆形；花冠蓝色，深裂，花冠裂片椭圆形，脉纹深蓝色，每个裂片基部两侧有腺窝。

散布于中国西南部西藏、四川、云南、青海和陕西。生长于高山草甸、草地和灌丛中。海拔 3000~4500 m。花果期 8~10 月。

Perennial, 10-20 cm, rhizomes much divided to rosette and flowering stems; flowering stems black purple, erect; rosette and lower stem leaves petiolate, blade spathulate; flowers solitary or few in cymes terminal; flowers 5-merus; calyx deeply divided, calyx lobes narrowly elliptic; corolla deeply divided, blue, corolla lobes elliptic, veins dark blue, each lobe with 2 nectaries at base. Scattered in Xizang, Sichuan, Yunnan, Qinghai, and Shaanxi provinces of China. In alpine meadows, grassland and bushes. Altitude: 3000-4500 m. Flowering & fruiting: August-October.

A-D. 宿根肋柱花 *Lomatogonium perenne*
中国四川省雀儿山
Queershan Mt., Sichuan, China. 21.08.2006

E. 生长在蚤缀上的宿根肋柱花 *Lomatogonium perenne*
中国四川省乡城县
Growing on Arenaria cushion, Xiangcheng, Sichuan, China. 26.08.1990

D

E

辐状肋柱花

Lomatogonium rotatum

一年生草本，高 15~40 cm；茎紫红色，直立，单一或基部以上分枝，分枝紧密；茎生叶无柄，狭长披针形到线形；聚伞花序排列成紧缩总状或圆锥状；花 5 基数，花萼深裂至基部，裂片稍不整齐，线状披针形；花冠淡蓝色，深裂，花冠裂片椭圆状披针形，脉纹深蓝色，每个裂片基部两侧有腺窝。

广布于中国西部和部及俄罗斯、日本等地。生长在高山草地和灌丛边缘。海拔 1500~3500 m。花果期 8~10 月。

Annual, 15-40 cm; stems reddish purple, erect, simple or branched above base, branches densely arranged; stem leaves sessile, narrowly long lanceolate to linear; cymes in dense raceme or panicle; flowers 5-merus; calyx divided to base, lobes subequal, linear-lanceolate; corolla deeply divided, pale blue, corolla lobes elliptic-lanceolate, veins dark blue, each lobe with 2 nectaries at base.

Widespread in western to northern China, Russian and Japan. In alpine grassland and bush margins. Altitude: 1500-3500 m. Flowering & fruiting: August-October.

A–C. 中国四川省松潘县 Songpan, Sichuan, China. 14.09.1994

C

中甸肋柱花

Lomatogonium zhongdianense

一年生草本，高 15~20 cm；茎紫红色，茎从基部起多分枝，铺散或斜升；茎生叶无柄，线形；聚伞花序总状，松散；花 5 基数，花萼深裂，萼筒明显，花萼裂片稍不整齐，线形，渐尖；花冠淡蓝色到蓝紫色，深裂，花冠裂片卵形，脉纹深蓝色，每个裂片基部两侧有腺窝。

本种于合萼肋柱花接近，但叶片和萼裂片为线形而不同。特产于中国云南西北部。生长在高山草地和灌丛边缘等开阔地。海拔 ~3500 m。花果期 8~10 月。

Annual, 15-20 cm; stems reddish purple, much branched from base, prostrate or ascending; stem leaves sessile, linear; cymes terminal racemelike, loose; flowers 5-merus; calyx deeply divided, calyx tube conspicuous, calyx lobes subequal, linear, acuminate; corolla pale blue to blue-purple, deeply divided, corolla lobes ovate, veins dark blue, each lobe with 2 nectaries at base.

This species highly resembles to *Lomatogonium gamosepalum*, key differences are linear leaves and calyx lobes. Endemic to northwestern Yunnan of China. In open places of alpine grassland and bush margins. Altitude: ~3500 m. Flowering & fruiting: August-October.

中国云南省香格里拉市（香格里拉原名“中甸”） Xianggelila (original name was “Zhongdian”), Yunan, China. 30.09.1992

大钟花 *Megacodon stylophorus*
中国云南香格里拉大雪山 Grand Snow Mountain, Xianggelila, Yunnan, China. 04.08.2011

4.2.11 大钟花属 *Megacodon*

多年生高大草本；根粗壮，稍肉质；莲座叶丛发达，叶椭圆形；花枝高大，高 50~150 cm；聚伞圆锥花序；花大型，5 基数；花萼筒钟状；花冠钟状，冠筒短，无褶；雄蕊 5 枚，着生冠筒基部，与花冠裂片互生；子房一室，花柱粗壮，柱头二裂；蜜腺轮状围绕子房基部。

本属 2 种，特产中国喜马拉雅地区。

Gigantic perennial; root stout, slightly fleshy; rosette leaves well developed, elliptic; flower stems stout, erect, 50-150 cm high; inflorescences a thyrse; flowers large, 5-merus; calyx campanulate; corolla tube campanulate, short, without plicae; stamens inserted basal of corolla tube, alternative with lobes; ovary 1-celled. Style stout; stigma 2-lobed; nectaries in a whorl on gynophore.

2 species, endemic to the Sino-Himalayan areas.

大钟花

Megacodon stylophorus

高大多年生草本，根粗大，肉质；营养叶莲座状丛生，叶大型，椭圆形；花枝直立，高 50~150 cm；茎生叶数对，无柄，基部者大型，椭圆形，光滑，5~7 条粗壮弧形脉；聚伞圆锥花序；花大，具短柄，下垂；花萼绿色，钟状；花冠钟状，黄绿色，深裂，5 裂片长圆状匙形，先端圆钝；蒴果，圆锥状。

较广泛分布于喜马拉雅东部山区，印度、尼泊尔、锡金、不丹到中国的西藏、云南和四川省。生长在高山林缘及林间空地。海拔 3000~4000 m。花果期 6~9 月。

Gigantic perennial, root stout, fleshy; vegetative shoots as rosette, leaves large, elliptic; flower stem erect, 50-150 cm high; cauline leaves sessile, basal larger, blades elliptic, glabrous, with 5-7 strong longitudinal veins; thyrse with large flowers, pedicellate, pendulous; calyx green, campanulate; corolla campanulate, greenish yellow, deeply divided into 5 lobes, lobes oblong-spathulate, apex rounded; Capsule conical.
Widely spread in eastern Himalayas, in India, Nepal, Sikkim, Bhutan, and Xizang, Yunnan and Sichuan provinces of China. In margins and open places in forest. Altitude: 3000-4000 m. Flowering & fruiting: June-September.

A

B

A-B. 中国云南省香格里拉大雪山
Grand Snow Mountain, Yunnan, China. 04.08.2011

C-D. 中国云南省香格里拉市
Xianggelila, Yunnan, China. 22.08.2002

C

D

E. 中国云南省香格里拉市 摄影 /Photo: Zhekun Zhou
Xianggelila, Yunnan, China. 07.1998

F. 中国四川省乡城县
Xiangcheng, Sichuan, China. 06.08.2002

川东大钟花

Megacodon venosus

与大钟花较相似，主要区别是它的叶显著较窄，线状椭圆形，花更小；分布区独立，仅见于中国中部的湖北、重庆和四川等局部山区；生态环境也不同，为海拔较低（1000~2000 m）的砾石山坡，矮灌丛等生境；花果期较晚，9~10 月。

Highly resemble *Megacodon stylophorus*; major differences include narrower, linear-elliptic leaves, smaller flowers, disjunct distribution in central China, limited to mountains in Hubei, Chongqing and Sichuan, different habitat, at lower altitude (1000-2000 m), drier gravel hill slopes, scrub, and later flowering and fruiting season in September to October.

中国重庆市城口县 Chengkou, Chongqing, China. 21.08.2004

翼萼蔓 *Pterygocalyx volubilis*

中国甘肃省康县 Kangxian Gansu, China. 02.10.2010

4.2.12
翼萼蔓属
Pterygocalyx

一年生缠绕草本；叶对生，全缘，叶片椭圆状披针形；聚伞花序腋生；花 4 基数，花蕾稍扁压；花萼筒钟状，具 4 个宽翅；花冠筒状，4 裂，无褶；雄蕊 4 枚，着生冠筒基部，与花冠裂片互生。

本属仅 1 种，特产于东亚地区。

Annual, stems twisting; leaves opposite, entire, blade elliptic-lanceolate; cymes axillary; flowers 4-merus, flower buds somewhat flattened; calyx campanulate, with 4 wide wings; corolla tube tubular, 4 lobes without plicae; stamens 4 inserted basal of corolla tube, alternative with lobes.

1 species, endemic to eastern Asian countries.

翼萼蔓

Pterygocalyx volubilis

一年生缠绕草本；叶对生，全缘，椭圆状披针形；聚伞花序腋生；花具短柄；花萼绿色，萼筒钟状，具 4 宽翅，裂片披针形；花冠管状，淡蓝紫色，4 裂，裂片长圆形，先端圆钝。

较广泛分布于东亚到东北亚地区，从中国西南部、中部到东北，及俄罗斯、朝鲜和日本。生长在高山林缘及灌丛中。海拔 1000~3000 m。花果期 8~10 月。

Annual twisting; leaves opposite, entire, elliptic-lanceolate; cymes axillary; flower shortly petiolate; calyx green, campanulate, with 4 conspicuous wings, calyx lobes lanceolate; corolla tubular pale blue-purple, divided into 4 lobes, lobes oblong, apex obtuse.

Widely spread in eastern Asian, from southwestern, central to northeastern China, Russian, Korea Peninsula, and Japan. In margins of forest and bushes. Altitude: 1000-3000 m. Flowering & fruiting: August-October.

A–B. 中国云南省德钦县 Deqin, Yunnan, China. 26.09.1999

C. 中国四川省九寨沟 Jiuzhaigou, Sichuan, China. 13.09.1994

C

中国甘肃省夏河县景观 Scenery in Xiahe, Gansu, China. 12.07.1990

4.2.13
异型花属
Sinoswertia

异型花属 *Sinoswertia* 为单种属，仅异型花 *Sinoswertia tetraptera* 1 种，特产于中国青藏高原东北部的西藏、四川、青海和甘肃。生长于牧场、路边、农田边等扰动过的生境。海拔 2000~4000 m。

一年生草本，分枝和花异型。直立枝正常，叶对生，全缘，叶片卵状披针形；聚伞花序排列成总状；花较大，异花受粉，4 基数，花萼 4 深裂，裂片卵状披针形；花冠绿色，4 深裂，无褶，裂片基部具 2 腺窝；雄蕊 4 枚，着生冠筒基部，与花冠裂片互生。基部分枝纤细，横卧地面；横卧枝叶细小，线状披针形，二歧聚伞花序，花小，闭合，闭花受粉，4 基数，无腺体。花果期 7~9 月。

Annual, dimorphic plants with dimorphic flowers. Main plant stem erect, leaves opposite, entire, blade ovate-lanceolate; cymes arranged racemelike; flowers bigger, open pollination, 4-merus, calyx deeply divided into 4 lobes, lobes ovate lanceolate; corolla green, deeply divided into 4 lobes, without plicae, lobes with 2 nectaries at base; stamens 4, inserted at base of corolla, alternative with corolla lobes. Basal branches slender, prostrate on ground, leaves minor, linear-lanceolate, bifurcate cyme, flowers minor, closed, cleistogamous, 4-merus without nectary. Flowering and Fruiting: July-September.

Endemic to northeastern part of the Qinghai-Xizang Plateau of China, in Xizang, Sichuan, Qinghai and Gansu provinces. Often growing in disturbed habitats such as pastures, roadside, and margins of farming land. Altitude: 2000-4000 m.

A

B

A. 中国甘肃省榆中县 Yuzhong, Gansu, China. 12.07.1990

B. 中国青海省河南蒙古族自治县 Henan, Qinghai, China. 09.09.2023

C–D. 中国甘肃省碌曲县 Luqu, Gansu, China. 08.09.2023

中国新疆维吾尔自治区天山 Tianshan Mountain, Xinjiang, China. 04.09.2004

4.2.14
獐牙菜属
Swertia

一年或多年生草本；常有明显主根；茎粗壮或纤细，或花葶状；叶对生，稀互生或轮生；聚伞花序或单花；花 4 或 5 基数，辐状；花萼深裂至基部；花冠深裂，冠筒短，无褶，裂片基部或中部有腺窝或腺斑，腺窝常有缘毛；雄蕊着生冠基部，与花冠裂片互生；子房 1 室，花柱短，柱头 2 裂；朔果 2 瓣裂，种子小而多。

本属约 135 种，分布于亚洲、非洲、欧洲和北美，亚洲种类较为丰富。

Annual or perennial; tap root well developed; stem stout, slender or scapiform; leaves opposite, occasionally alternative or whorled; inflorescences cymose, flowers rarely solitary; flowers 4- or 5-merus, rotate; calyx deeply divided to base; corolla deeply divided, tube short, without plicae, corolla lobes with conspicuous nectaries at middle or base, nectary often fimbriate; stamens inserted basal of corolla tube, alternative with lobes; ovary 1-celled. Style short; stigma 2-lobed. Capsule 2-valved; seeds small, numerous.

About 135 species, widespread in Asia, Africa, Europe and North America.

狭叶獐牙菜

Swertia angustifolia

一年生草本，茎直立，四棱形，高 20~60 cm；茎生叶无柄，披针形到椭圆状披针形，全缘；聚伞花序圆锥状；花艳丽，4 基数；花萼绿色，深裂，裂片线状披针形；花冠白色到黄白色，具蓝紫色纹路，5 深裂，裂片卵形，中下部具 1 个袋状腺窝，被圆形绿色鳞片覆盖。

散布于中国喜马拉雅地区各国，在中国南部和西南部较常见。生长在山坡草地、荒地和农田边缘。海拔 500~3000 m。花果期 8~10 月。

Annual; stem erect, quadrangular, 20-60 cm; stem leaves sessile, lanceolate to elliptic-lanceolate, entire; cymes a panicle; flower showery, 4-merus; calyx green, deeply divided, calyx lobes linear-lanceolate; corolla white to yellowish white with blue-purple stripes, divided into 5 lobes, lobes ovate, each with 1 nectary covered by a green scale.

Widespread in the Sino-Himalayan areas; in southern and southwestern provinces in China. In hill slopes, grassland, wasteland, and margins of farming fields. Altitude: 500-3000 m. Flowering & fruiting: August-October.

A

B

A–B. 中国云南省哀牢山 摄影 /Photo: Adrien Favre
Ailaoshan Mt., Yunnan, China. 09.09.2004

C–D. 中国云南省宾川县 Binchuan, Yunnan, China. 05.10.1999

C

D

二叶獐牙菜

Swertia bifolia

多年生草本，具短根状茎；花枝直立，花葶状，高 10~30 cm；基生叶 1~2 对，具柄，狭椭圆形，全缘，先端圆钝，基部楔形，渐狭成柄；茎生叶生顶部，苞片状；聚伞花序顶生；花 5 基数；花萼绿色，深裂，裂片披针形，稍不整齐；花冠蓝紫色，5 深裂，裂片长圆形，基部具 2 个腺窝，边缘被柔毛状流苏。

散布于中国西部到西南部的西藏、四川、青海、甘肃和陕西。生长在高山草地、草甸和灌丛边缘。海拔 2800~3500 m。花果期 7~9 月。

Perennial with short rhizome; stem erect, scapiform, 10-30 cm; basal leaves 1-2 pairs, petiolate, narrowly elliptic, entire, apex obtuse, base narrowed to petiole, stem leaves on top only, scale-like; cymes terminal; flower 5-merus; calyx green, deeply divided, calyx lobes lanceolate; corolla blue-purple, deeply divided into 5 lobes, lobes oblong, each with 2 nectary at base, nectary margins with pilose fimbriae.

Scattered in the western and southwestern provinces in China, Xizang, Sichuan, Qinghai, Gansu and Shaanxi. In alpine grassland, meadows, and bush margins. Altitude: 2800-3500 m. Flowering & fruiting: July-September.

A–B. 中国四川省德格县 Dege, Sichuan, China. 21.08.2006

獐牙菜

Swertia bimaculata

一年生草本；茎直立，高 30~130 cm；茎生叶无柄，椭圆形至卵状披针形，全缘；聚伞花序顶生和腋生，圆锥状，疏松开展；花 5 基数，偶尔 4 基数；花萼绿色，深裂，裂片狭椭圆形；花冠黄白色，具紫色斑点，5 深裂，裂片长圆形，中部具 2 个黄绿色腺斑，腺斑光滑，无附属物。

广布于中国 - 喜马拉雅及东南亚各地。生长在山坡草地、林下和灌丛中。海拔 300~3000 m。花果期 7~10 月。

Annual; stem erect, 30-130 cm; stem leaves sessile, elliptic to ovate-lanceolate, entire; cymes terminal and axillary forming a panicle, loose and spreading; flower 5-merus, occasionally 4-merus; calyx green, deeply divided, calyx lobes narrowly elliptic; corolla yellowish white with purple dots and patches, deeply divided into 5 lobes, lobes oblong, each with 2 nectaries at middle, nectaries yellow-green, orbicular, reduced to a naked gland patch without raised margin.

Widespread in the Sino-Himalayan and southeastern Asian countries. In hill slope, grassland, forest and bushes. Altitude: 300-3000 m. Flowering & fruiting: July-October.

A–B. 中国浙江省龙泉市 Longquan, Zhejiang, China. 17.10.2022

C–E. 中国湖北省神农架 Shennongjia Mt., Hubei, China. 18.08.2004

F–G. 中国甘肃省康县 Kangxian, Gansu, China. 03.10.2010

宾川獐牙菜

Swertia binchuanensis

一年生草本；茎直立，高 20~40 cm，四棱形；茎生叶无柄，披针形，全缘；聚伞花序顶生和腋生，圆锥状，疏松开展；花 5 基数；花萼绿色，深裂，裂片披针形；花冠黄色，5 深裂，裂片卵状披针形，近基部两侧具 2 个黄色条形腺窝，腺窝边缘具长流苏。

特产于中国云南西部。生长在较干燥的向阳山坡草地和灌丛边缘。海拔 1800~2200 m。花果期 9~10 月。

Annual; stem erect, 20-40 cm, quadrangular; stem leaves sessile, lanceolate, entire; cymes terminal and axillary forming a panicle, loose and spreading; flower 5-merus; calyx green, deeply divided, calyx lobes lanceolate; corolla yellow, deeply divided into 5 lobes, lobes ovate-lanceolate, each with 2 nectaries at base, nectaries linear, with long fimbriae along margins.
Endemic to western Yunnan of China. In dry and sunny hill slope, grassland, and bush margins. Altitude: 1800-2200 m. Flowering & fruiting: September-October.

A–B. 中国云南省宾川县 Binchuan, Yunnan, China. 04.10.1999

叶萼獐牙菜

Swertia calycina

多年生草本，具短根状茎；花枝直立，花葶状，高 10~30 cm；基生叶具柄，线状长圆形，全缘；茎生叶生顶部，苞片状；聚伞花序顶生；花 5 基数；花萼绿色，深裂，裂片卵状披针形；花冠淡黄色，5 深裂，裂片卵状长圆形，基部具 2 个褐色腺窝，边缘被柔毛状流苏。

特产于中国西南部的云南和四川。生长在高山草地、草甸和灌丛边缘。海拔 2800~3500 m。花果期 7~9 月。

Perennial with short rhizome; stem erect, scapiform, 10-30 cm; basal leaves petiolate, linear-oblong, entire, stem leaves on top only, scale-like; cymes terminal; flower 5-merus; calyx green, deeply divided, calyx lobes ovate-lanceolate; corolla pale yellow, deeply divided into 5 lobes, lobes ovate-oblong, each with 2 nectaries at base, nectary margins with pilose fimbriae.

Endemic to Sichuan and Yunnan provinces in southwestern China. In alpine grassland, meadows, and bush margins. Altitude: 2800-3500 m. Flowering & fruiting: July-September.

A-C. 中国云南省丽江 Lijiang, Yunnan, China. 03.10.1992

西南獐牙菜

Swertia cincta

一年生草本，茎直立，高 30~100 cm；茎生叶具短柄，椭圆状披针形；聚伞圆锥花序顶生，松散；花 5 基数；花萼绿色，深裂，裂片卵状披针形，稍长于花冠，边缘具长睫毛；花冠黄绿色，5 深裂，裂片卵形，基部具连续的紫色环带，每裂片基部具 2 个绿色裸露腺斑，无附属物；花丝基部扩大扁平联合成短筒状包围子房。

特产于中国西南部的云南、四川和贵州。生长在潮湿山坡、林缘和灌丛边缘。海拔 2000~3500 m。花果期 8~9 月。

Annual, stem erect, 30-100 cm; stem leaves shortly petiolate, elliptic-lanceolate; cymes as a panicle, terminal; flower 5-merus; calyx green, deeply divided, calyx lobes ovate-lanceolate, slightly longer than corolla, margin ciliate; corolla yellowish green, deeply divided into 5 lobes, lobes ovate with a continuous purple ring at base, each lobe with 2 green naked nectary patches at base, without any appendage; filaments flattened and connate at base to tube-like surrounding pistil.

Endemic to Sichuan, Yunnan and Guizhou provinces in southwestern China. In moist slopes, forest and bush margins. Altitude: 2000-3500 m. Flowering & fruiting: August-September.

A. 中国云南省昆明市 unming，Yunnan, China. 19.09.1999

B. 中国云南省云龙县 Yunlong, Yunnan, China. 22.09.2023

C. 中国云南省云龙县 Yunlong, Yunnan, China. 22.09.2023

楔叶獐牙菜

Swertia cuneata

多年生草本，有短根状茎，直立，高10~30 cm；基生叶具柄，长圆状匙形，全缘，基部渐狭成叶柄；茎生叶较少，狭椭圆形；聚伞花序顶生；花5基数；花萼绿色，深裂，裂片匙形；花冠淡蓝紫色，5深裂，裂片卵状椭圆形，基部具2个深色腺窝，边缘被柔毛状流苏。

特产于中国喜马拉雅山中部的印度、尼泊尔和中国西藏西南部。生长在高山草地、草甸和灌丛边缘。海拔3500~4500 m。花果期9~10月。

Perennial with short rhizome; stem erect, 10-30 cm; basal leaves petiolate, oblong-spathulate, entire, base cuneate, narrowed to petiole; stem leaves few, narrowly elliptic; cymes terminal; flower 5-merus; calyx green, deeply divided, calyx lobes spathulate; corolla pale blue-purple, deeply divided into 5 lobes, lobes ovate-elliptic, each with 2 nectary at base, nectary margins with pilose fimbriae.

Limited to central Himalayas in India, Nepal and southwestern Xizang of China. In alpine grassland, meadows, and bush margins. Altitude: 3500-4500 m. Flowering & fruiting: September-October.

A-B. 中国西藏自治区聂拉木县
Nielamu, Xizang, China. 16.09.1992

观赏獐牙菜

Swertia decora

一年生草本；茎直立，高 5~15 cm，四棱形，基部以上多分枝；茎生叶近无柄，披针形，全缘；花单生分枝顶端，5 基数；花萼绿色，深裂，裂片线状披针形；花冠粉紫色，5 深裂，裂片卵状长圆形，近基部两侧具 2 个黄色条形腺窝，腺窝边缘具短流苏。

特产于中国西南部云南和四川。生长在较干燥的向阳山坡草地和灌丛边缘。海拔 1800~2500 m。花果期 8~10 月。

Annual; stem erect, 5-15 cm, quadrangular, much branched from base and above; stem leaves subsessile, lanceolate, entire; flower solitary on terminal, 5-merus; calyx green, deeply divided, calyx lobes linear-lanceolate; corolla pink-purple, deeply divided into 5 lobes, lobes ovate-oblong, each with 2 nectaries at base, nectaries linear, with short fimbriae along margins.

Endemic to Yunnan and Sichuan in southwestern China. In dry and sunny hill slope, grassland, and bush margins. Altitude: 1800-2500 m. Flowering & fruiting: August-October.

A–C. 中国云南省大理白族自治州

Dali, Yunnan, China. 06.10.1999

丽江獐牙菜

Swertia delavayi

一年生草本；茎直立，高 10~30 cm，四棱形，具狭翅，基部以上多分枝；茎生叶近无柄，椭圆状披针形，全缘；花单生分枝顶端，5 基数；花萼绿色，深裂，裂片卵形，3 大 2 小；花冠粉紫色，5 深裂，裂片卵状长圆形，近基部两侧具 2 个黄色条形腺窝，腺窝边缘具短流苏。

特产于中国西南部云南和四川。生长在较干燥的向阳山坡草地和灌丛边缘。海拔 1800~2500 m。花果期 8~10 月。

Annual; stem erect, 10-30 cm, quadrangular with narrow wings, much branched from base and above; stem leaves subsessile, elliptic-lanceolate, entire; flower solitary on terminal, 5-merus; calyx green, deeply divided, calyx lobes ovate, 3 bigger and 2 smaller; corolla pink-purple, deeply divided into 5 lobes, lobes ovate-oblong, each with 2 nectaries at base, nectaries linear, with short fimbriae along margins.

Endemic to Yunnan and Sichuan in southwestern China. In dry and sunny hill slope, grassland, and bush margins. Altitude: 1800-2500 m. Flowering & fruiting: August-October.

A–C. 中国云南省丽江市 Lijiang, Yunnan, China. 09.10.1998

歧伞獐牙菜

Swertia dichotoma

一年生草本，茎纤细，四棱形，平铺或斜升，二歧分枝；茎生叶具短柄，匙形到卵状披针形；花细小，单生顶端和叶腋，开展；花 4 基数；花萼绿色，深裂，裂片卵形；花冠黄白色，4 深裂，裂片卵形，每裂片基部具 2 个腺窝，下陷，形成类似于短距的结构。

较广泛散布于中国北方及俄罗斯、蒙古和日本。生长在较潮湿的山坡、林缘和灌丛边缘的开阔地。海拔 1200~3000 m。花果期 6~7 月。

Annual, stem slender, quadrangular, prostrate or ascending, branched dichotomously; stem leaves shortly petiolate, spathulate to ovate-lanceolate; flowers small, 4-merus; calyx green, deeply divided, calyx lobes ovate; corolla yellowish white, deeply divided into 4 lobes, lobes ovate, each lobe with 2 green nectaries at base, sunk into a short spur-like structure.

Widespread in northern China, Russia, Mongolia, and Japan. In open places of moist slopes, forest and bush margins. Altitude: 1200-3000 m. Flowering & fruiting: June-July.

中国甘肃省民乐县 Minle, Gansu, China. 20.07.1995

北方獐牙菜

Swertia diluta

一年生草本；茎直立，高 10~50 cm，四棱形；茎生叶无柄，狭披针形，全缘；聚伞花序圆锥状，花多数；花 5 基数；花萼绿色，深裂，裂片线形到狭披针形；花冠黄白色到淡蓝色，脉深蓝色，花冠裂片卵状披针形，近基部两侧具 2 个黄色条形腺窝，腺窝边缘具长流苏。

习见于东亚的俄罗斯、蒙古、朝鲜半岛、日本及中国的西北、华北到东北各省。生长在较干燥向阳山坡草地和灌丛边缘。海拔 500~2500 m。花果期 8~10 月。

Annual; stem erect, 10-50 cm, quadrangular; stem leaves sessile, narrowly lanceolate, entire; inflorescence a panicle with many flowers; flower 5-merus; calyx green, deeply divided, calyx lobes linear to narrowly lanceolate; corolla yellowish white to pale blue, veins dark blue, deeply divided into 5 lobes, lobes ovate-lanceolate, each with 2 nectaries at base, nectaries yellow, linear, with long fimbriae along margins.

Common in eastern Asia in Russia, Mongolia, Korean peninsula, Japan and China (northwestern, northern to northeastern provinces). In dry and sunny hill slope, grassland, and bush margins. Altitude: 500-2500 m. Flowering & fruiting: August-October.

A-B. 中国甘肃省宕昌县 Tanchang, Gansu, China. 02.10.2010

叉序獐牙菜

Swertia divaricata

多年生草本；花枝粗壮，直立，有条棱，高 50~70 cm；基生叶不发达或无，长圆状椭圆形，全缘；茎生叶椭圆形，下部叶有短柄，上部叶无柄；聚伞圆锥花序，松散，二叉分枝；花 5 基数；花萼绿色，深裂，裂片披针形；花冠黑紫色，5 深裂，裂片椭圆形，中部具 2 个褐色裸露腺窝，边缘被短毛状流苏。

特产于中国云南西北部。生长山谷林缘。海拔 2400 m 以下。花果期 8~9 月。

Perennial; flowering stem robust, erect, striate, 50-70 cm; basal leaves few or absent, petiolate, oblong-elliptic, entire, stem leaves elliptic, low ones petiolate, upper ones sessile; inflorescence a sparse panicle, dichotomously branched; flower 5-merus; calyx green, deeply divided, calyx lobes lanceolate; corolla blackish purple, deeply divided into 5 lobes, lobes elliptic, each with 2 connected nectaries at middle, nectary margins with short pilose fimbriae.

Endemic to northwestern Yunnan province in China. In woodland margins in valleys. Altitude: ~2400 m. Flowering & fruiting: August-September.

A–B. 中国云南省贡山独龙族怒族自治县
摄影 /Photo：付鹏程
Gongshan, Yunnan, China. 26.08.2018

高獐牙菜

Swertia elata

多年生草本，具短根状茎；花枝直立，高 40~100 cm 不分枝；叶多基生，具柄，线状长圆形至狭披针形，全缘；茎生叶少，较小；聚伞圆锥花序；花 5 基数；花萼绿色，深裂，裂片卵状披针形；花冠黄绿色，具紫色斑点和条纹，5 深裂，裂片卵状长圆形，基部具 2 个褐色腺窝，边缘被柔毛状流苏。

特产于中国西南部的云南和四川。生长在高山草地、草甸和灌丛边缘。海拔 3000~4000 m。花果期 7~9 月。

Perennial with short rhizome; stem erect, 40-100 cm; leaves mostly basal, petiolate, linear-oblong to lanceolate, entire, stem leaves few, smaller; inflorescence a panicle; flower 5-merus; calyx green, deeply divided, calyx lobes ovate-lanceolate; corolla yellow-green, deeply divided into 5 lobes, lobes ovate-oblong, each with 2 nectaries at base, nectary margins with lone pilose fimbriae.

Endemic to Sichuan and Yunnan provinces in southwestern China. In alpine grassland, meadows, and bush margins. Altitude: 3000-4000 m. Flowering & fruiting: July-September.

A-B. 中国四川省理塘县 Litang，Sichuan, China. 08.08.2002

C-E. 中国四川省理塘县 Litang，Sichuan, China. 08.08.2002

红直獐牙菜

Swertia erythrosticta

多年生草本，具短根状茎；茎直立，高 20~50 cm 不分枝，紫红色；叶多基生，具柄，长圆状椭圆形，全缘，具紫黑色斑块；茎生叶少，较小，卵状椭圆形；聚伞圆锥花序；花 5 基数，常下垂；花萼绿色，深裂，裂片卵状披针形；花冠黄绿色，具紫色斑点，5 裂片卵状披针形，基部具 1 个圆形褐色腺窝，边缘被柔毛状流苏。

散布于中国西南到华北的四川、湖北、青海、甘肃、陕西、山西、河北到内蒙古及朝鲜。生长在高山草地、溪流边和灌丛边缘。海拔 1500~4000 m。花果期 8~10 月。

Perennial with short rhizome; stem erect, 20-50 cm, reddish purple; leaves mostly basal, petiolate, oblong-elliptic, entire, with black spots, stem leaves few, smaller; inflorescence a panicle; flowers 5-merus, often pendulous; calyx green, deeply divided, calyx lobes ovate-lanceolate; corolla yellow-green with purple spots, deeply divided into 5 lobes, lobes ovate-lanceolate, each with 1 round nectary at base, nectary margins with lone pilose fimbriae.

Spread from SW to N China, Sichuan, Hubei, Qinghai, Gansu, Shaanxi, Shanxi, Hebei, and Inner Mongolia, and Korea. In alpine grassland, streamside, and bush margins. Altitude: 1500-4000 m. Flowering & fruiting: August-October.

A–B. 中国四川省松潘县 Songpan，Sichuan, China. 15.09.1994

C. 中国四川省松潘县 Songpan，Sichuan, China. 15.09.1994

抱茎獐牙菜

Swertia franchetiana

一年生草本；茎直立，高 10~50 cm，四棱形；茎生叶无柄，披针形，全缘，叶片基部耳状半抱茎，稍下延成茎的翅；聚伞花序圆锥状，疏松开展；花 5 基数；花萼绿色，深裂，裂片披针形；花冠黄白色到淡蓝色，脉深蓝色，花冠裂片卵状披针形，近基部两侧具 2 个蓝色条形腺窝，腺窝边缘具长流苏。

特产于中国西部西藏、四川、青海和甘肃。生长在较干燥向阳山坡草地和灌丛边缘。海拔 2200~3500 m。花果期 8~11 月。

Annual; stem erect, 10-50 cm, quadrangular; stem leaves sessile, lanceolate, entire, base auriculate and subamplexicaul, decurrent into narrow wings on stem; inflorescence a panicle, loose and spreading; flower 5-merus; calyx green, deeply divided, calyx lobes lanceolate; corolla yellowish white to pale blue, veins dark blue, deeply divided into 5 lobes, lobes ovate-lanceolate, each with 2 nectaries at base, nectaries blue, linear, with long fimbriae along margins.

Endemic to western China, Xizang, Sichuan, Qinghai and Gansu. In dry and sunny hill slope, grassland, and bush margins. Altitude: 2200-3500 m. Flowering & fruiting: August-November.

A–B. 中国四川省康定市 Kangding, Sichuan, China. 14.09.2023

C-D. 中国西藏自治区林芝市 Linzhi, Xizang, China. 21.09.2003

矮獐牙菜

Swertia handeliana

多年生矮小垫状草本，具短根状茎和营养枝；花枝直立，花葶状，高 3~5 cm；叶基生，具柄，莲座状，叶片卵圆形，全缘；茎生叶苞片状；花单生分枝顶端；4 基数；花萼绿色，深裂，裂片卵形；花冠常闭合状，蓝色到蓝紫色，裂片卵圆形，基部具 1 个腺斑，无流苏。

特产于中国的云南和西藏。生长在高山草甸。海拔 3500~4500 m。花果期 8~10 月。

Perennial with short rhizome and vegetative shoots, cushion-like; stem erect, scapiform, 3-5 cm; leaves mostly basal, petiolate, rosette, blade ovate-orbicular, stem leaves few, scale-like; flowers solitary on terminals; 4-merus; calyx green, deeply divided, calyx lobes ovate; corolla often closed, blue to blue-purple, deeply divided into lobes ovate-orbicular, each with 1 nectary at base without fimbriae.

Endemic to Xizang and Yunnan provinces of China. In alpine meadows. Altitude: 3500-4500 m. Flowering & fruiting: August-October.

A–B. 中国西藏自治区多雄拉山
Duoxiongla Mt., Xizang, China. 24.09.2003

C-D. 中国西藏自治区多雄拉山 Duoxiongla Mt., Xizang, China. 24.09.2003

浙江獐牙菜

Swertia hickinii

一年生草本；茎直立，高 15~30 cm，四棱形，有窄翅；茎生叶无柄，披针形，全缘；聚伞花序圆锥状，疏松开展；花 5 基数；花萼绿色，深裂，裂片披针形；花冠黄白色到淡蓝色，脉深蓝紫色，花冠裂片卵形，近基部两侧具 2 个腺窝，腺窝边缘具白色长流苏。

特产于中国中部、南部到东部。生长在向阳山坡草地和灌丛边缘。海拔 200~1500 m。花果期 9~11 月。

Annual; stem erect, 15-30 cm, quadrangular, narrowly winged; stem leaves sessile, lanceolate, entire; inflorescence a panicle, loose and spreading; flower 5-merus; calyx green, deeply divided, calyx lobes lanceolate; corolla white to pale blue, veins dark blue purple, deeply divided into 5 lobes, lobes ovate, each with 2 nectaries at base, nectaries round, with long white fimbriae along margins.

Spread in central, southern to eastern China. In sunny hill slope, grassland, and bush margins. Altitude: 200-1500 m. Flowering & fruiting: September-November.

A–B. 中国浙江省龙泉市 Longquan, Zhejiang, China. 17.10.2022

C–E. 中国浙江省龙泉市 Longquan, Zhejiang, China. 17.10.2022

毛萼獐牙菜

Swertia hispidicalyx

一年生草本；茎直立，高 5~20 cm，四棱形，紫色，从基部分枝，侧枝开散；茎生叶无柄，披针形，全缘；聚伞花序圆锥状，疏松开展；花 5 基数；花萼绿色，深裂，裂片卵状披针形；花冠粉紫色，花冠裂片卵形，近基部两侧具 2 个腺窝，腺窝被覆盖圆形鳞片和短流苏覆盖，朝花冠裂片基部开口；花丝基部扁平扩大。

特产于中国西藏。生长在向阳的河滩、空地和灌丛边缘。海拔 3500~4500 m。花果期 8~10 月。

Annual; stem erect, 5-20 cm, quadrangular, purple, branched from base, branches loose, spreading; stem leaves sessile, lanceolate, entire; inflorescence a panicle, loose and spreading; flower 5-merus; calyx green, deeply divided, calyx lobes ovate-lanceolate; corolla pink-purple, deeply divided into 5 lobes, lobes ovate, each with 2 nectaries at base, nectaries round, covered by an orbicular scale and short fimbriae, open toward base of lobe; filaments flatten and enlarged toward base.

Endemic to Xizang of China. In sunny river banks, open places, and bush margins. Altitude: 3500-4500 m. Flowering & fruiting: August-October.

A

B

C

A–G. 中国西藏自治区林芝市 Linzhi, Xizang, China. 22.09.2003

D
E
F
G

粗壮獐牙菜

Swertia hookeri

多年生草本，具短根状茎；茎直立，粗壮，花葶状，高 20~80 cm，紫红色；叶多数基生莲座状，具柄，匙状椭圆形，全缘；茎生叶少，较小，椭圆形；聚伞圆锥花序；花 4 基数；花萼绿色，深裂，裂片卵形；花冠红紫色，脉深褐色，裂片卵状椭圆形，基部具 1 个褐色圆形裸露的腺窝。

特产于喜马拉雅山尼泊尔、不丹、锡金和中国西藏。生长在高山草地和灌丛中。海拔 3800~4200 m。花果期 7~9 月。

Perennial with short rhizome; stem erect, stout, scapiform, 20-80 cm, reddish purple; leaves mostly basal, rosette like, petiolate, spathulate-elliptic, entire, stem leaves few, smaller; inflorescence a panicle; flowers 4-merus, often pendulous; calyx green, deeply divided, calyx lobes ovate; corolla reddish purple, veins darker, 4 lobes ovate-elliptic, each with 1 round naked nectary at base, no fimbriae.

Endemic to the Himalayas, in Nepal, Bhutan, Sikkim and Xizang of China. In alpine grassland and bushes. Altitude: 3800-4200 m. Flowering & fruiting: July-September.

A–D. 中国西藏自治区错那市 Cuona, Xizang, China. 14.08.2006

C

D

黄花獐牙菜

Swertia kingii

多年生草本，具短根状茎；花枝粗壮，直立，高 50~100 cm；基生叶具柄，宽椭圆形；茎中上部叶多对，无柄，卵形；聚伞圆锥花序；花 5 基数；花萼绿色，深裂，裂片卵状披针形；花冠黄绿色，具蓝紫色条纹，5 深裂，裂片长圆状卵形，基部具 2 个圆形腺窝，边缘被短柔毛状流苏。

特产于喜马拉雅山尼泊尔、锡金、不丹和中国西藏。生长在高山灌丛中。海拔 3500~3800 m。花果期 8~9 月。

Perennial with short rhizome; flowering stem stout, erect, 50-100 cm; basal leaves petiolate, wide elliptic, middle and upper stem leaves numerous pairs, sessile, ovate, smaller; Inflorescence a panicle; flower 5-merus; calyx green, deeply divided, calyx lobes ovate-lanceolate; corolla yellow-green, deeply divided into 5 lobes, lobes oblong-ovate, each with 2 orbicular nectaries at base, nectary margins with short pilose fimbriae.

Endemic to Himalayas, in Nepal, Sikkim, Bhutan,and Xizang of China. In alpine bushes. Altitude: 3500-3800 m. Flowering & fruiting: August-September.

A

B

A-B. 中国西藏自治区聂拉木县
Nielamu, Xizang, China. 19.09.1992

大籽獐牙菜

Swertia macrosperma

一年生草本；茎直立，高 30~100 cm，四棱形，紫色，基部以上分枝；茎生叶无柄，披针形，全缘；聚伞花序圆锥状，疏松开展；花 5 基数；花萼绿色，深裂，裂片披针形；花冠黄白色到淡粉色，脉深紫色，花冠裂片卵状椭圆形，近基部两侧具 2 个蓝色条形腺窝，腺窝边缘具长流苏；种子 3~4 粒，较其他种大。

广布于中国喜马拉雅地区。生长在较干燥向阳山坡草地、灌丛边缘、牧场及扰动过的空地。海拔 1500~3500 m。花果期 8~11 月。

Annual; stem erect, 30-100 cm, quadrangular, purple, branched above base; stem leaves sessile, lanceolate, entire; inflorescence a panicle, loose and spreading; flower 5-merus; calyx green, deeply divided, calyx lobes lanceolate; corolla yellowish white to pale pink, veins dark purple, deeply divided into 5 lobes, lobes ovate-elliptic, each with 2 nectaries at base, nectaries linear, with long fimbriae along margins; seeds 3-4, big comparing to other species.

Widespread in the Sino-Himalayan areas. In dry and sunny hill slope, grassland, bush margins, pastures and disturbed open places. Altitude: 1500-3500 m. Flowering & fruiting: August-November.

A–B. 中国云南省丽江市 Lijiang, Yunnan, China. 24.08.2002

C

D

C. 中国云南省大理白族自治州 Dali, Yunnan, China. 06.10.1999

D. 中国四川省盐源县 Yanyuan, Sichuan, China. 16.09.2023

E. 中国云南省德钦县 Deqin, Yunnan, China. 20.09.2023

E

膜边獐牙菜

Swertia marginata

多年生草本，具短根状茎；花枝粗壮，直立，高 20~40 cm，黄绿色；基生叶具柄，狭椭圆形；茎中上部叶无柄，卵状椭圆形；聚伞圆锥花序，密集；花 5 基数；花萼绿色，深裂，裂片卵状披针形，具膜质边缘；花冠黄绿色，具蓝紫色条纹，5 深裂，裂片长圆状卵形，基部具 2 个圆形腺窝，边缘被短柔毛状流苏。

特产于克什米尔地区、俄罗斯、蒙古及中国新疆。生长在高山灌丛中。海拔 2500~3000 m。花果期 8~9 月。

Perennial with short rhizome; flowering stem stout, erect, 20-40 cm, yellowish green; basal leaves petiolate, narrowly elliptic, middle and upper stem leaves sessile, ovate-elliptic, smaller; inflorescence a panicle, dense; flower 5-merus; calyx green, deeply divided, calyx lobes ovate-lanceolate, margins membranous; corolla yellow-green, deeply divided into 5 lobes, lobes oblong-ovate, each with 2 orbicular nectaries at base, nectary margins with short pilose fimbriae.

Endemic to the Kashmir region, Russia, Mongolia, and Xinjiang of China. In alpine bushes. Altitude: 2500-3000 m. Flowering & fruiting: August-September.

A

B

A–C. 中国新疆维吾尔自治区巩乃斯

Gongnaisi, Xinjiang, China. 07.09.2004

中国新疆维吾尔自治区和静县巴音布鲁克，远处为天山
Tianshan seen from distance, Bayinbuluk, Hejing, Xinjiang, China. 09.09.2004

川西獐牙菜

Swertia mussotii

一年生草本；茎直立，高 20~60 cm，四棱形，绿色，基部起多分枝，松散；茎生叶无柄，卵状披针形，全缘；大型聚伞花序圆锥状，疏松开展；花小，4 基数；花萼绿色，深裂，裂片披针形；花冠暗紫红色，花冠裂片披针形，近基部两侧具 2 个腺窝，腺窝边缘具长流苏。

原产于中国西藏、青海、四川和云南。生长在向阳山坡草地、河谷和灌丛边缘。海拔 2000~3500 m。花果期 8~10 月。

Annual; stem erect, 20-60 cm, quadrangular, green, branched from base, branches spreading; stem leaves sessile, ovate-lanceolate, entire; inflorescence gigantic panicle, loose and spreading; flowers small, 4-merus; calyx green, deeply divided, calyx lobes lanceolate; corolla dark reddish purple, 4 lobes lanceolate, each with 2 nectaries at base with long fimbriae along margins.

Native to Xizang, Qinghai, Sichuan and Yunnan provinces of China. In sunny hill slope, grassland, river banks, and bush margins. Altitude: 2000-3500 m. Flowering & fruiting: August-October.

A

B

A. 中国西藏自治区察隅县 Chayu, Xizang, China. 09.10.2003

B-C. 中国西藏自治区昌都市 摄影 /Photo：付鹏程

Changdu, Xizang, China. 16.08.2017

C

显脉獐牙菜

Swertia nervosa

一年生草本；茎直立，高 30~100 cm，四棱形，有翅，基部起多分枝，松散；茎生叶无柄，狭椭圆形至披针形；大型聚伞花序圆锥状，疏松开展；4 基数；花萼绿色，深裂，裂片披针形，长约花冠的 2 倍；花冠黄绿色，花冠裂片长圆状披针形，先端一半具暗紫红色花纹，近基部具 1 个圆形腺窝，被圆形鳞片覆盖，边缘具短流苏。

较广泛分布于中国 - 喜马拉雅地区较温暖各地区。生长在向阳山坡、草地、河谷和灌丛边缘。海拔 600~2500 m。花果期 8~11 月。

Annual; stem erect, 30-100 cm, quadrangular with wings, branched from base, branches spreading; stem leaves sessile, narrowly elliptic to lanceolate; inflorescence gigantic panicle, loose and spreading; flower 4-merus; calyx green, deeply divided, calyx lobes lanceolate, about twice as long as corolla; corolla yellow-green, 4 lobes oblong-lanceolate, upper half with dark purple stripes, each with 1 orbicular nectary at base, covered by a scale, shortly fimbriate along margins.

Widespread in the warm areas in Sino-Himalayan region. In sunny hill slope, grassland, river banks, and bush margins. Altitude: 600-2500 m. Flowering & fruiting: August-November.

A

B

A-B. 中国甘肃省康县 Kangxian, Gansu, China. 03.10.2010

C. 中国云南省鹤庆县 Heqin, Yunnan, China. 08.10.1999

C

互叶獐牙菜

Swertia obtusa

多年生草本，直立，高 15~40 cm；叶全部互生，基生叶具长柄，长圆形，全缘，基部渐狭成叶柄，茎生叶较少，狭椭圆形，无柄；聚伞花序顶生成狭窄的圆锥状；花 5 基数，偶尔 6 基数；花萼绿色，深裂，裂片披针形；花冠蓝紫色，5 深裂，裂片椭圆形，基部具 2 个深色腺窝，边缘被柔毛状流苏。

特产于阿尔泰山地区的中国新疆及俄罗斯和蒙古。生长在高山草地、草甸和灌丛边缘。海拔 2100~2500 m。花果期 8~9 月。

Perennial; stem erect, 15-40 cm; all leaves alternative, basal leaves petiolate, oblong, entire, base cuneate, narrowed to petiole; stem leaves few, narrowly elliptic; cymes terminal as narrow thyrse; flower 5-merus occasionally 6-merus; calyx green, deeply divided, calyx lobes lanceolate; corolla blue-purple, deeply divided into 5 lobes, lobes elliptic, each with 2 nectaries at base, nectary margins with pilose fimbriae. Endemic to the Altai Mountain, in Xinjiang of China, Russia, and Mongolia. In alpine grassland, meadows, and bush margins. Altitude: 2100-2500 m. Flowering & fruiting: August-September.

A

B

A-B. 中国新疆维吾尔自治区喀纳斯
Hanasi, Xinjiang, China. 09.08.1995

开展獐牙菜

Swertia patula

一年生草本;茎直立,高5~15 cm,四棱形,紫色,自基部多分枝;茎生叶近无柄,披针形,全缘;花单生分枝顶端,4基数;花萼绿色,深裂,裂片线状披针形,长于花冠;花冠白色到淡紫色,具密集的深紫色脉纹,裂片匙状长圆形,近基部两侧具2个绿色圆形腺窝,腺窝被鳞片覆盖,边缘具长流苏。

特产于中国西南部云南和四川。生长在较干燥的向阳山坡草地和灌丛边缘。海拔1500~3000 m。花果期7~9月。

Annual; stem erect, 5-15 cm, quadrangular, much branched from base; stem leaves subsessile, lanceolate, entire; flower solitary on terminal, 4-merus; calyx green, deeply divided, calyx lobes linear-lanceolate; corolla white to purple, veins dark purple, deeply divided into 4 lobes, lobes spathulate-oblong, each with 2 green nectaries at base, nectaries orbicular covered by scale, with long fimbriae along margins.

Endemic to Yunnan and Sichuan in southwestern China. In dry and sunny hill slope, grassland, and bush margins. Altitude: 1500-3000 m. Flowering & fruiting: July-September.

A–D. 中国云南省丽江市 Lijiang, Yunnan, China. 10.10.1999

祁连獐牙菜

Swertia przewalskii

多年生草本，具短根状茎；花枝直立，黄绿色，花葶状，高 10~25 cm；基生叶 1~2 对，具柄，椭圆形，全缘，先端圆钝，基部楔形，渐狭成柄；茎生叶 1~2 对，生顶部，苞片状；聚伞花序顶生；花 5 基数；花萼绿色，深裂，裂片披针形，不整齐；花冠黄绿色，5 深裂，裂片披针形，基部具 2 个腺窝，边缘被柔毛状流苏。

特产于中国西部青海和甘肃。生长在高山草地、草甸和灌丛边缘。海拔 3000~4000 m。花果期 7~9 月。

Perennial with short rhizome; stem erect, greenish yellow scapiform, 10-25 cm; basal leaves 1-2 pairs, petiolate, elliptic, entire, apex obtuse, base narrowed to petiole, stem leaves 1-2 pairs on top only, scale-like; cymes terminal; flower 5-merus; calyx green, deeply divided, calyx lobes unequal, lanceolate; corolla greenish yellow, deeply divided into 5 lobes, lobes lanceolate, each with 2 nectary at base, nectary margins with pilose fimbriae.

Endemic to western China, in Qinghai and Gansu. In alpine grassland, meadows, and bush margins. Altitude: 3000-4000 m. Flowering & fruiting: July-September.

中国青海省泽库县 Zeku, Qinghai, China. 31.08.1994

毛獐牙菜

Swertia pubescens

一年生草本，全株密被短柔毛；茎直立，高 30~80 cm；茎生叶无柄，卵状披针形；聚伞圆锥花序，松散；花 5 基数；花萼绿色，裂片不等大，2 大，卵形，3 小，披针形；花冠白色，5 深裂，裂片阔卵形，基部具连续的紫色环带，每裂片基部具 2 个绿色裸露腺斑，无附属物；花丝基部扩大扁平联合成短筒状包围子房。

特产于中国云南。生长在潮湿山坡、溪边、林缘和灌丛边缘。海拔 2800~4000 m。花果期 8~10 月。

Annual, whole plant densely puberulent; stem erect, 30-80 cm; stem leaves sessile, ovate-lanceolate; cymes as a panicle, loose; flower 5-merus; calyx green, calyx lobes unequal, 2 bigger, ovate, 3 smaller, lanceolate; corolla white, deeply divided into 5 lobes, lobes broad ovate with a continuous purple ring at base, each lobe with 2 green naked nectary patches at base, without any appendage; filaments flattened and connate at base to tube-like surrounding pistil. Endemic to Yunnan province of China. In moist slopes, streamside, forest and bush margins. Altitude: 2800-4000 m. Flowering & fruiting: August-October.

A

B

A–C. 中国云南省丽江市 Lijiang, Yunnan, China. 11.10.1999

紫红獐牙菜

Swertia punicea

一年生草本；茎直立，高 15~60 cm，四棱形，有窄翅，中部以上分枝；茎生叶无柄，披针形；聚伞花序圆锥状，疏松开展；5 基数；花萼绿色，深裂，裂片披针形，较花冠短；花冠暗紫红色，花冠裂片长圆状披针形，近基部具 2 个条形腺窝，边缘具短流苏。

较广泛分布于中国中部和西南部。生长在向阳山坡、草地、河谷和灌丛边缘。海拔 600~3000 m。花果期 8~11 月。

Annual; stem erect, 15-60 cm, quadrangular with narrow wings, branched from middle; stem leaves sessile, lanceolate; inflorescence a panicle, loose and spreading; flower 5-merus; calyx green, deeply divided, calyx lobes lanceolate, shorter than corolla; corolla dark reddish-purple, 5 lobes oblong-lanceolate, each with 2 linear nectaries at base, shortly fimbriate along margins.

Common in central to southwestern China. In sunny hill slope, grassland, river banks, and bush margins. Altitude: 600-3000 m. Flowering & fruiting: August-November.

A

B

A–B. 中国云南省丽江市 Lijiang, Yunnan, China. 01.10.1998

莲座獐牙菜

Swertia rosularis

多年生小草本，有细弱的根状茎；茎直立，高 5~15 cm，中部以上分枝；叶多数基生，莲座状，具柄，匙形；茎生叶无柄，匙形到倒披针形；聚伞花序圆锥状，疏松开展；5 基数；花萼绿色，深裂，裂片钻形，较花冠短；花冠白色，花冠裂片长圆状披针形，近基部具 1 个圆形绿色腺窝，边缘具长流苏。

特产于中国四川峨眉山。生长在潮湿滴水的岩壁上。海拔 1300 m 以下。花果期 8~9 月。

Small perennial with slender rhizomes; stem erect, 5-15 cm, branched from middle; leaves mostly basal as rosette, petiolate, blade spathulate; stem leaves sessile, spathulate to oblanceolate; inflorescence a panicle, loose and spreading; flower 5-merus; calyx green, deeply divided, calyx lobes subulate, much shorter than corolla; corolla white, 5 lobes oblong-lanceolate, each with 1 round green nectaries at base, with long fimbriae along margins.

Endemic to Emeishan Mountain in Sichuan province of China. On moist rocks dripping water. Altitude: -1300 m. Flowering & fruiting: August-September.

A-C. 中国四川省峨眉山

Emeishan Mt., Sichuan, China. 25.08.2006

细瘦獐牙菜

Swertia tenuis

一年生草本，茎直立，细瘦，中部有分枝，高 10~30 cm；茎生叶线形或线状披针形，全缘；聚伞花序圆锥状；花细小，5 基数；花萼绿色，深裂，裂片线形；花冠白色到淡蓝色，5 深裂，裂片卵形，中下部具 1 个腺窝，边缘具长柔毛状流苏。

较特产于中国云南和四川。生长在高山林缘及灌丛中。海拔 1500~3000 m。花果期 8~10 月。

Annual; stem erect, slender, branched from middle part, 10-30 cm; stem leaves linear or linear-lanceolate, entire; cymes a panicle; flower small, 5-merus; calyx green, deeply divided, calyx lobes linear; corolla white to pale blue, divided into 5 lobes, lobes ovate, each with 1 nectary covered by few long fimbriae.

Endemic to Yunnan and Sichuan provinces of China. In margins of forest and bushes. Altitude: 1500-3000 m. Flowering & fruiting: August-October.

中国云南省丽江市 Lijiang, Yunnan, China. 01.10.1998

苇叶獐牙菜

Swertia wardii

多年生草本，具短根状茎；花枝粗壮，直立，高 30~80 cm；基生叶具柄，披针形至狭椭圆形；茎中上部叶数对，无柄，披针形；聚伞圆锥花序；花 5 基数；花萼绿色，深裂，裂片披针形；花冠淡蓝到蓝色，具深蓝紫色脉纹，裂片长圆状披针形，基部具 2 个圆形紫色腺窝，边缘被长柔毛状流苏。

特产于喜马拉雅山锡金、不丹和中国西藏。生长在高山草甸。海拔 3500~4500 m。花果期 7~9 月。

Perennial with short rhizome; flowering stem stout, erect, 30-80 cm; basal leaves petiolate, lanceolate, middle and upper stem leaves several pairs, sessile, lanceolate, smaller; inflorescence a panicle; flower 5-merus; calyx green, deeply divided, calyx lobes lanceolate; corolla pale blue to blue, veins dark blue-purple, deeply divided into 5 lobes, lobes oblong-lanceolate, each with 2 orbicular purple nectaries at base, nectary margins with long pilose fimbriae.

Endemic to Himalayas, in Sikkim, Bhutan, and Xizang of China. In alpine meadows. Altitude: 3500-4500 m. Flowering & fruiting: July-September.

A. B. D. 中国西藏自治区错那市
Cuona, Xizang, China. 14.08.2006
C. 中国西藏自治区林芝市
Cuona, Xizang, China. 17.08.2006

C

D

华北獐牙菜

Swertia wolfangiana

多年生草本，具短根状茎；花枝直立，有条棱，高 10~50 cm；基生叶 1~2 对，具柄，矩圆形至椭圆形；茎中部无叶，上部 1-2 对叶，极小无柄，苞片状，卵形；聚伞圆锥花序 2~7 花；花 5 基数；花萼绿色，深裂，裂片披针形；花冠黄绿色，背部中央蓝色，裂片椭圆形，基部具 2 个圆形黄色腺窝，边缘被长柔毛状流苏。

产于中国西藏、四川、青海、甘肃、湖北和山西等省。生长在高山草甸、沟谷及灌丛边缘潮湿处。海拔 1500~4500 m。花果期 7~9 月。

Perennial with short rhizome; flowering stem erect, striate, 10-50 cm; basal leaves 1-2 pairs, petiolate, blade oblong to elliptic, middle stem without leaf, upper stem with 1-2 pairs much smaller leaves, bract-like, sessile, ovate; Inflorescence a panicle with 2-7 flowers; flower 5-merus; calyx green, deeply divided, calyx lobes lanceolate; corolla greenish yellow, outside blue in middle, deeply divided into 5 lobes, lobes elliptic, each with 2 orbicular yellow nectaries at base, nectary margins with long pilose fimbriae. Endemic to Xizang, Sichuan, Qinghai, Gansu, Hubei and Shanxi provinces of China. In alpine meadows woodland margins. Altitude: 1500-4500 m. Flowering & fruiting: July-September.

A

B

A-C. 中国青海玉树 摄影 /Photo：付鹏程
Yushu, Qinghai, China. 12.08.2017

少花獐牙菜

Swertia younghusbandii

多年生草本，具短根状茎；花枝直立，紫色，花葶状，高 5~20 cm；基生叶 1~2 对，具柄，线状椭圆形，全缘，先端圆钝，基部楔形，渐狭成柄；茎生叶 1~2 对，生顶部，苞片状；花单生分枝顶端，1~3 朵；花 5 基数；花萼绿色，深裂，裂片披针形；花冠黄色，5 深裂，裂片椭圆形，基部具 2 个腺窝，边缘被柔毛状流苏。

特产于中国西藏。生长在高山草甸和灌丛中。海拔 3800~4500 m。花果期 7~9 月。

Perennial with short rhizome; Stem erect, purple, scapiform, 5-20 cm; basal leaves 1-2 pairs, petiolate, linear-elliptic, entire, apex obtuse, base narrowed to petiole, stem leaves 1-2 pairs on top only, scale-like; flowers solitary on terminals of branches, 1-3 flowers; flower 5-merus; calyx green, deeply divided, lobes lanceolate; corolla yellow, deeply divided into 5 lobes, lobes elliptic, each with 2 nectaries at base, nectary margins with pilose fimbriae.

Endemic to Xizang of China. In alpine meadows and bushes. Altitude: 3800-4500 m. Flowering & fruiting: July-September.

中国西藏自治区当雄县 Dangxiong, Xizang, China.
22.09.1992

云南獐牙菜

Swertia yunnanensis

一年生草本；茎直立，高5~30 cm，四棱形，紫色，基部起多分枝；茎生叶无柄，线状披针形，全缘；聚伞花序圆锥状，疏松开展；花5基数；花萼绿色，深裂，裂片披针形；花冠淡蓝色，脉深蓝紫色，花冠裂片卵形，近基部两侧具2个条形不太明显的腺窝，腺窝边缘具白色长流苏。

特产于中国西南部的云南、四川和贵州。生长在较干燥向阳山坡草地和灌丛边缘。海拔1200~3500 m。花果期8~11月。

Annual; stem erect, 5-30 cm, quadrangular, purple, branched from base; stem leaves sessile, linear-lanceolate, entire; Inflorescence a panicle, loose and spreading; flower 5-merus; calyx green, deeply divided, calyx lobes lanceolate; corolla pale blue, veins dark blue-purple, deeply divided into 5 lobes, lobes ovate, each with 2 nectaries at base, nectaries linear, inconspicuous, with long white fimbriae along margins.

Endemic to Yunnan, Sichuan, and Guizhou in southwestern China. In dry and sunny hill slope, grassland, and bush margins. Altitude: 1200-3500 m. Flowering & fruiting: August-November.

A–B. 中国云南省丽江市 Lijiang, Yunnan, China. 24.08.2002

黄秦艽 *Veratrilla baillonii*

中国四川省稻城县 Daocheng, Sichuan, China. 08.07.2014

4.2.15
黄秦艽属
Veratrilla

花单性，雌雄异株；多年生草本；营养枝的叶呈莲座状；花枝较粗壮，花葶状；聚伞花序；花4基数，辐状；花萼深裂至基部；花冠深裂，冠筒短，无褶，裂片基部有腺斑；雄蕊着生花冠裂片的弯缺处，与花冠裂片互生；子房1室，花柱短，柱头2裂；蒴果2瓣裂，种子小，周边具翅。

本属2种，特产于中国－喜马拉雅地区的不丹、中国西南部、印度东北部及锡金。

Dioiceous; perennial; leaves of vegetative shoots as rosette; flower stem stout, scapiform; inflorescences cymose; flowers 4-merus, rotate; calyx deeply divided to base; corolla deeply divided, tube short, without plicae, corolla lobes with nectaries at base; stamens inserted at sinus between corolla lobes, alternative with lobes; ovary 1-celled. Style short; stigma 2-lobed. Capsule 2-valved; seeds small, discoid winged.

2 species, in the Sino-Himalayan regions, in Bhutan, SW China, NE India, and Sikkim.

黄秦艽

Veratrilla baillonii

多年生草本，具短根状茎；茎直立，粗壮，花葶状，高 30~50 cm，绿色；叶多数基生莲座状，具柄，长圆状匙形，全缘；茎生叶较小，卵状椭圆形；聚伞圆锥花序；雌株花较少，花序狭窄，花小，4 基数；花萼绿色，裂片披针形；花冠黄绿色，裂片匙形，基部具 2 个紫色腺斑，雄蕊退化；雄株花序较大，花数量多，密集，花的构造形态与雌花相似，雌蕊退化。

特产中国西藏、云南和四川。生长在高山草地、森林边缘和灌丛中。海拔 3000~4500 m。花果期 5~9 月。

Perennial with short rhizome; stem erect, stout, scapiform, 30-50 cm, green; leaves mostly basal as rosette, petiolate, oblong-spathulate, entire, stem leaves smaller, ovate-elliptic; inflorescence a panicle; female plant with few flowers, flowers small, 4-merus; calyx green, lobes lanceolate; corolla yellow-green, 4 lobes spathulate, each with 2 naked nectaries at base, stamens reduced; male plant with large inflorescence with numerous flowers densely arranged, flowers similar to female flowers, pistil reduced.

Endemic to Xizang, Yunnan and Sichuan of China. In alpine grassland, forest margins and bushes. Altitude: 3000-4500 m. Flowering & fruiting: May-September.

A

B

A–B. 雌株；中国云南省大理白族自治州
Female plant, Dali, Yunnan, China. 02.08.2011

C. 雄株；中国云南省大理白族自治州
Male plant, Dali, Yunnan, China. 02.08.2011

D. 中国云南省香格里拉县
Xianggelila, Yunnan, China. 01.10.1999

C

D

加那利群岛拉帕玛岛景观 La Palma, Canary Islands, Spain. 25.05.1997

4.3
绮皇花属 [绮龙花族]
Ixanthus [Tribe *Chironieae*]

单种属，仅绮皇花 *Ixanthus viscosus* 一种，特产于大西洋的加那利群岛。

在较早期的多数植物学文献中，该属一直被包括在龙胆族 *Gentianeae*，但从生物地理学的角度看，它与该族其他植物没有关联。近来的基于形态解剖和 DNA 分子序列的系统发育分析揭示了它其实与分布于地中海地区的绮莲花属 *Blackstonia* 近缘，应该归属于绮龙花族 *Chironieae*。收录在此，仅供比较参考。

Monotypic genus, with only one species, *Ixanthus viscosus*, endemic to the Canary Islands in the Atlantic Ocean.

In most early literatures, this genus was included in tribe *Gentianeae*, however, it was difficult to understand its relationships with other members of the tribe from the point of biogeography. Recent phylogenetic studies based on morphology, anatomy and DNA molecular sequences revealed that it is in fact closely related to the genus *Blackstonia* which is endemic to the Mediterranean Basin, thus, should be included in the tribe *Chironieae*. It is shown here for information and comparison only.

绮皇花

Ixanthus viscosus

多年生草本；茎直立，粗壮，基部木质化，高可达 200 cm；叶无柄，基部半抱茎，叶片卵状椭圆形或长圆状披针形，全缘，分泌胶质，有黏性；大型聚伞圆锥花序；花 5 基数，偶尔 4 或 6 基数，基部有一对基部合生的苞片；花萼筒短小；花冠亮黄色，裂片椭圆形，雄蕊与雌蕊向相反方向弯曲。

特产于加那利群岛，生长在月桂林下及边缘。花果期 4~5 月。

Perennial; stem erect, stout, woody at base, up to 200 cm, high; leaves sessile, subamplexicaul, blades ovate-elliptic to oblong-lanceolate, entire, producing viscous gums; inflorescence a thyrse; flowers mostly 5-merous, occasionally 4- or 6-merous, each flower with a pair of basally connate bracts; calyx minor; corolla bright yellow, lobes elliptic. Stamens and pistil curved to opposite directions.

Endemic to the Canary Islands in Atlantic Ocean. In laurel forest and margins. Flowering & fruiting: April-May.

A

B

A–F. 加那利群岛拉帕玛岛

La Palma, Canary Islands, Spain. 25.05.1997

C

D

E

F

线叶龙胆 *Gentiana lawrencei* var. *farreri* + 蓝玉簪龙胆 *Gentiana veitchiorum*

中国青海省河南蒙古族自治县 Henan, Qinghai, China; 09.09.2023

麻花艽 *Gentiana straminea*

中国甘肃省卓尼县 Zhuoni, Gansu, China. 08.09.2023

5

龙胆与人类生活

Gentians and Human Life

龙胆科植物不仅在大自然中野生，自在自然，多姿多彩，维持着生态系统的稳定、美丽与多样，为世界各地人民所喜爱，而且许多种类与人类生活密不可分，具有重要的社会和经济价值。有关龙胆族植物的应用及其价值，可以概括为以下四个方面。

Gentianaceae plants are diverse and colorful when wild in the nature, playing critical roles in sustaining the stability, beauty, and diversity of the ecosystems on the earth. They are loved by people around the world, and many of their species are closely related to human life, possessing important social and economic value. The applications and value of gentians can be discussed from the following four aspects.

龙胆类植物的药用具有悠久的历史，其价值自古就被中华民族所认识。

5.1 药用

世界现存最早的药物学专著《神农本草经》成书年代不详，一般认为是秦汉时期或更早，至少有 2200 多年的历史了。《神农本草经》中记载了植物药 252 种，分上（有益）、中（无害）、下（有毒）三品。其中就包括了龙胆族植物 2 种，即上品龙胆和中品秦艽。“龙胆”和“秦艽”这些中文植物名称也是《神农本草经》最早记载下来并沿用至今的。

龙胆作为植物（药物）名称，一般认为是描述其苦味的。中国另一本中药学经典著作《开宝本草》（公元 973—974 年）就这样解释龙胆："叶似龙葵，味苦如胆，因此为名。"也有解释认为，中国古人以"龙"作为最高级、最厉害的象征，"龙胆"则意味着最苦了。大多数龙胆科植物的口感的确是非常苦的。中医药常有"苦口良药"或"良药苦口"的说法，也许说的就是龙胆或龙胆科其他植物。据不完全统计，龙胆族作为药用的植物就有 65 种之多，详见表 5-1。

每一种不同的龙胆族植物药材，甚至同一植物的不同器官（根、茎、叶、花等）都可能有不同的药性和应用，国内外都有较多专门研究，中国的传统和现代中药材专著，如《本草纲目》《晶珠本草》《中国药典》《中华本草》等都有比较系统的描述。龙胆族药材基本都有着比较类似的功效，主要为清热解毒、清热利湿、调经活血和健胃等，可以主治风湿痹痛、黄疸型肝炎、扁桃腺炎、胆囊炎、气管炎、咳嗽、痢疾、感冒发烧、咽喉肿痛、风湿性关节炎、消化不良、毒蛇咬伤等。

由于龙胆族植物以高山植物为主，尤其被我国的藏族医学所重视，应用历史悠久，积累了丰富的经验，藏族医学经典《四部医典》和《晶珠本草》中就记载了多种龙胆类植物。如图 5-1 所示，《四部医典》的传统挂图里就包括了多种龙胆，包括乌奴龙胆、岷县龙胆、蓝玉簪龙胆、云雾龙胆等，其中对乌奴龙胆的表现就惟妙惟肖，非常准确地展现了它的花和叶的形态特征（图 5-2）。现代藏医药也以龙胆类植物为基本原料，开发出了许多新药品，在一些病症如肝胆病变的治疗方面有比较显著的疗效。

虽然龙胆和秦艽都被列为"上品"和"中品"，对身体无毒无害，但必须强烈提醒本书读者，不要尝试私自食用龙胆类药材（其他类药材也一样），若有需要使用，使用前必须咨询专业医师，听从医师建议。

表 5-1　中国龙胆族药用植物

Table 5-1　Medicinal Plants of the tribe *Gentianeae* from China

原植物 Species	中药材 TCM	药用部位　Organs used
Crawfurdia campanulacea	云南蔓龙胆	全草　whole plant
Gentiana apiata	茱苓草	全草　whole plant
Gentiana aristata	完布	全草　whole plant
Gentiana crassicaulis	吉解完保（兰花秦艽）、秦艽	花、根　flower, root
Gentiana dahurica	秦艽	根　root
Gentiana davidii	落地荷花	带花全草　whole plant with flower
Gentiana filistyla	蓝花龙胆	带根全草　whole plant with root
Gentiana flavomaculata	黄斑龙胆	全草　whole plant
Gentiana georgei	旁间那保（黑花龙胆）	花　flower
Gentiana hexaphylla	六叶龙胆	根　root
Gentiana lhassica	莪代哇	全草　whole plant
Gentiana loureirii	龙胆地丁	带根全草　whole plant with root
Gentiana macrophylla	秦艽	根　root
Gentiana manshurica	龙胆	根和根茎　root, rhizome
Gentiana napulifera	菊花参	根　root
Gentiana nubigena	旁间完保（兰花龙胆）、杂花龙胆	花、根、根茎　flower, root, rhizome
Gentiana officinalis	吉解完保（兰花秦艽）	花、根　flower, root
Gentiana pubigera	矮脚龙胆	全草　whole plant
Gentiana purdomii	旁间嘎保、白花龙胆、高山龙胆	花、带根全草　flower, whole plant with root
Gentiana rhodantha	红花龙胆	全草　whole plant
Gentiana rigescens	龙胆	根和根茎　root, rhizome
Gentiana rubicunda	小儿血参	带根全草　whole plant with root
Gentiana scabra	龙胆	根和根茎　root, rhizome
Gentiana sinoornata	旁间察吾（杂花龙胆）	花　flower
Gentiana siphonantha	吉解完保（兰花秦艽）	花、根　flower, root

原植物 Species	中药材 TCM	药用部位 Organs used
Gentiana spathulifolia	奥拉毛	带根全草 whole plant with root
Gentiana squarrosa	石龙胆	全草 whole plant
Gentiana straminea	吉解嘎保、（白花秦艽）、秦艽	花、根 flower, root
Gentiana triflora	龙胆	根和根茎 root, rhizome
Gentiana urnula	冈嘎琼	全草 whole plant
Gentiana veitchiorum	旁间察吾、双色龙胆、杂花龙胆	花、带根全草、根 flower, whole plant with root, root
Gentianopsis barbata	加蒂嘎布、扁蕾	全草 whole plant
Gentianopsis paludosa	蒂达（代用品）	全草 whole plant
Halenia corniculata	花锚	全草 whole plant
Halenia elliptica	蒂达（代用品）、黑及草、甲蒂然果	全草、地上部分 whole plant, aerial part
Latouchea fokiensis	匙叶草	全草 whole plant
Lomatogonium rotatum	肋柱花	全草 whole plant
Metagentiana rhodantha	小龙胆草	根及全草 root, whole plant
Sinoswertia tetraptera	蒂达（代用品）	全草 whole plant
Swertia angustifolia	青叶胆	全草 whole plant
Swertia bifolia	代哇、乌金草	全草 whole plant
Swertia bimaculata	獐牙菜	全草 whole plant
Swertia chirata	甲蒂	全草 whole plant
Swertia ciliata	蒂达（正品）	全草 whole plant
Swertia davidi	鱼胆草	全草 whole plant
Swertia dichotoma	蒂达（代用品）	全草 whole plant
Swertia diluta	淡花当药	全草 whole plant
Swertia erythroisticta	蒂达（代用品）、红直当药	全草 whole plant
Swertia kingii	黄花獐牙菜	根 root
Swertia kouitchensis	贵州獐牙菜	全草 whole plant

（续表 continued）

原植物 Species	中药材 TCM	药用部位 Organs used
Swertia mileensis	青叶胆	全草 whole plant
Swertia mussotii	蒂达（正品）、藏茵陈、桑蒂	全草 whole plant
Swertia muticaulis	赛保古椎窍	根 root
Swertia nervosa	翼梗獐牙菜	全草 whole plant
Swertia patens	小儿腹痛草	全草 whole plant
Swertia phragmitiphylla	赛尔蒂	全草 whole plant
Swertia pseudochinensis	当药	全草 whole plant
Swertia punicea	山飘儿草	全草 whole plant
Swertia yunnanensis	青叶胆	全草 whole plant
Tripterospermum chinense	肺形草	幼嫩全草 whole young plant
Tripterospermum cordatum	青鱼胆草	全草 whole plant
Tripterospermum nienkui	黄金线	全草 whole plant
Tripterospermum taiwanense	台湾肺形草	全草 whole plant
Tripterospermum volubile	小筋骨藤	全草 whole plant
Veratrilla baillonii	黄秦艽	全草、根 whole plant, root

图 5-1　成书于公元 8 世纪的我国藏族医学经典《四部医典》的挂图第 28 图。注意第 1 排最右侧的乌奴龙胆，第 3 排最右侧的“白花龙胆”（即岷县龙胆），和第 4 排最左侧的 2 个分别为“蓝花龙胆”（蓝玉簪龙胆）和“杂花龙胆”（云雾龙胆）（感谢马溪芮先生供图）

Fig. 5-1　Illustration No. 28 of the ancient Tibetan classic work “The Four Treatises of Tibetan Medicine”, which was written in 8th century AD. Note the gentians, *Gentiana urnula* at the right most of line 1, *Gentiana purdomii* at the right most of line 3, *Gentiana veitchiorum* and *Gentiana nubigena* at left most of line 4 (photo courtesy from Mr. Xirui Ma)

图 5-2　传统中药和藏药用龙胆科植物

Fig. 5-2　Gentians used as traditional Chinese and Tibetan medicines

A. 秦艽，根入药
B. 龙胆（三花龙胆）及药材用根
C. 乌奴龙胆
D. “白花龙胆”（岷县龙胆）
E. “蓝花龙胆”（蓝玉簪龙胆）
F. “杂花龙胆”（云雾龙胆）（C-F: 花或全草入药）

A. *Gentiana macrophylla*, root as medicine
B. *Gentiana triflora*, root as medicine
C. *Gentiana urnula*
D. *Gentiana purdomii*
E. *Gentiana veitchiorum*
F. *Gentiana nubigera* (C-F: flower or whole plant as medicine)

The medicinal use of gentian plants has a long history and their value has been well recognized by Chinese since ancient times.

5.1 Medicines

The earliest pharmacology monograph in the world, "Shen Nong Ben Cao Jing", was written at an unknown time. It is generally believed to be written in the Qin or Han dynasties or earlier, with a history of at least 2200 years. It is a well recognized classic work of traditional Chinese medicine, which recorded 252 plant medicines, divided them into three categories: superior (beneficial), medium (harmless), and inferior (toxic). Among them, two species belong to the Gentianaceae family, namely the superior beneficial Long Dan, *Gentiana scabra*, and the medium harmless Qin Jiao, *Gentiana macrophylla*. The Chinese names of those species were also recorded in the "Shen Nong Ben Cao Jing" for the first time and have been used to this day.

"Long Dan" (meaning Dragon gallbladder), as a plant (drug) name, is generally considered to describe its bitter taste. Another Chinese classic work on traditional Chinese medicine, "Kaibao Materia Medica" (973-974 AD), explains "Long Dan" in this way: "The leaves are like eggplant, the taste is as bitter as gallbladder, so it is named". There are also explanations that suggest that ancient Chinese people used "Long" (dragon) as the highest and most powerful symbol, thus "Long Dan (dragon gallbladder)" means the most bitter material. The taste of most Gentianaceae plants is indeed very bitter. Traditional Chinese medicine often has the expressions "bitter medicine are better medicines" or "the bitter medicine the better", which perhaps referring to plant medicines of gentians. According to incomplete statistics, as many as 65 species of the tribe *Gentianeae*, are used as medicines in China, as shown in Table 5-1.

Each different species of medicinal herb of the tribe *Gentianeae*, and even different organs of the same species (roots, stems, leaves, flowers, etc.), may have different medicinal properties and applications. There are many specialized studies both domestically and internationally. Traditional and modern Chinese herbal medicine monographs, such as "Compendium of Materia Medica", "Jingzhu Materia Medica", "Pharmacopoeia of China", "Materia Medica Sinica", and so on, have systematic descriptions. Nevertheless, medicinal herbs of the tribe *Gentianeae* basically have relatively consistent effects: mainly for fever removing and detoxifying, clearing heat and relieving dampness, regulating menstruation and promoting blood circulation, and strengthening the stomach functions. It can be used to treat rheumatic pains, jaundice hepatitis, tonsillitis, cholecystitis, tracheitis, cough, dysentery, cold and fever, sore throat, rheumatoid arthritis, indigestion, snake bites, and so on.

Due to the fact that the plants of the tribe *Gentianeae* are mainly alpine plants, they are particularly valued by Tibetan medicine in China. They have a long history of application and accumulated rich experience. The Tibetan medical classics "The Four Treatises of Tibetan Medicine", and "Jingzhu Materia Medica" recorded various gentian plants. As shown in Fig. 5-1, the traditional thangka paintings of the "The Four Treatises of Tibetan Medicine" included various types of gentians, including *Gentiana urnula*, *Gentiana purdomii*, *Gentiana veitchiorum* and *Gentiana nubigera*, among which the presentation on *Gentiana urnula* is vivid, accurately depicting the morphological characteristics of its flowers and leaves (Fig. 5-2). Modern Tibetan medicine also uses gentian plants as the basic raw material and has developed many new drugs, which have significant therapeutic effects on some diseases such as liver and gallbladder diseases.

Although both "Long Dan" and "Qin Jiao" are classified as "beneficial" and "harmless" to the body, readers of this book are strongly reminded not to try to consume any gentians (other plants as well) medicinal materials without permissions or advice from doctors. It is important to consult a professional physician or doctor and follow their advice before any consumption.

5.2 花卉

龙胆花如此美丽，自然而然应该是观赏植物的首选。然而，现实中，只有在西方发达国家的高海拔或高纬度地区才偶尔能看到栽培的龙胆花。放眼整个龙胆科，也只有个别种类是较为常见的大众花卉。藻百年（*Exacum affine*），也被称为紫芳草，原产自也门的索科特拉岛，它的植株小巧玲珑，花色鲜艳秀气，很耐看，时常作为小盆栽栽培，供桌面摆设，在欧美较常见，中国近几年也有引进（图 5-3：A-C）。另一个常见的龙胆科花卉就是常常作为鲜切花的洋桔梗（*Eustoma grandiflora*），也被称为草原龙胆，它原产于墨西哥，现在作为切花在世界各地较为广泛地种植（图 5-3：D）。典型的龙胆族植物作为花卉栽培的并不常见，仅有少数几个例子。

据报道，野生于中国青藏高原的类华丽龙胆（*Gentiana sinoornata*）和线叶龙胆（*Gentiana lawrencei* var. *farreri*）在 20 世纪初被首次引种到英国时曾引起轰动；爱丁堡皇家植物园等机构尝试栽培和品种选育，这方面的努力历经百年不衰，经反复杂交筛选，已经命名注册的（或品牌商标）就有 150 多个华丽龙胆品种，如“蓝鸟”“蓝色帝国”“蓝色国王”“蓝色天国”“蓝色火焰”“蓝钻”“极品尤金”（重瓣花）等。然而，类华丽龙胆和线叶龙胆及其人工繁育品种的栽培并不普及，100 多年过去了，它的栽培还仅限于欧美高纬度地区的植物园和个别园艺爱好者的私家花园，并没有走进寻常百姓家，这主要是由于这些龙胆花的栽培比较困难，尤其不耐热，难以适应低纬度和低海拔地区的气候。

另一组栽培较多的龙胆花是龙胆属大花组的狭叶龙胆（*Gentiana angustifolia*）和与它近缘的西部龙胆（*Gentiana occidentalis*），它们都是欧洲本土原产的，非常相似，区别主要在于基生叶的长宽比例，狭叶龙胆的叶较细长（图 5-4：A），而西部龙胆的叶较短宽（图 5-4：B）。也许它们是该组比较耐热的种类，尤其狭叶龙胆的耐热性可能是最好的，在欧洲的一些植物园和私家花园时有栽培，也有一些栽培品种，但总的来说栽培也并不普及。

新近流行栽培的龙胆花在国际上一般称之为“日本龙胆”，由日本岩手县最早培育栽培，据说是用日本本土野生的龙胆草（*Gentiana scabra*）和三花龙胆（*Gentiana triflora*）杂交培育而来。早期的杂种是高杆的，花色也有蓝色和紫红，主要用来做切花，一度非常流行，一花难求。欧美好多园艺学家也加入培育，近几年矮生型品种被培育出来，耐热性更好，花量更大，有“海洋之心”（图 5-4：C）及改进商标版“岩间钻石海洋之心”等（图 5-4：D）。这些品种深受世界各地龙胆迷们的喜爱，很快流传开来，在中国也较为流行，是新晋网红花卉。有丰富的野生资源作为遗传资源，园艺学家们一定会培育出更多更好的龙胆花新品种，值得我们期待。有了耐热性能更好的品种，配以工业化的大规模扦插繁殖和自动化栽培管理（图 5-5），龙胆花走进千家万户的日子必将到来。

最后，必须强调的是，请大家爱护野生龙胆花资源，在野外时敬请大家仅仅观看欣赏和拍照，绝不要随意采摘野生龙胆花，更不要尝试移栽。基本可以肯定的是野生龙胆花移栽是很难成活的，正如美国诗人 Amy Lowell 写诗告诫大家的：龙胆花是很恋家的，它离开家的当天就会死去。

图 5-3　龙胆科最常见的两种花卉

Fig. 5-3　The two most common ornamental plants from Gentianaceae

A–C. 藻百年　A–C. *Exacum affine*
D. 洋桔梗　D. *Eustoma grandiflora*

图 5-4　龙胆属常见的几种花卉品种

Fig. 5-4　The common ornamental cultivars of gentians

A. 狭叶龙胆“自由”
B. 西部龙胆（品种不详）
C. 日本龙胆“海洋之心”
D. 日本龙胆“岩间钻石海洋之心”

A. *Gentiana angustifolia* “Frei”
B. *Gentiana occidentalis* (cultivar unknown)
C. *Gentiana scabra* “Blue Heart”
D. *Gentiana scabra* “Rocky Diamond Blue Heart”

图 5-5 观赏用日本龙胆花的商业化大规模生产

Fig. 5-5 Commercial large scale production of ornamental *Gentiana scabra*

A. 生根扦插苗
B-E. 不同生长期的花苗及自动化浇灌
F. 即将发售的开花苗

A. Rooted cutting
B-E. Potted plants at different growing stage and automatic irrigation system
F. Flowering plant ready to market

5.2 Ornamentals

However, in reality, cultivated gentian flowers can only be seen occasionally in high-altitude or high-latitude areas of developed western countries. Looking at the entire Gentianaceae family, only a few species are common popular flowers. One best case is *Exacum affine*, which is native to the Socotra Island in Yemen. Its plants are small and delicate, with bright and delicate flowers, which are very durable and often cultivated as small potted plants for table top uses. It is common in Europe and America, and has also been introduced in China in recent years (Fig. 5-3: A-C). Another common flower in the Gentianaceae family is the *Eustoma grandiflora*, also known as the pasture gentian, which is often used as a fresh cut flower. It is native to Mexico and is now widely cultivated as a cut flower worldwide (Fig. 5-3: D). Other most typical gentian plants are not commonly cultivated as flowers, with only a few examples as exceptions.

According to reports, *Gentiana sinoornata* and *Gentiana lawrencei* var. *farreri*, both of which are wild in the Qinghai-Xizang Plateau of China, caused a sensation when they were first introduced to the UK in the early 20th century. Institutions such as the Royal Botanical Garden Edinburgh have attempted cultivation and new variety breeding, and such effort has been enduring for a century. After repeated hybridization and screening, over 150 magnificent varieties have been named or branded under these species, such as "Blue Bird", "Blue Empire", "Blue King", "Blue Heaven", "Blue Flame", "Blue Diamond", "Eugens Allerbester" (with double corolla), and so on. However, the cultivation of these two gentians and their artificially bred varieties is not widespread. More than 100 years have passed, and their cultivation is still limited to botanical gardens and private gardens of individual horticultural enthusiasts in high latitude regions of Europe and America, and has not entered ordinary people's homes. This is mainly because the cultivation of these gentian flowers is relatively difficult, especially they are not heat-tolerant, and difficult to adapt to the climate of low latitude and low altitude regions.

Another group of commonly cultivated gentian flowers includes *Gentiana angustifolia*, which has rather narrow leaves, and a closely related gentian *Gentiana occidentalis*, both belonging to sect. *Ciminalis* of the genus *Gentiana*. They are both native to Europe and are very similar, with the main difference being the length to width ratio of the basal leaves. The leaves of *Gentiana angustifolia* are thinner and longer (Fig. 5-4: A), while the leaves of *Gentiana occidentalis* are shorter and wider (Fig. 5-4: B). Perhaps they are the more heat-resistant species in this group, especially *Gentiana angustifolia*, which may have the best heat resistance. It has been cultivated in some botanical gardens and private gardens in Europe, and some cultivated varieties have also been claimed, but overall, cultivation is not widespread.

The newly popular cultivated gentian flower, commonly known as "Japanese gentian", was firstly cultivated in Iwate Prefecture in Japan. It is said to be a hybrid between the two wild Japanese species Gentiana scabra and *Gentiana triflora*; Early hybrids were tall, with blue and magenta flower colors. They were mainly used for cutting flowers and were once very popular. Many horticulturists in Europe and America joined in the cultivation and breeding. In recent years, dwarf varieties have been cultivated with better heat resistance and larger flower numbers, including the so-called “Blue Heart" (Fig. 5-4: C) and the improved and branded version “Rocky Diamond Blue Heart“ (Fig. 5-4: D), which are loved by gentian fans around the world and quickly spread worldwide. They are also popular in China and are a new popular flower on the internet. With good varieties that are heat-resistant, coupled with industrialized large-scale cutting propagation and automated cultivation management (Fig. 5-5), the day when gentian flowers enter millions of households will come sooner or later. There are abundant wild resources as genetic resources, and horticulturists will definitely bring us more and better new varieties of gentian flowers, which we look forward to.

Finally, it must be emphasized that everyone should cherish the resources of wild gentian flowers. When you are in the wild nature, please watch only, appreciate, and take photos, and never pick any wild gentian flower, let alone try transplanting. It is generally certain that transplanting wild gentian flowers is not possible to survive. As American poet Amy Lowell wrote a poem warning everyone, gentian flowers are very homely and will die on the day they leave their home.

5.3 香料

龙胆味苦又安全无害，可作良药，亦可调味，其中最有名的例子应该属欧洲的黄龙胆了。

黄龙胆是多年生草本，植株寿命可超过 50 年。它有很粗壮的根，富含淀粉、糖分及丰富的黄酮类化合物。黄龙胆的根粗壮又深，采挖黄龙胆根既是很重的力气活，也是一个技术活，在法国甚至有专门的采挖工具，叫作魔鬼叉（fourche du diable），是一把很重很坚固的钢叉，有 2 m 多高，重达 12 kg（图 5-6：A-F）。

早期在欧洲，黄龙胆根曾作为健胃草药，后来广泛用来制作龙胆酒，在法国、瑞士、德国和奥地利一带比较流行，尤其是在比利牛斯山、汝拉山和阿尔卑斯山等山区，形成了一个传统的特色产业。龙胆酒的种类很多，主要有两大类：即酒精含量超过 40% 的烈性酒和低于 20% 的开胃酒（图 5-6：G-I）。

与普通白酒的生产类似，黄龙胆的根由于富含糖分和淀粉，可以直接发酵。清洁后的新鲜根或干燥根经过适当粉碎就可以发酵，发酵后经过蒸馏就可以生产高度烈酒。黄龙胆根给酵母提供养分的同时，也给产品提供了独特的风味，有草药和柑橘的香味，微苦中带有回甘，十分爽口。据说还有较好的安神和助消化功效，因此，常常用来作为餐后的助消化酒，也可以用来调制鸡尾酒。这类烈性龙胆酒在德国、奥地利和瑞士较为流行。

法国和瑞士法语区则比较流行低度开胃酒，品类和品牌非常多，其中以法国的苏士（Suze）龙胆酒最著名。苏士龙胆酒的历史非常悠久，最早可以追溯到 1885 年，在法国的 Maisons-Alfort 首创，以后作为开胃酒被发扬光大。它以黄龙胆的根为原料，采集生长 20 年以上的野生或栽培的黄龙胆根后，首先进行精选和清洗处理，适当粉碎后浸入 50% 左右的食用酒精中搅拌浸泡；浸泡几个月后再压榨过滤，滤液经过特殊工艺蒸馏，蒸馏出的粗酒再经过与其他香草提取物进行调配，就制成了口味独特的开胃酒。这种开胃酒的酒精度一般在 15%~20%，散发着柔和的柑橘气息和星星点点泥土的芬芳，也有些许花香，在微苦的基调中甘甜若隐若现。它可以纯饮，也可以调制鸡尾酒。据说，毕加索就十分喜爱苏士龙胆酒，1912 年还专门为这款酒创作了一幅画，名为 Verre et bouteille de Suze（酒杯和一瓶苏士龙胆酒）。

啤酒花为啤酒带来了爽口的苦味，它的清香味也为啤酒带来了灵魂；黄龙胆为龙胆酒带来的不仅仅有灵魂，还有肉体，既有香味与口感，还有发酵产物酒精和糖分。龙胆酒真的是好东西，既能帮你开胃，又能帮你消化，让你吃得又多又好。另一方面，人生也有如一杯龙胆酒，微甜有苦，让人兴奋，也让人沉醉。

图 5-6　黄龙胆根的采挖及欧洲市面上不同的龙胆酒例子（A-F: 感谢刘倩菲女士供图）

Fig. 5-6　Harvesting process of the roots of *Gentiana lutea*, and examples of alcoholic drinks made from the roots on Europe market (A-F: photos courtesy from Ms Qianfei Liu)

A-F. 黄龙胆根的采挖及其专用工具魔鬼叉
G-H. 高度烈性助消化龙胆酒
I. 低度开胃龙胆酒

A-F. Harvesting process and specially designed tool for this purpose
G-H. Digestive liquors with alcohol >40%
I. Aperitif with alcohol <20%

Gentians are safe and have strong bitter tastes used as good medicines, so they can be used as flavors for bitter tasting as well.

5.3 Flavors

Gentians are safe and have strong bitter tastes used as good medicines, so they can be used as flavors for bitter tasting as well. The most famous of these is probably the European Yellow Gentian, *Gentiana lutea*.

Yellow Gentian is a perennial herb with a lifespan of over 50 years. It has thick and robust roots which contains lots of starches, sugars, and flavonoids. The roots of Yellow Gentian are thick and deep, and digging the roots is not only a heavy physical task, but also a technical task. In France, there is even a specialized digging tool called the "fourche du diable" (Evil Garden Fork), which is a heavy and strong steel fork that is over 2 meters high and over 12kg weight (Fig. 5-6: A-F).

In the early time of Europe, the root of Yellow Gentian was used as a stomach building herb, and later widely used to make gentian wine/liquor. It was rather popular in France, Switzerland, Germany, and Austria, especially in mountainous areas such as the Pyrenees, the Jura, and the Alps, even forming a traditional specialty industry. There are many types of gentian wines, and can be divided into two categories: strong spirits with an alcohol content of over 40% and aperitifs with an alcohol content of less than 20% (Fig. 5-6: G-I).

Similar to the production of ordinary liquor, the root of yellow gentian can be directly fermented because it is rich in sugar and starch. Clean fresh or dry roots can be fermented by crushing them appropriately, and after fermentation, they can be distilled to produce stronger liquor. The roots of Yellow Gentian not only provide nutrients for yeast, but also provide a unique flavor to the product. It has a herbal and citrus aroma, with a slight bitterness and a sweet aftertaste, making it very refreshing. It is said to have good calming and digestive effects as well, so it is often used as a digestive aid after meals and can also be used to make cocktails. This type of strong gentian liquor is more popular in Germany, Austria, and Switzerland, France and the French speaking regions of Switzerland are more popular for low alcohol aperitifs, with a wide collection of varieties and brands, among which French Suze is the most famous.

The history of Suze is very long, dating back to 1885, first created in Maisons Alfort, France, and later promoted as an aperitif. It uses the roots of Yellow Gentian as the raw material. After collecting wild or cultivated roots of Yellow Gentian that have been growing for more than 20 years, it is first selected and cleaned, and then appropriately crushed and immersed in about 50% edible alcohol to stir and soak; After soaking for a few months, it is pressed and filtered, and the filtrate is distilled through a special process. The distilled raw wine is then blended with other aromatic herb extracts to make a unique appetizer. This aperitif typically has a alcohol content of 15%-20% with a soft citrus aroma, a hint of earthy aroma, as well as a hint of floral fragrance. It also provides a slight sweetness and a slightly bitter aftertaste. It can be consumed pure or mixed with others as cocktails. It is said that Picasso had a great love for Suze, and in 1912, he created a painting specifically for this wine called "Verre et bouteille de Suze".

Hops bring refreshing bitterness and aroma to beer, thus bring soul to the beer; Yellow Gentian brings not only the soul, the aroma and the taste, but also the fleshy body to the gentian wines, the alcohol and the sugars, because it provides both the nutrients for fermentations and the flavors to the products. Gentian wine/liquor is really a good thing, it can help you both appetize and digest, let you to eat more and better. Life, on the other hand, is also like a cup of gentian wine/liquor, slightly sweet and bitter, making people excited and intoxicated.

中国云南省德钦县白马雪山的钟花龙胆
这里现在通隧道了，老公路上车少了，人退花进，龙胆花又回到了路边

Gentiana nannobella on the Baima Snow Mountain in Yunnan of China.
A new tunnel was built so there are much less traffic on the old road,
thus the gentian flowers could return to the roadside.

5.4 文化艺术

在一般中国人的生活里，除了药用能接触到龙胆和秦艽外，鲜有龙胆花的身影，文化艺术中也见不到龙胆花元素。然而，世界的另一边，却是另一番景象。除了做酒，用龙胆花来装点生活，进行文艺创作，甚至印在邮票（图 5-7）和钱币（图 5-8）上。欧洲人，尤其是阿尔卑斯山的山民们把龙胆花画在画里，做成工艺品，绣在衣服上，甚至作为商品的商标或食品的包装，在生活中随处可见；龙胆花与火绒草和杜鹃花一起，构成了山里的“三原色”红白蓝组合，被阿尔卑斯山人无限宠爱。在瑞士，如果你去阿尔卑斯山区旅游，时常可以看到“龙胆花酒店”“龙胆花饭馆”和“龙胆花巴士”；在阿彭策尔的小路上你还可以见到盛装的大人小孩，衣服上绣满了雪绒花和龙胆花；甚至在酒店的餐桌上，隔热垫和餐巾布上也会绣上龙胆花；喝咖啡的奶油或抹面包的黄油外包装上自然也少不了龙胆花，龙胆花就是山里的生活。龙胆花也是摄影和绘画等艺术创作的良好素材，在欧洲时常能见到龙胆花主题的墙报、挂图、明信片和绘画作品（图 5-9）。中国著名植物画家曾孝濂也曾创作一幅中国龙胆花头花龙胆的绘画，也非常漂亮。

中国历史悠久，文化积淀深厚，有大量赞美各种风光景色和鲜花的诗词。诗经《关雎》里国人皆知的荇菜（*Nymphoides peltata*），“参差荇菜，左右流之。窈窕淑女，寤寐求之”。荇菜是爱情和浪漫的象征，早期就曾经被放在龙胆科，新近研究证明它并不属于龙胆科。除此之外，诗经里没有别的龙胆，唐诗和宋词里也没有龙胆花，想必是龙胆花离我们的诗人们太遥远了吧。

英国小说家和诗人戴维·赫伯特·劳伦斯（David Herbert Lawrence，1885—1930），在他生命接近终点的 1929 年，创作了最后一首题为《巴伐利亚龙胆花》的诗，作品情感强烈而忧郁，表达了他对死亡和重生的看法（图 5-10）。当他接近死亡时，他将一朵龙胆花比作他的火炬，作为生命的象征，也许劳伦斯要传达的信息是死亡本身就是生命的永恒，就像中国诗人描述的“轻舟已过万重山”！

劳伦斯的《巴伐利亚龙胆花》关乎生死，充满了忧郁和压抑。蓝色是深邃的，充满了忧郁，但蓝色也可以是广阔的，充满亮丽；龙胆花的火炬可以引领劳伦斯到暗界永生，与冥王冥后对话；更可以引领一个人到名山大川，逾越艰难险阻，到达宽阔的草地高原，看牛羊撒欢，看百花尽妍，看碧天如洗，看云卷云舒。常说爱是蓝色的，也就是说爱就是龙胆花，是不是这个道理？

除了劳伦斯，美国女作家艾米·洛威尔（1874—1925）也写过有关龙胆花的诗——《流苏龙胆》，带给我们的则是清新明快和淡淡的乡愁，还有对龙胆花的挚爱。

这两首诗歌都附在了后面，让我们期待会有更多、更好的作品面世。

图 5-7　邮票上的龙胆花（感谢袁方提供邮票实物及图片）

Fig. 5-7　Gentian flowers on stamps (Thanks to Fang Yuan for providing the stamps and photos)

图 5-8　钱币上的龙胆花

Fig. 5-8　Gentian flowers on money

A. 罗马尼亚发行的 1 列伊流通纸币（袁方供图）；
B. 瑞士发行的 10 瑞郎纪念硬币（孙瑜供图）；
C. 奥地利发行的 1 欧分流通硬币（孙瑜供图）。

A. 1 LEU Romanian banknotes in circulation from Romania(Photo by Fang Yuan);
B. 10 Swiss Franc commemorative coin from Switzerland (photo by Yu Sun);
C. 1 Euro Cent coin in circulation from Austria (photo by Yu Sun).

图 5-9　瑞士生活中随处可见的龙胆花装饰元素

Fig. 5-9　Decorations using gentian flowers found everywhere in Switzerland

Bavarian Gentians

D. H. Lawrence

Not every man has gentians in his house
in Soft September, at slow, Sad Michaelmas.

Bavarian gentians, big and dark, only dark
darkening the daytime torchlike with the smoking blueness of Pluto's gloom,
ribbed and torchlike, with their blaze of darkness spread blue
down flattening into points, flattened under the sweep of white day
torch-flower of the blue-smoking darkness, Pluto's dark-blue daze,
black lamps from the halls of Dis, burning dark blue,
giving off darkness, blue darkness, as Demeter's pale lamps give off light,
lead me then, lead me the way.

Reach me a gentian, give me a torch!
Let me guide myself with the blue, forked torch of this flower
down the darker and darker stairs, where blue is darkened on blueness
even where Persephone goes, just now, from the frosted September
to the sightless realm where darkness is awake upon the dark
and Persephone herself is but a voice
or a darkness invisible enfolded in the deeper dark
of the arms Plutonic and pierces with the passion of dense gloom,
among the splendour of torches of darkness, shedding darkness on the lost bride and her groom.

[Text cited from The New Oxford Book of English Verse, edited by Gardner H., Oxford University Press, 1972]

巴伐利亚龙胆花

D. H. 劳伦斯

不是每个人的家里都摆放着龙胆花，
在柔和的九月，闲散而伤感的圣米迦勒节，

巴伐利亚龙胆，硕大而深邃，唯有深邃，
火炬般浸染着白天，似冥王的忧郁般冒烟的幽蓝，
就像带着棱纹的火炬，用它们黑暗的火焰散播幽蓝，
白日扫荡挤压，挤压聚集成一个个光点，
火炬一样的花朵，蓝得冒烟的黑暗，似冥王的深蓝色晕眩，
冥府厅堂里的黑灯，燃着黑暗的蓝，
黑暗流泻，深蓝色的黑暗，就如同德米特尔暗淡的灯光，
来引领我吧，给我引路。

递我一支龙胆花，有如给我一把火炬，
让我用这蓝色的，光焰四射的花朵之炬，
来引领自己，
走下越来越暗的台阶，在那里蓝色被黑暗染得更蓝，
甚至到冥后的去处，就是现在，霜冻的九月，
前往没有视觉的国度，在那里黑暗把黑暗唤醒，
冥后也只是一个声音，
只有看不见的黑暗被更暗所包裹，
在冥王的怀里，
被浓密忧暗的激情穿透，
在黑暗火炬的辉煌里，为迷失的新娘和她的新郎，
挥洒黑暗。

［本书作者译］

图 5-10　劳伦斯和巴伐利亚龙胆花

Fig. 5-10　Lawrence and *Bavarian gentians*

A. 矗立在英国诺丁汉大学校园里的劳伦斯雕像，手里捧着一朵龙胆花，在思考人生的死亡和重生（感谢英国诺丁汉张林博士提供照片）
B. 现实中生长的巴伐利亚龙胆花，它就是劳伦斯心中那蓝光四射的火炬吗
C. 巴伐利亚的新天鹅堡

A. Statue of D H Lawrence in the university campus of Nottingham in UK (Photo courtesy from Dr Lin ZHANG, Nottingham UK)
B. The living Bavarian gentian, *Gentiana bavarica*, a forked blue-smoking torch in mind of Lawrence
C. The Schloss Neuschwanstein in Bavaria

德国巴伐利亚贝奇嘎登的山水景观及鹰巢峰，这里不但有美丽的风景和著名而神秘的城堡，更有巴伐利亚龙胆 *Gentiana bavarica*、匈牙利龙胆 *Gentiana pannonica* 和德国假龙胆 *Gentianella germanica*。

The Eagle's Nest peak in Berchtesgaden, Bavaria, Germany, where you find not only beautiful scenery and a mysterious castle, but also Bavarian genian *Gentiana bavarica*, Hungarian gentian *Gentiana pannonica*, and German gentian *Gentianella germanica*. 19.08.2005.

Fringed Gentians

Amy Lowell

Near where I live there is a lake
As blue as blue can be, winds make
It dance as they go blowing by
I think it curtseys to the sky
It's just a lake of lovely flowers
And my Mamma says they are ours
But they are not like those we grow
To be our very own, you know
We have a splendid garden, there
Are lots of flowers everywhere
Roses, and pinks, and four o'clocks
And hollyhocks, and evening stocks
Mamma lets us pick them, but never
Must we pick any gentians — ever
For if we carried them away
They'd die of homesickness that day

流苏龙胆

艾米·洛威尔

我家的旁边有一片湖
湛蓝湛蓝到蓝的极致
微风轻拂
它就起舞
向天空道万福
多么可爱的鲜花湖
妈妈说它就是我们的
可它却不像是我们种的
完全为我们所属
我们有个繁盛的花园
花团锦簇
玫瑰、石竹、紫茉莉
还有锦葵和夜香紫罗兰
妈妈说我们可以随意采摘园子里的花
但不要去采任何龙胆花，绝不
因为带走它们的当天
它们就会由于想家
而死

[本书作者译]

5.4 Inspiration of Culture & Art

In the daily lives of ordinary Chinese people, besides being able to come into contact with Long Dan (*Gentiana scabra*) and Qin Jiao (*Gentiana macrophylla*) for medicinal purposes, there is rarely the presence of gentian flowers in cultivation, and the element of gentian flowers is also not seen in Chinese culture and art. However, on the other side of the world, there is a different scene. In addition to making wine/liquor, the gentian flowers there have become culture, art, used to decorate life, and for artistic creation, and even to illustrate stamp(Fig. 5-7) and money(Fig. 5-8). Europeans, especially the people in the Alps, draw gentian flowers in their paintings, make them into handicrafts, embroider them on clothes, and even use them as trademarks or food packaging, which can be seen everywhere in daily life. Gentians, along with Edelweiss and Rhododendron, forms the "three primary colors" of red, white, and blue in the mountains, which are infinitely favored by the people of the Alps. In Switzerland, if you travel in the Alps, you can often see "Gentian Hotel", "Gentian Restaurant", and "Gentian Bus". On the paths in Appenzell, you can also see adults and children brightly dressed in costumes, embroidered with edelweiss and gentian flowers. Even on hotel dining tables, insulation pads and napkins are embroidered with gentian flowers. The butter for coffee or bread cannot be served without gentian flowers on the packing; what a wonderful gentian life in the mountains!

Gentians are also good materials for photography, painting and other artistic creation. In Europe, we often see posters, postcards and paintings with the theme of gentians (Fig. 5-9). Prof. Xiao-Lian Zeng, a well-known Chinese artist for plant painting, had also created a nice portrait for a Chinese gentian, *Gentiana cephalantha*.

China has a long history and profound cultural heritage, with a large number of poems praising various scenery and flowers. In the "Book of Songs, Guan Ju", there is a well-known Chinese aquatic flower *Nymphoides peltata*, which reads, "*Nymphoides peltata* flowers growing here and there in flowing water, graceful and beautiful girls seeking them constantly and hard." This poem was considered as highly romantic, and the flower as a symbol of love and romance. The species *Nymphoides peltata* was once placed in the Gentianaceae family, but recent research has proven that it does not belong to this family. Besides that, there is no other gentian in the Book of Songs, and there is no gentian flower in any poetry of the Tang and Song dynasties. It must be that the gentian flower is too far away from our ancient Chinese poets, right?

British novelist and poet David Herbert Lawrence (September 11, 1885 ~ March 2, 1930) wrote his last poem titled "Bavarian Gentians" in 1929, near the end of his life. The emotions in the poem were so strong and melancholic, expressing his views on death and rebirth (Fig. 5-10). When he was close to death, he used a gentian flower as a torch, as a symbol of life. Perhaps Lawrence wanted to convey the message that death itself is the eternal essence of life, that is, "the small boat has passed all the gorges", as Chinese poet said!

Lawrence's gentian poetry is full of melancholy and oppression, concerning life and death. Blue is deep, full of melancholy, but it can also be broad, full of brightness. Believing that the Gentian Torch can lead Lawrence to eternal life in the dark world and engage in dialogue with the underworld king and queen. It can also lead a person to come over mountaintops, to the vast plateau and grassland full of vitality, to view cows and sheep grazing happily, crowded flowers blooming crazy one after another, blue sky shining as washed, and white clouds rolling down and up across the horizon. It is often said that love is blue, so love is also gentian, isn't it?

In addition to Lawrence, American writer Amy Lowell (1874-1925) has also written poetry about gentian flowers. Her "Fringed Gentians" brings us a fresh, lively, and faint nostalgia, as well as a deep love for gentian flowers.

These two poems on gentians are attached bellow, hoping to attract more and better works, which we look forward to.

线叶龙胆 *Gentiana lawrencei* var. *farreri*
中国青海省河南蒙古族自治县 Henan, Qinghai, China. 09.09.2023

飞翔在龙胆国

Flying across Gentianland

龙胆花的引领

给我一片艳阳天，给我一片龙胆蓝，
循着梦中彩虹路，领我上高原，
任野风轻拂，看百花尽妍，
看天宽地阔，看牛羊撒欢。
许我身轻如飞燕，许我翼薄似蝶蝉，
心有灵犀向龙胆，领我上山巅，
任霜寒露冷，看雾起雾散，
看日升月落，看星移斗转。
——作者自勉

Guidance of Gentians

Find me a sunny blue sky, find me a carpet of gentian flowers,
Following the trails in my dream, bring me to the highlands,
Wild wind blows, thousand flowers dance,
Vast meadow extends, yaks and sheep graze;
Allow me to fly as a swallow, with wings as transparent as butterfly's,
Following the direction of gentians, bring me to mountaintops,
Frost grows on gentian petals, heavy fog arising in valleys,
Sun comes, gone moon, stars changing their tracks.

- the author to encourage himself

【摄影 /Photo：马惠】

参考文献
References

ANCHISI E, BERNINI A, CARTASEGNA N, et al., 1989. Gentianes d' Europe[M]. Italy: Gruppo Naturalistico Oltrepo' Pavese.

CHASSOT P, NEMOMISSA S, YUAN Y-M, et al., 2001. High paraphyly of Swertia L. (Gentianaceae) in the Gentianella-lineage as revealed by nuclear and chloroplast DNA sequence variation[J]. Plant System Evol, 229: 1-21.

CHEN S Y, XIA T, WANG Y J, et al., 2005. Molecular systematics and biogeography of Crawfurdia, Metagentiana and Tripterospermum (Gentianaceae) based on nuclear ribosomal and plastid DNA sequences[J]. Ann Bot, 96: 413-424. DOI:10.1093/aob/mci188.

ENGLER A, PRANTL K(eds), 1928. Die naturlichen Pflanzenfamilien, 4(2): 50-62[M]. Leipzig: Verlag von Wilhelm Engelmann.

FAVRE A, YUAN Y-M, KÜPFER P, et al., 2010. Phylogeny of subtribe Gentianinae (Gentianaceae): biogeographic inferences despite limitations in temporal calibration points[J]. Taxon, 59: 1701-1711.

FAVRE A, MATUSZAK S, SUN H, et al., 2014. Two new genera of Gentianinae (Gentianaceae): Sinogentiana and Kuepferia supported by molecular phylogenetic evidence[J]. Taxon, 63: 342-354.

FAVRE A, MICHALAK I, CHEN C H, et al., 2016. Out-of-Tibet: the spatio-temporal evolution of Gentiana (Gentianaceae)[J]. Journal of Biogeography, 43: 1967-1978.

FAVRE A, et al., 2020. Phylogenetic relationships and sectional delineation within Gentiana (Gentianaceae)[J]. Taxon, 69: 1221-1238. https://doi.org/10.1002/tax.12405.

FU P C, SUN S S, TWYFORD A D, et al., 2021. Lineage-specific plastid degradation in subtribe Gentianinae (Gentianaceae)[J]. Ecology and Evolution, 11: 3286-3299. DOI: 10.1002/ece3.7281.

FU P C, CHEN S L, SUN S S, et al., 2022. Strong plastid degradation is consistent within section Chondrophyllae, the most speciose lineage of Gentiana[J]. Ecology and Evolution, 12: e9205. DOI: 10.1002/ece3.9205.

FU P C, et al., 2021. Recurrent hybridization underlies the evolution of novelty in Gentiana (Gentianaceae) in the Qinghai-Tibetan Plateau[J]. AoB PLANTS, Vol. 13: No. 1. DOI:10.1093/aobpla/plaa068.

HALDA J J, 1997. The genus Gentiana[M]. Sen, Dobre.

HO T N, CHEN S, 2002. Metagentiana, a new genus of Gentianaceae[J]. Bot Bull Academia, 43:83-91.

HO T N, LIU S W, 1990. Infrageneric classification of Gentiana (Gentianaceae)[J]. Bulletin of the British Museum of Natural History (Botany Series), 20: 169-192.

HO T N, LIU S W, 2001. A worldwide Monograph of Gentiana[M]. Beijing: Science Press.

HO T N, LIU S W, 2015. A worldwide Monograph of Swertia and its allies[M]. Beijing: Science Press.

HO T N, PRINGLE J S, 1995. Gentianaceae[M]. In: Wu ZY, Raven PH (eds) Flora of China, vol 16. Beijing: Science Press, and St. Louis: Missouri Botanic Garden Press.

KADEREIT J W, VON HAGEN K B, 2003. The evolution of flower morphology in Gentianaceae-Swertiinae and the roles of key innovations and niche width for the diversification of Gentianella and Halenia in South America[J]. Interl J Plant Sci, 164: S441-S452.

KÜPFER P, YUAN Y M, 1996. Karyological studies on Gentiana sect. Chondrophyllae Bunge (Gentianaceae) from China[J]. Plant Systematics and Evolution, 200: 161-176.
NISHIHARA M, TASAKI K, SASAKI N, et al., 2018. Development of basic technologies for improvement of breeding and cultivation of Japanese gentian[J]. Breeding Science, 68(1): 14-24.
RYBCZYNSKI J J, MICHAEL R, MIKUŁA D A(eds.), 2014. The Gentianaceae, Volume 1: Characterization and ecology [M]. Berlin: Springer.
RYBCZYNSKI J J, MICHAEL R, MIKUŁA D A(eds.), 2015. The Gentianaceae, Volume 2: Biotechnology and Applications [M]. Berlin: Springer.
STRUWE L, ALBERT V A, 2002. Gentianaceae: systematics and natural history[M]. Cambridge: Cambridge University Press.
TAKAMURA Y, ASANO C, HIKAGE T, et al., 2019. Production of interspecific hybrids between Japanese gentians and wild species of Gentiana[J]. Breeding Science, 69: 680-687.
THE ANGIOSPERM PHYLOGENY GROUP, 2016. An update of the Angiosperm Phylogeny Group classification for the orders and families of flowering plants: APG IV[J]. Bot J Linn Soc, 181: 1-20.
TUTIN T G, et al., (eds). 1972. Flora Europaea Volume 3[M]. Cambridge: Cambridge University Press.
VON HAGEN K B, KADEREIT J W, 2001. The phylogeny of Gentianella (Gentianaceae) and its colonization of the southern hemisphere as revealed by nuclear and chloroplast DNA sequence variation[J]. Org Divers Evol, 1: 61-79.
VON HAGEN K B, KADEREIT J W, 2002. Phylogeny and flower evolution of the Swertiinae (Gentianaceae-Gentianeae): homoplasy and the principle of variable proportions[J]. Syst Bot, 27: 548-572.
VON HAGEN K B, KADEREIT J W, 2003. The diversification of Halenia (Gentianaceae): ecological opportunity versus key innovation[J]. Evolution, 57: 2507-2518.
YUAN Y M, 1993. Seedcoat micromorphology and its systematic implications for Gentianaceae of Western China[J]. Botanica Helvitica, 103: 73-82.
YUAN Y M, 1993. Karyological studies on Gentiana sect. Cruciata (Gentianaceae) from China[J]. Caryologia, 46: 99-114.
YUAN Y M, KÜPFER P, 1993. Karyological studies of Gentianopsis Ma and some related genera of Gentianaceae from China[J]. Cytologia, 58: 115-123.
YUAN Y M, KÜPFER P, 1993. Karyological studies on Gentiana sect. Frigida sensu lato and sect. Stenogyne (Gentianaceae) from China[J]. Bulletin de la Societe Neuchateloise des Sciences Naturelles, 116: 65-78.
YUAN Y M, KUEPFER P, 1995. Molecular phylogenetics of the subtribe Gentianinae (Gentianaceae) inferred from the sequences of internal transcribed spacers (ITS) of nuclear ribosomal DNA[J]. Plant Systematics and Evolution, 196: 207-226.
YUAN Y M, KÜPFER P, DOYLE J, 1996. Infrageneric phylogeny of the genus Gentiana (Gentianaceae) inferred from nucleotide sequences of the internal transcribed spacers (ITS) of nuclear ribosomal DNA[J]. American Journal of Botany, 83(5): 641-652.
YUAN Y M, KÜPFER P, 1997. The monophyly and rapid evolution of Gentiana sect. Chondrophyllae Bunge s. l. (Gentianaceae): evidence from the nucleotide sequences of the internal transcribed spacers of nuclear ribosomal DNA[J]. Botanical Journal of the Linnean Society, 123: 25-43.

YUAN Y M, KÜPFER P, ZELTNER L, 1998. Chromosomal evolution of Gentiana and Jaeschkea (Gentianaceae), with further documentation of chromosome data for 35 species from Western China[J]. Plant Systematics and Evolution, 210: 231-247.

YUAN Y M, WOHLHAUSER S, MÖLLER M, et al., 2003. Monophyly and relationships of the tribe Exaceae (Gentianaceae) inferred from nuclear ribosomal and chloroplast DNA sequences[J]. Mol Phylogene Evol, 28: 500-517. DOI:10.1016/S1055-7903(03)00068-X.

YUAN Y M, WOHLHAUSER S, MÖLLER M, et al., 2005. Phylogeny and biogeography of Exacum (Gentianaceae): a disjunctive distribution in the Indian Ocean Basin resulted from long distance dispersal and extensive radiation[J]. Syst Biol, 54:21-34. DOI:10.1080/10635150590905867.

ZHANG X L, WANG Y J, GE X J, et al., 2009. Molecular phylogeny and biogeography of Gentiana sect. Cruciata (Gentianaceae) based on four chloroplast DNA datasets [J]. Taxon, 58: 862-870.

ZHANG X L, YUAN Y M, GE X J. 2007. Genetic structure and differentiation of Gentiana atuntsensis W. W. Smith and G. striolata T. N. Ho (Gentianaceae) as revealed by ISSR markers[J]. Botanical Journal of the Linnean Society,154: 225-232.

国家中医药管理局《中华本草》编委会, 1999. 中华本草[M]. 上海: 上海科学技术出版社.

国家中医药管理局《中华本草》编委会, 2002. 中华本草 藏药卷[M]. 上海: 上海科学技术出版社.

何廷农, 1988. 中国植物志 第六十二卷[M]. 北京: 科学出版社.

何廷农, 刘尚武, 刘建全, 2013. 青藏高原一新特有属: 异型株属及其传粉模式[J]. 植物分类与资源学报, 35: 393-400.

中华人民共和国卫生部药典委员会, 2020. 中华人民共和国药典（2020 年版）[M]. 北京: 中国医药科技出版社.

中文名称索引
Index of Chinese Names

学名索引
Index of Scientific Names

H

I

J

K

L

M

龙胆花开天涯近，痴心远时地不偏；
风景独在这边好，眼界穷处有大川。

When gentians bloom, skyline is touchable; When hearts pursue, one reaches ends of the earth;
Scenery is better here, streams run through horizon.

中国青海省河南蒙古族自治县景观；Scenery in Henan, Qinghai, China. 09.09.2023.

后记：一切遇到过的美丽都值得被纪念

对龙胆花的野外观察和拍摄有如膜拜，有如约会，无比神圣，也令人无比激动。而本书的编写过程则是又一个再学习和再认识的过程，内心同样忐忑，充满惊喜和新发现。你以为你已经很了解很熟悉一个种了，但每次回顾和推敲，会发现并没有。像每一个富有魅力的人一样，龙胆花们值得被研究、被学习，继续学习，不停地学习，因为你总能看到以前或多或少忽视了的一些细节；只有不断端详、不断琢磨、才会不断发现、不断前行。书能成稿，的确值得庆贺，终于可以如释重负，了却一番心愿，虽未必尽善尽美，却已尽心尽力，尽了一份责任。唯有如此，才能让自己卸下心里的重担，迈开更轻快的步伐，继续在追寻龙胆花的路上徜徉。所以，书稿的完成并不是终点，而只是新的起点，我会继续努力，希望自己在追寻龙胆花的路上走得更高，走得更远，看到更美。

最后，在付梓之际，我愿意借此机会感谢那些引领我、成就我、帮助我研究龙胆花的人和机构。首先要感谢的是瑞士纳沙泰尔大学植物研究所的 Philippe Kuepfer 教授，是他接收我去他的实验室攻读博士学位，并以龙胆花作为研究对象，为我打开了一个全新的世界，重塑了我的整个人生，他指导我完成了学业，让我走上了热爱龙胆这条没有回程的快乐单行道。其次，要感谢我的家人，尤其是我的夫人马惠女士，除了给我坚定的持久的各种支持外，还亲自陪同我在阿尔卑斯山和横断山等欧亚各地的山野里跋涉、寻找龙胆。还要感谢华南植物园的葛学军教授，他几乎参与了我一半以上的龙胆花野外工作，多次与我一起专程考察龙胆花，一起走遍了多半个中国，足迹遍布东南西北。正是 30 多年来的那些艰苦快乐的野外考察让我见识了龙胆的美丽和多样，奠定了本书的基础。还要感谢何廷农教授和刘尚武教授的支持与指点，他们赠送给我的他们自己编写的专著和植物志，几乎成了我的教科书；我虽尚未掌握它们的全部精髓，书却已被我翻得稀烂。另外，彭泽祥教授、陈家瑞教授、洪德元院士、Louis Zeltner 博士、Egidio Anchisi 先生等前辈也为我的龙胆花研究提供了大量的帮助和指导。

非常感谢 Adrien Favre 博士和付鹏程博士提供了大量他们拍摄的龙胆花照片，使得本书更加完整、系统，为本书增色不少。此外，马惠、葛学军、曾佑派、詹雪菁、刘彬彬、刘源、贾留坤、张林、张粤、马溪芮、孟德昌、孙瑜、袁方、周浙昆、Christian Bravard、Elizabeth Byers、Peter Connolly、Philippe Kuepfer、Cheng-Tao LIN、Regula Naef、Erica Simonson 等人也提供了照片或授权我使用他们上传到不同图库的照片（主要是 PPBC 和 iNaturalist）。

刘运笑女士为本书绘制了两幅线描图，为本书增添了亮色，特此致谢。另外，也要特别感谢本书的责任编辑、美术编辑及其他编辑人员，他们的精心编排完美呈现了本书内容，也为本书加分不少。

还有许多值得感谢的专家和朋友，在本书资料收集和整理过程中提供了各种各样的帮助，包括但不限于柏国清、陈东方、陈昊、冯虎元，李攀，李思锋，刘建全、刘倩菲、刘志学、吴振海、辛晓萍、赵桂芳、张丽兵、张思宇、朱相云，等等，在此一并致谢。

承蒙 Philippe Kuepfer 教授、李思锋教授、刘志学教授、葛学军教授、陈东方博士、付鹏程博士和 Adrien Favre 博士审阅初稿并提供宝贵意见，也一并致谢。

野生植物的摄影，尤其是在边远山区的专类野生植物的摄影，有诸多不易。首先必须在野外正确的地方，找到要找的植物，正确地识别它，更必须是在正确的时刻，在它的花期或果期，天气好、不刮风、不下雨、最好还有好阳光的时刻为它拍照。虽竭尽全力，一些照片仍不尽如意，甚至有些比较重要的种类依然没有拍到，难免遗憾。

幸好，随着交通条件的更加便利和人们生活水平的不断提高，看到现在越来越多的人们成为自然爱好者和发烧友，他们走川西，进西藏，上更高的山，跨更广阔的高原，龙胆花也得到更多的关注与爱戴；而且，地球这颗蓝色星球上，大海蓝天依然湛蓝，龙胆花们还在继续绽放、继续争艳，愿所有的美好都得以继续、灿烂如初。

本书由拙耕园（上海）野生植物科技开发有限公司资助出版。

Postscript:

All the beauties encountered are worth remembering

Observing and photographing wild gentian flowers are a kind of worshiping and dating, which is incredibly sacred, and also incredibly exciting. The process of writing a picture illustration book on gentians is another process of re-learning and re-understanding, with the same anxiety in the heart, full of surprises and also new discoveries. You think you already knew and are familiar with a gentian species, but every time you review and reflect it when writing about it, you find that you do not know it. Like every charming person, gentian flowers are worth studying, learning, continuing to learn, and constantly learning, because you can always see some details that were more or less overlooked in the past; Only by constantly examining and pondering, can we constantly discover and move forward. The completion of the book is indeed worth celebrating. Finally, I can feel relieved and fulfill my wish. Although it may not be perfect, I have done my best and fulfilled my responsibility. Only in this way can I unload the heavy burden in my heart, take lighter steps, and continue to wander lightly on the path of pursuing more gentian flowers. So, the completion of the manuscript is not the end point, but just a new starting point. I will continue to work hard, hoping to go higher, farther, and see more beautiful scenery on the path of pursuing gentian flowers.

Finally, on the occasion of completion of the book, I would like to take this opportunity to express my sincere gratitude to those who have led me, accomplished me, and helped me with my research on gentian flowers. The first to thank is Professor Philippe Kuepfer from the Institute of Botany at the University of Neuchatel in Switzerland. He accepted me to pursue a doctoral degree in his laboratory and set up the subject to study gentian flowers. He opened up a new world for me, reshaped my entire life, guided me through my studies, and embarked me on the one-way street of loving gentian, which has no return journey, happily. Secondly, I would like to express my gratitude to my family, especially my wife, Ms. Hui Ma, for not only providing me with unwavering and lasting support, but also personally participating in accompanying me on my treks in the mountains of various regions of Europe and Asia, such as the Alps and the Hengduan Mountains, in search of gentians. I would also like to thank Professor Xue-Jun Ge from South China Botanical Garden, who has participated in more than half of my field trips searching gentian flowers. He has made special trips to investigate gentian flowers with me for many times across nearly the entire China, from the south to north and from the east to the west. It was those hard but joyful trips made this book possible. I would like to thank Professor Ting-Nong Ho and Professor Shang-Wu Liu for their support and guidance for me to start with gentians. They have given me their excellent monographs, flora and other works, which have almost become my textbooks, which I have thoroughly read again ad again. Professor Ze-Xiang Peng, Professor Jia-Rui Chen, Academician De-Yuan Hong, Dr. Louis Zeltner, and Mr. Egidio Anchisi also provided a lot of helps and guidance for my research on gentians.

I am very much grateful to Dr. Adrien Favre and Dr. Peng-Cheng Fu for providing me a large number of photos they took, which made this book more complete and systematic, and thus more valuable. In addition, Hui Ma, Xue-Jun Ge, You-Pai Zeng, Xue-Jing Zhan, Bin-Bin Liu, Yuan Liu, Liu-Kun Jia, Lin Zhang, Yue Zhang, Xi-Rui Ma, De-Chang Meng, Yu Sun, Fang Yuan, Zhe-Kun Zhou, Christian Bravard, Elizabeth Byers, Peter Connolly, Philippe Kuepfer, Cheng-Tao LIN, Regula Naef, and Erica Simonson have provided me photos or authorized me to use their photos that they uploaded to public databases (here mainly PPBC and iNaturalist).

Special thanks to Ms Yun-Xiao Liu, who has kindly drawn 2 nice drawing illustrations for this book, which added special highlight to the book. In addition, special thanks should be given to the responsible editor, art editor, and other editors. Their careful, stylish and creative lay out designs have added great value to this book.

Many experts and friends deserve my thanks for providing me various helps in the process of data collection, processing and writing of this book. They are Guo-Qing Bo, Dong-Fang Chen, Hao Chen, Hu-Yuan Feng, Pan Li, Si-Feng Li, Jian-Quan Liu, Qian-Fei Liu, Zhi-Xue Liu, Zhen-Hai Wu, Xiao-Ping Xin, Gui-Fang Zhao, Li-Bing Zhang, Si-Yu Zhang, Xiang-Yun Zhu, and many others.

Prof. Philippe Kuepfer, Prof. Si-Feng Li, Prof. Zhi-Xue Liu, Prof. Xue-Jun Ge, Dr. Dong-Fang Chen, Dr. Peng-Cheng Fu and Dr. Adrien Favre kindly reviewed the book draft and provided valuable comments and suggestions, for which I am very appreciated.

Photographing of wild plants, especially when targeting specific species in remote mountainous areas, has many challenges. You must be in the right place in the wild nature, find the right plant you are looking for, identify it correctly, and at the right time, during its flowering or fruiting period, encounter good weather, no wind, no rain, and preferably with good sunlight. Despite making every effort, Some photos I got are still not satisfactory, and even some species are still regrettably missing.

Fortunately, with the more convenient transportation conditions and the continuous improvement of people's living standards, more and more people are becoming nature lovers and enthusiasts. They travel to western Sichuan, Xizang, to climb higher mountains, and visit broader plateaus. Gentians are attracting more attentions and love; Moreover, on the blue planet Earth, the sea and sky are still azure, and the gentian flowers are continuously blooming, year after year. May all goodness be lasting, as good as, for ever.

Publishing of the book was funded by Dragon Yuan (Shanghai) Wild Plant Research & Development Co. Ltd.

类华丽龙胆 *Gentiana sinoornata*

中国西藏自治区林芝市 Linzhi, Xizang, China. 21.09.2003